Modern Thermodynamics

Carl S. Helrich

Modern Thermodynamics
with Statistical Mechanics

 Springer

Prof. Dr. Carl S. Helrich
Goshen College
1700 S. Main St.
Goshen IN 46526
USA
carlsh@goshen.edu

ISBN: 978-3-642-09909-0   e-ISBN: 978-3-540-85418-0

DOI 10.1007/978-3-540-85418-0

© Springer-Verlag Berlin Heidelberg 2010

This work is subject to copyright. All rights are reserved, whether the whole or part of the material is concerned, specifically the rights of translation, reprinting, reuse of illustrations, recitation, broadcasting, reproduction on microfilm or in any other way, and storage in data banks. Duplication of this publication or parts thereof is permitted only under the provisions of the German Copyright Law of September 9, 1965, in its current version, and permission for use must always be obtained from Springer. Violations are liable to prosecution under the German Copyright Law.

The use of general descriptive names, registered names, trademarks, etc. in this publication does not imply, even in the absence of a specific statement, that such names are exempt from the relevant protective laws and regulations and therefore free for general use.

Cover design: eStudio Calamar S.L.

Printed on acid-free paper

9 8 7 6 5 4 3 2 1

springer.com

Preface

Thermodynamics is not the oldest of sciences. Mechanics can make that claim. Thermodynamics is a product of some of the greatest scientific minds of the 19th and 20th centuries. But it is sufficiently established that most authors of new textbooks in thermodynamics find it necessary to justify their writing of yet another textbook. I find this an unnecessary exercise because of the centrality of thermodynamics as a science in physics, chemistry, biology, and medicine. I do acknowledge, however, that instruction in thermodynamics often leaves the student in a confused state. My attempt in this book is to present thermodynamics in as simple and as unified a form as possible.

As teachers we identify the failures of our own teachers and attempt to correct them. Although I personally acknowledge with a deep gratitude the appreciation for thermodynamics that I found as an undergraduate, I also realize that my teachers did not convey to me the sweeping grandeur of thermodynamics. Specifically the simplicity and the power that James Clerk Maxwell found in the methods of Gibbs were not part of my undergraduate experience. Unfortunately some modern authors also seem to miss this central theme, choosing instead to introduce the thermodynamic potentials as only useful functions at various points in the development.

I introduce the combination of the first and second laws and then the compete set of the potentials appears in chapter four. The remainder of the text is then built on the potentials. Before discussing modern laboratory measurements, for example, I show that the fundamental quantities sought in the laboratory are those which are required for determining the potentials.

The question of how to present our microscopic understanding of matter in a thermodynamics course confronts the author of any text. Presently the subjects of thermodynamics and statistical mechanics are a seamless whole [cf. [154, 155]]. I believe that we should try to convey that to our students without recourse to probability arguments. And so I have elected to present statistical mechanics as an integral part of the text. I begin with a chapter that takes the reader as far as we can go with simple models of molecules. Then I present ensemble theory as succinctly and simply as I am able, with the seamless connection to thermodynamics. Because of the modern work in Bose Einstein Condensation and in astrophysics, I then added a chapter on quantum statistical mechanics.

Because of its importance in modern applications I have chosen to treat irreversibility and the ideas of Ilya Prigogine. This provides a logical transition into the application of thermodynamics to chemical reactions. And irreversibility is at the center of all of biophysics.

I have used irreversibility as a step to considerations of stability and then to chemical equilibrium, solutions, and the equilibrium of heterogeneous systems. As a physicist I have also looked at chemical reaction rates and transition state theory. TST is a very interesting branch of theoretical physics. I encourage any physicist considering this text to not disregard this chapter.

This text is intended to be used as an introduction to modern thermodynamics and statistical mechanics. I believe it has the depth that can be appreciated without the extreme length to which many textbooks have gone. Consistent with this I have limited the exercises at the end of chapters. The exercises I have used are not intended to teach methods of solution nor are they intended as drill. Some are even intended as vehicles for investigation.

I suspect that my interests as a physicist will be apparent. My original training as an engineer, however, has led me to believe that applications only follow from understanding. Thermodynamics is subtle.

As an author I owe deep debts of gratitude to many that I can never hope to repay. I encountered the beauty and subtlety of thermodynamics first from Jerzy R. Moszynski. I was privileged then to work with David Mintzer and Marvin B. Lewis, from whom I gained an understanding of statistical mechanics and kinetic theory. I am also very grateful to generations of students who have helped this text emerge from my efforts to introduce them to thermodynamics. In an evaluation one student remarked that you have to work to hang on, but if you do the ride is worth it. I am also grateful to those who have been personally involved in the writing of this book. My wife, Betty Jane, has provided patience, understanding, and support, without which this book never would have been written. And Thomas von Foerster read and commended on the first drafts of almost all of the chapters. His critique and insistence have been invaluable.

Goshen, IN   *Carl Helrich*
August, 2008

Contents

Chapter 1
Beginnings

Willst du ins Unendliche Schreiten,
Geh nur im Endlichen nach allen Seiten.
If you want to reach the infinite,
explore every aspect of the finite.
Johann Wolfgang von Goethe

1.1 Introduction

Thermodynamics is the science that first dealt with heat and the motion caused by heat. Initially that meant the conversion of heat into motive power using an engine and a working substance like steam. Such things were an integral part of the industrial revolution and the wealth of nations. But as the foundations of the science emerged it became clear that we were not dealing only with motive power. We were dealing with matter and energy at a very fundamental level.

Max Planck's use of thermodynamic entropy and his eventual acceptance of Ludwig Boltzmann's formulation of a statistical entropy led to the interpretation of the black body radiation spectrum and the birth of the quantum theory [94, 131]. Thermodynamic arguments were the basis of Albert Einstein's paper on Brownian motion that closed the arguments about the existence of atoms [48]. And Einstein's thermodynamic treatment of radiation in a cavity produced the photon [47, 94].

We will treat systems of atoms and molecules with the methods of statistical mechanics. This includes extreme situations such as high temperature plasmas and studies of the behavior of matter near absolute zero of thermodynamic temperature. The roots of statistical mechanics are in thermodynamics. And there remains a seamless connection between the two.

We will also study the foundations of the theory of chemical reactions. In the language of the chemist, the combination of thermodynamics and statistical mechanics is physical chemistry. And as the American chemist G.N. Lewis[1] once remarked, physical chemistry includes everything that is interesting.

---

[1] Gilbert Newton Lewis (1875–1946) is one of the greatest of American scientists. He left an indelible mark on American teaching as well as research in physical chemistry.

C.S. Helrich, *Modern Thermodynamics with Statistical Mechanics*,
DOI 10.1007/978-3-540-85418-0_1, © Springer-Verlag Berlin Heidelberg 2009

For such a broad study we must carefully lay the foundations. And so we begin with heat and motion.

1.2 Heat and Motion

That motion can cause heat through friction is a part of common experience. To reverse the process and produce motion from heat required imagination and ingenuity. Then later it became evident that ingenuity alone did not suffice. A deeper understanding of matter and of energy was required. By the end of the 19th century this was beginning to crystallize as the sciences of thermodynamics and kinetic theory began to emerge.

That heat could cause motion occurred to people as they noticed the production of vast amounts of steam from the boiling of water. Hero (Heron) was a Greek mathematician (geometer) and writer on mechanics who lived in Alexandria in the first century of the common era. Among the known works of Hero is the record of an engine which used heat from a fire, and steam as an intermediary, to produce motion.

Hero's engine was a glass globe pivoted on glass tubes, which fed steam from a cauldron to the globe. The steam escaped through tubes located circumferentially around the globe. The tubes were bent so that the steam escaped tangentially and the globe rotated on its axis. This is a reaction engine. The force driving the engine was generated by the steam in the same sense as the thrust of a rocket engine is generated by the escaping gases. But the source of the motive power was the fire that produced the steam.

A steam pump developed by Thomas Savery in 1698 used heat to do work with steam as an intermediary. This pump was used to raise water for supplying reservoirs and in pumping water out of coal mines. Mines became flooded when excavation intersected ground water sources. A modification of Savery's pump, which actually included a cylinder, was produced by Thomas Newcomen in 1705. The action was slow and probably required the operator to manage the valves by hand. James Watt, an instrument maker in Glasgow, when employed to repair a Newcomen engine, observed that the action could be made more efficient. By 1782 Watt was producing an engine with a flywheel and continuous operation.

Much of the effort in steam engine development in the 18th century was of the sort of practical nature that required only a rudimentary concept of forces and a good dose of ingenuity. The engineer simply did not have the scientific tools necessary to design efficient engines.

At the beginning of the 19th century there were two competing concepts of heat, but really only one theory. One concept was that heat is a substance, referred to as caloric . The other concept was that heat is actually the motion of the constituents of matter. There was a theory for caloric; but there was no well-structured theory for the kinetic concept of heat. Therefore, even though some of the bases of the caloric theory were refuted experimentally at the end of the 18th century, there was no competing theory providing an alternative.[2]

In 1824 a remarkable memoir *Reflexions sur la puissance motrice du feu et sur les machines propre à développer cette puissance,* usually translated simply as *Reflections on the motive power of fire,* was published by a young French military engineer, Sadi Carnot,[3] outlining the concepts of the motive power of heat [21].

Unfortunately Carnot's memoir, which was written in non-mathematical language so that practical engineers would find it accessible, was essentially ignored by both engineers and scientists until its republication by Émile Clapeyron[4] two years after Carnot's death.

In the memoir Carnot made three primary points: (1) heat engines work because caloric flows through them much as water flows over a water wheel, (2) an efficient heat engine should operate in a cycle, in which the working substance[5] remains enclosed in a cylinder fitted with a piston, and (3) the most efficient engine should be reversible in the sense that in each step there is a very small difference between the forces and temperatures within the cylinder of the engine and those outside [21]. Very small changes in external conditions could then reverse the direction of the process and the heat engine could operate as a heat pump, pumping the caloric upwards across the temperature difference as a water pump.

Carnot explicitly cited, with praise, the British and Scottish engineers who had invented and developed the steam engine [[55], p. 63]. But Carnot's ideas put to an end the implicit belief that efficient steam engines could be built on the basis of ingenuity alone. Carnot had presented a new concept in which the heat itself was the important quantity, not the force of the steam on the cylinder.

The diagram in Fig. 1.1 illustrates the Carnot idea. In Fig. 1.1 the heat engine is represented by a filled circle.

Theoretically caloric was a conserved quantity. So a continuous arrow of uniform width is used to represent the caloric flowing through the heat engine. Work is produced as the caloric falls through the temperature difference, so the arrow representing work is separate from the caloric arrow. Carnot showed that the efficiency of the engine, which is the work done divided by the caloric input, must be independent of the working substance in the engine and is a function only of the height of fall, i.e. temperature difference. He was, however, unable to obtain a functional form for the efficiency.

Carnot later rejected the caloric theory. But he published nothing after the memoir. We know of Carnot's rejection of the caloric theory from his scientific notes. These had been in his brother's possession for forty-six years after Carnot's death, and were finally published in 1878 [37].

In the notes Carnot writes "... we may state as a general proposition, that the quantity of motive power in nature is fixed and that, strictly speaking, motive power

---

[2] These classic experiments are those of Benjamin Thompson (Count Rumford) and Humphry Davy.

[3] Nicolas Léonard Sadi Carnot (1796–1832) was a French physicist and military engineer.

[4] French engineer Benoît Paul Émile Clapeyron (1799–1864).

[5] Carnot used air in his example.

Fig. 1.1 Carnot's concept
of flowing caloric. Work is
done as caloric flows from
a high temperature to a low
temperature

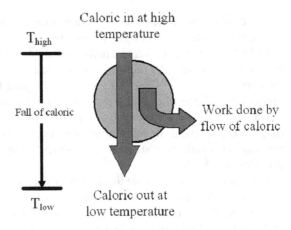

is neither produced nor destroyed. It is true that it changes its form; that is it sometimes produces one kind of motion, sometimes another. But it is never annihilated."
[[55], p. 30] We may easily substitute the modern term "energy" for "motive power."
Then this is not only a rejection of the caloric theory, it is a clear, verbal statement
of conservation of energy, which is the essence of the first law of thermodynamics.

1.3 The Laws

1.3.1 First Law

The relationship between heat and work is traditionally credited to the amateur scientist, James P. Joule,[6] in England. This is intermingled with the search for a conservation of energy, which occupied the attention of particularly German and English
scientists in the 1840's [69, 75, 110, 128]. The difficulties were in part a consequence of the fact that the relationship between the commonly used term caloric
and the developing concept of energy was not clear. That there was a relationship
was evident. Hermann von Helmholtz[7] attempted to place the conservation of energy on firm grounds based on mechanics as well as the other sciences [69]. Von
Helmholtz's arguments were visionary and based on a firm belief in a mechanical
picture [[37] pp. 72–74].[8] But Joule performed the beautiful and critical experiments
resulting in a laboratory measured result.

---

[6] James Prescott Joule (1818–1889).

[7] Hermann Ludwig Ferdinand von Helmholtz (1821–1894) was a German physician and physicist.

[8] The story is more nuanced. In 1847 von Helmholtz was a 26 year old physician. He finally had
to arrange for a private publication after rejection by Poggendorff's Annalen.

Joule performed a series of experiments between 1840, when he was twenty one, and 1847. He was interested in equivalences among electrical, thermal, chemical, and mechanical effects. Many of the experiments he performed were technically very difficult. And his data often reflected the difficulty in the variation in results he obtained in the earlier experiments. In 1847 he began the series of experiments for which he is remembered, dealing with the relationship between mechanical work and heat.

It became quite clear to Joule that for the experiments he had planned a thermometer exceeding the accuracy of anything commercially available was necessary. He had thermometers capable of measurements of hundredths of a degree Fahrenheit made by J. B. Dancer, a Manchester instrument maker.

In Fig. 1.2 we illustrate the experiment for which Joule is famous. This is often simply called the Joule experiment, and we forget the previous six years of hard work.

In the experiment diagrammed in Fig. 1.2 Joule allowed a weight to fall through a measured distance. By the arrangement of a pair of pulleys this caused the shaft to turn and vertical vanes attached to the shaft churned the water in the vessel. The observed increase in temperature of the water in the vessel, which Joule measured with one of his specially constructed thermometers, was 0.563°F. This was a measurement of the heat transferred to the system.

Of course there was no heat actually transferred to the system because of the insulated vessel. An amount of work, proportional to the distance through which the weight fell, was done on the system. That was the point. Joule was interested in finding a mechanical equivalent of heat. In this experiment he obtained the value of 773.64 ft lbf Btu^{-1}. The modern accepted value is 778 ft lbf Btu^{-1} [[37], p. 63].

Joule presented these results at the 1847 Oxford meeting of the British Association for the Advancement of Science, where he was asked to keep his presentation as brief as possible. After Joule's presentation there was silence until a young man stood up and pointed out the importance of what Joule had done. The young man

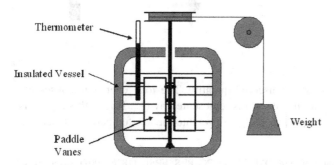

Fig. 1.2 One of the Joule experiments. The paddle wheels do work on the water in the insulated vessel as the weight falls. The thermometer measures the rise in temperature of the water

was Thomson,[9] who had just taken the position as professor of natural philosophy at Glasgow University, and would later become Lord Kelvin.

Thomson pointed out that in these experiments a certain amount of work was being done on an insulated system. For a given amount of work the result on the system, in this case the change in temperature, was always the same regardless of how the work was performed.[10] Doing work on an insulated system then produced a change in a property of the system, that is now called the internal energy.

The results of Joule's experiments may be taken as a statement of the first law [cf. [89], p. 152]. Heat is then introduced as a derived quantity. This approach has an attractive logic based entirely on experiment. We shall instead follow Rudolf Clausius' development, which introduces conservation of energy as a primary concept.

The first law of thermodynamics in its present form was published in 1850 by Clausius[11] [27] in Germany. But in 1850 Clausius referred to the internal energy as a "sensible heat." The confusion was resolved in 1865 with Clausius' great work on the second law [28–30], which was based on Carnot's work of 1824.

Clausius' ideas are represented in Fig. 1.3. In Fig. 1.3 the heat engine, is represented by a filled circle. The processes are cyclic. In (a) the heat Q_{in} entering the system from a high temperature source (reservoir) is separated into the heat Q_{out} flowing out of the system to a low temperature sink (reservoir) and the work done on the surroundings, W. The diagram (b) represents the reverse of (a). In (b) the heat taken in at the low temperature is combined with the work done on the system to produce an amount of heat transferred to the high temperature reservoir. This is possible if the heat engine is reversible, as Carnot required.

Fig. 1.3 Heat flow and work done in a cycle. Heat flow from high to low temperature can be made to produce work (**a**). Heat can be made to flow from low to high temperature if work is done on the system (**b**)

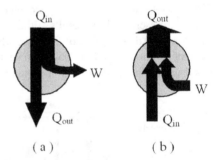

(a)   (b)

In Fig. 1.3 heat is not the conserved quantity. The total energy is conserved. The flow of this energy into the system is represented by Q_{in}. Part of this energy is converted into work W and part of it, Q_{out}, flows into the low temperature reservoir.

---

[9] William Thomson, 1st Baron Kelvin, OM, GCVO, PC, PRS, FRSE, (1824–1907) was a British mathematical physicist and engineer.

[10] Some of Joule's experiments also measured heat produced by frictional processes.

[11] Rudolf Julius Emanuel Clausius (1822–1888), was a German physicist and mathematician. Clausius formulated the second laws in terms of the entropy as a thermodynamic property. He was also instrumental in formulating atomic and kinetic theory of matter.

The amount of the heat which can be converted into work depends solely on the temperature difference.

Joule's equivalence of heat and work is evident in Fig. 1.3. But the engine is not insulated and there is no change in internal energy of the engine in a cycle. The cycle returns the engine, including the working substance contained in the engine, to its initial condition. There is then no change in any property of the engine. With the realization that energy is the conserved quantity we can combine Clausius' ideas in Fig. 1.3 with Thomson's interpretation of Joule's experiments.

Let us consider a small step in the cyclic process illustrated in Fig. 1.3. According to Thomson (and Joule) work done on an insulated system is equal to the increase in internal energy. According to Clausius, if the system is not insulated some of this internal energy is lost to the outside as an equivalent amount of heat.

During the small step in the cycle we are considering the engine is not insulated. Then the amount of work ΔW done on the system may initially give rise to a small increase in the internal energy ΔU of the system. But because the system is not insulated some of this internal energy is lost to the surroundings as an equivalent amount of heat ΔQ. That is, in the small step in the cycle

$$\Delta W \text{ (on the system)} = \Delta U \text{ (of the system)} + \Delta Q \text{(from the system)}. \quad (1.1)$$

In utilizing heat to perform work we consider that heat is the source of the motion resulting in work done on the surroundings. Then we choose heat entering the system as positive and work done by the system as positive. Heat leaving the system and work done on the system are then negative and the terms ΔQ (from the system) and ΔW (on the system) in (1.1) are negative quantities. With this convention (1.1) becomes

$$\Delta Q = \Delta U + \Delta W. \quad (1.2)$$

Equation (1.2) is a general statement of a conservation of energy applied to a system in which only energy, but no mass, is transferred. The energy is transferred in the form of heat or work. The process generally results in a change in the internal energy of the system.

Of the three terms appearing in (1.2) only ΔU is the change in a system property and depends solely on the initial and final states of the process. The heat transferred and the work done depend upon the process. For example, we may choose to insulate the system in which case $\Delta Q = 0$. That is the first law for a closed system requires that the difference in two process dependent quantities, the heat transferred and work done, is always equal to a term which is dependent only on end states of the system.

1.3.2 Second Law

The second law is traditionally given verbally in terms that are expressed visually in Fig. 1.3. These statements were formulated by Thomson (Lord Kelvin) and Clausius.

The Kelvin statement of the second law denies the possibility of a cycle that converts all of the heat transferred to it into work. That is, the Kelvin statement denies the possibility of Fig. 1.3(a) with $Q_{out} = 0$.

The Clausius statement of the second law denies the possibility of a cycle which transfers heat from a low to a high temperature reservoir without doing work on the system. That is, the Clausius statement denies the possibility of Fig. 1.3(b) with $W = 0$.

The second law does not deny Fig. 1.3(a) with $W = 0$. That is, heat flows naturally from high to low temperature without doing work. We may only transfer heat from a low to a high temperature in a cycle by performing work on the system during the cycle as shown in Fig. 1.3(b). The result is a refrigerator in which work is performed by the compressor.

There is nothing in the general statement of the first law that requires the process to run in one direction or the other. Energy is conserved whether we choose to obtain work from the flow of heat or to produce heat flow from work done. The second law, however, imposes a directionality. The natural flow of heat, when no work is done, is from high to low temperature.

The second law also imposes limitations on what we can do. Our engine has lost energy in the form of heat to a low temperature reservoir. In a sense this energy has been degraded. We can only use it in another cycle to obtain work if we have a still lower temperature to which we can exhaust heat. We have lost the possibility of using this low temperature energy in the same way that we used the energy stored in the reservoir at the high temperature.

If we allow a natural flow of heat from high to low temperature without producing useful work then we have lost an opportunity. This lost opportunity is irreversible.

In the 1865 paper Clausius [28] identified a new property, which he called the entropy from the Greek $\eta \tau \rho o \pi \eta$, which means transformation [[96], p. 80]. Entropy is a measure of irreversibility. In one of the great steps in theoretical physics, Clausius was able to show that entropy must increase whenever an irreversible process occurs in an isolated system. Specifically if any spontaneous process occurs inside the system, without any external cause, then there will be an increase in the entropy of the system. The entropy of Clausius is then a measure of the irreversible nature of a process occurring in a system.

Irreversible processes in a system may result from any of a number of causes. For example differences in temperature, electrical potential, concentrations will all result in irreversible motion of parts of a system. Such processes result in entropy increases, or entropy production within a system.

Clausius was able to obtain an expression for the infinitesimal change in the entropy of a closed system dS in a reversible process in terms of the infinitesimal amount of heat transferred reversibly during that process δQ_{rev} as

$$dS = \frac{1}{T} \delta Q_{rev},\tag{1.3}$$

where T the thermodynamic temperature, which can never be zero. This mathematical definition may be used for the calculation of the change in the entropy. But this definition is incomplete without the statement that the entropy must increase whenever an irreversible process occurs in the system.

The first and the second laws of thermodynamics are considered as the two great principles of the science. These were neatly tied together by J.W. Gibbs[12] at the end of the 19th century. We then possessed what was apparently a complete science of heat and motion. With that came the new discipline of physical chemistry.

Germany was particularly dependent on the chemical industry. And the next great step in thermodynamics came from Germany with roots in problems associated with the chemical industry. The third law was discovered by Walther Nernst[13] in Göttingen as he tried to resolve the problem of calculating equilibrium constants for reactions. The final formulation of the third law is in terms of the behavior of matter in the neighborhood of absolute zero of temperature. A consequence of the third law is the unattainability of absolute zero of temperature.

A specific reference value was required for the entropy in order to be able to calculate equilibrium constants for high temperature gas reactions. The third law provided this. In practical terms the third law has no effect on the Gibbs formulation. However, it has been indispensable for studies of chemical equilibrium and has motivated cryogenic studies (low temperature studies).

1.4   Modern Directions

Much of the more recent work in thermodynamics has been with nonequilibrium systems. This has resulted in no new laws, but has provided a deeper understanding of the second law. The behavior of nonequilibrium systems is based on the rate at which entropy is produced within the system. Particularly when coupled with the exploration of nonlinear dynamical systems and chaos, these ideas on entropy production provide great insight into some of the more fundamental questions related to the evolution of physical, chemical, and biological systems [49, 121, 133–135]. Such studies generally come under the blanket term of complexity or studies of complex systems. The term complexity has also found rather broad use outside of the scientific community. Nevertheless, it is one of the most active areas of interdisciplinary scientific study [122]. This is particularly because of the applications of these ideas in the neurosciences.

Modern work in thermodynamics also includes the application of statistical mechanics to nonequilibrium and to extreme conditions. Statistical mechanics cannot be separated from thermodynamics. It emerged logically from our attempts to understand the basic structure of matter and was motivated by an attempt to discover a microscopic (atomic and molecular) basis of the laws of thermodynamics. And the development of statistical mechanics engaged the same people who

---

[12] Josiah Willard Gibbs (1839–1903) was an American theoretical physicist, chemist, and mathematician. Gibbs was the first professor of Mathematical Physics at Yale College, now Yale University.

[13] Walther Hermann Nernst (1864–1941) was a German physicist.

were instrumental in the development of thermodynamics. particularly James Clerk
Maxwell,[14] Clausius, and Gibbs.

These modern applications of statistical mechanics include studies of high temperature plasmas, such as those present in thermonuclear fusion devices such as
the Tokamak, or encountered in astrophysics. Applications also include studies
near absolute zero, where statistical mechanics provides an understanding of such
phenomena as Bose-Einstein condensation (see Sect. 10.3.4).

1.5 Summary

In this chapter we have outlined the historical origins of the basic principles of thermodynamics in terms of the first, second, and third laws. The first law is a general
statement of conservation of energy. We cannot produce something out of nothing.
The second law denies us the possibility of using all of the energy we have. At every
step we lose energy to a lower temperature. A result of the third law we cannot attain
a temperature of absolute zero. Everything that we can imagine cannot always be
accomplished.

In Chap. 2 we shall formulate these fundamental laws of thermodynamics in a
way that makes them understandable in terms of the laboratory and in terms of
application.

Exercises

1.1. In the textbook situation a scientific revolution occurs when there are two
competing paradigms and one triumphs over the other. A paradigm is a pattern
of thinking about some aspect of the universe. We saw that caloric appeared
to be untenable but that there was no competing theory (paradigm). Try to
think of another example, out of any branch of human knowledge, in which a
paradigm shift has occurred based on the existence of competing theories. Try
to think of an instance in which the first paradigm was not working, but there
was no alternative.

1.2. Describe in your own words the points Carnot considered critical in developing a theory of the motive power of heat.

1.3. Figure 1.2 is a diagram of one of Joule's experiments. Design another experiment in which work can be used to heat water. Use your imagination, but
recognize that your design must be possible.

---

[14] James Clerk Maxwell (1831–1879) was a Scottish physicist. Maxwell is most remembered for
his synthesis of electromagnetism. He was also a pioneer in the development of atomic physics and
kinetic theory. Most of 19th century physics bears the imprint of Maxwell.

1.4. You want to carry out a Joule experiment to measure heat transfer in the nona-diabatic case. See Fig. 1.2. You construct two identical systems: one with an adiabatic boundary the other with a diathermal boundary. You arrange your pulley system so that you can drop the weights a large distance, i.e. out the window in a three storey building. For the adiabatic system you drop the mass, M, a height, H_0, and record a temperature change, ΔT. You notice that to get the same temperature rise in the diathermal case requires a distance, H, for the mass to fall.

(a) What is the heat transferred in the diathermal case?
(b) What assumptions have you made in the diathermal case? Consider the time.

1.5. A small explosion takes place inside a balloon with radius R and skin thick enough to remain in tact. The energy released in the explosion is $\mathscr{E}_{exp}$. To a reasonable approximation the balloon skin can be assumed to be adiabatic and the pressure of the balloon skin on the gas inside P_b may be assumed constant. The explosion is sufficient to increase the balloon volume by 10%.
What is the change in the internal energy of the gas inside the balloon?

1.6. You have studied Carnot's ideas carefully and built a very efficient engine too. In each cycle your engine extracts an amount of heat Q_1 from a high temperature reservoir at the temperature T_1 and exhausts Q_2 to a low temperature reservoir at T_2. In each cycle your engine does a useful amount of work W. You can relate the heats transferred and the work done using the first law. Because you know the heats transferred and the (thermodynamic) temperatures you can also calculate the entropy changes ΔS_1 and ΔS_2 for the reservoirs. You obtain a relationship among the total entropy change $\Delta S_T = \Delta S_1 + \Delta S_2$, the work done, and the heat input as $W = Q_1 (1 - T_2/T_1) - T_2 \Delta S_T$ (show this). You then notice that your engine has a measured efficiency $\eta = (1 - T_2/T_1) - \varepsilon$ where ε is not very large. Using the definition of work done in terms of efficiency $W = \eta Q_1$ you find that $\varepsilon Q_1 = T_2 \Delta S_T$ (show this). This intrigues you because it relates ε to ΔS_T. Comment on the result.

1.7. Place a wood cube at temperature T_1 on the wood laboratory table, which has a temperature $T_2 < T_1$. Neglect heat transfer from the cube to the room air. The heat transfer to or from a substance during a temperature change ΔT can be written as $C\Delta T$ where C is the heat capacity of the substance.

(a) What is the total entropy change in the laboratory during a (small) change in the cube temperature of ΔT?
(b) Is this entropy change positive or negative?
(c) You concluded in exercise 1.6 that positive entropy change means a loss. Does that concept hold here? Explain.

1.8. In Exercise 1.7 you found an expression for total increase in entropy during a temperature change ΔT. That is entropy increases when there is heat flow in the presence of a temperature difference. Relate this finding to the third point in Carnot's memoire.

1.9. Without imposing an engine between high and low temperatures heat will naturally flow from the high to the low temperature. It will not reverse and flow the other way. Show that heat flow in the presence of a temperature gradient is then an irreversible process resulting in a positive production of entropy.

1.10. With the first law in the form (1.2) the change in system entropy can be written as $dS = (dU + \delta W)/T$ for a very small process. So we do not need to think of entropy as only related to heat transfer. Anything that changes internal energy or is related to work can change the entropy. And, provided our conclusions in the previous exercises are correct, irreversible processes will increase the entropy. What are some irreversible processes that can occur in a system that are candidates for producing entropy?

1.11. The entropy change of the sun is negative. The earth absorbs some of the energy from the sun at a much lower temperature. So entropy is absorbed by the earth. Where is the entropy coming from?

Chapter 2
Formulation

*A theory is more impressive the greater
the simplicity of its premises is, the more
different kinds of things it relates, and the
more extended its area of applicability.
Therefore the deep impression which
classical thermodynamics made upon
me. It is the only physical theory of
universal content concerning which I am
convinced that, within the framework of
the applicability of its basic concepts, it
will never be overthrown.*

Albert Einstein

2.1 Introduction

Einstein's conviction, expressed in the quotation above, is based in part on the care
with which thermodynamics is constructed as a science. In our discussions in the
preceding chapter we have not required precise definitions of terms and concepts.
Indeed these may have hindered the discussion. As we begin to formulate the science
in such a way that we can have confidence in the results of our formulation, we must,
however, define terms and concepts carefully.

In his chapter we will introduce thermodynamic systems, states and processes.
We will also present the basic principles or laws of thermodynamics. These laws
govern the behavior of material substances in transformations associated with thermodynamic processes. The goal of this chapter is to present a brief, but sufficiently
complete discussion that will result in the basic formulation of the thermodynamics
of closed systems in terms of a single mathematical statement.

2.2 Systems and Properties

2.2.1 Systems

In a thermodynamic analysis we always consider a system. The system is a particular amount of a substance contained within definable boundaries. The substance

C.S. Helrich, *Modern Thermodynamics with Statistical Mechanics*,
DOI 10.1007/978-3-540-85418-0_2, © Springer-Verlag Berlin Heidelberg 2009

we may consider can be quite general. In most of what we do this will be matter composed of particles, i.e. atoms and/or molecules. However, we do not exclude systems containing radiation (photons). The substance need not be a single element or compound. And we shall treat substances containing particles that may undergo chemical reactions. And the substance or substances present may be in any phase or phases: solid, liquid or gas.

2.2.2 Boundaries

Boundaries separate the system under consideration from the immediate surroundings. These may be real material boundaries or imagined. We define the properties of the boundary so that the interaction between the system and the surroundings is specified in terms necessary for the analysis. Whether physically real or imaginary, the system boundary is real as far as the system properties are concerned.

Generally systems may be either impervious or open to the exchange of substances (matter) with the surroundings. Systems which are impervious to the exchange of matter with the surroundings are closed systems. Otherwise the system is open. In either case the system boundary may be fixed, such as a solid wall, or movable, such as a piston.

If we are interested in the heating of a solid block, for example, the boundary is most conveniently considered to be the outer surface of the block itself. The boundary is then closed and movable. The boundary is closed because the block neither gains nor loses mass. The boundary is movable because the block expands on heating. The boundary is physically real because the surface of the block is a real physical surface.

The analysis of a uniformly flowing fluid provides an example of an open system with a fixed boundary. A portion of the system boundary may be determined by a curved surface defined by the streamlines of the flowing fluid. The streamlines are defined by the velocity of the flow and there is no macroscopic flow across streamlines. In some circumstances we may be able to identify a real solid boundary that is parallel to the streamlines, such as for flow through a pipe or a nozzle. We complete the system by closing the ends of our system by open, imaginary surfaces at the ends, through which the fluid flows.

System boundaries are also classified according to whether or not they are permeable to heat. An adiabatic boundary is impermeable to the flow of heat. Heat may flow across a diathermal boundary. A system placed within a rigid adiabatic enclosure is then completely isolated from its surroundings. Work may be done on a system in a rigid adiabatic boundary in a Joule experiment using paddle wheels to stir the fluid. But an adiabatically enclosed system can do work on the surroundings only if the boundary is movable. If two systems are separated by a diathermal boundary, heat will flow from one system to the other until the temperatures of the two systems are equal. In a state of equilibrium, then, the temperatures of two systems separated by a diathermal boundary are the same.

System boundaries may then be open or closed, rigid or movable, and adiabatic or diathermal.

2.2.3 Properties and States

Pressure, P, volume, V, and temperature, T, are examples of thermodynamic properties of a system. The thermodynamic state of a system is determined if the thermodynamic properties of the system are known. We consider, for example, a system containing a certain amount of gas in a cylinder with a piston that is free to move vertically. The volume of the system is known from the position of the piston and the pressure of the system is determined by the weight of the piston and the weight of any mass that may be added to the piston. If the cylinder is diathermal the temperature of the system is the temperature of the room. The values of P, V, and T specify the thermodynamic state, or simply state of the system. We implicitly assume that P and T are uniform throughout the system so that we can speak of a single pressure and temperature of the system.

If there is no change in the state of the system with time we define the state as one of thermodynamic equilibrium.

2.2.4 Surfaces and States

Experiment shows that for any substance the three properties P, V, and T are not independent, but are related in a particular way that depends upon the identity of the substance. Specification of any two of these properties determines the third. For example, the pressure, P, is a function of the temperature, T, and the volume, V. The function $P = P(T,V)$ is then a surface in (P,V,T) space. An example is shown in Fig. 2.1. Each point on the surface in Fig. 2.1 is the representation of a thermodynamic state.

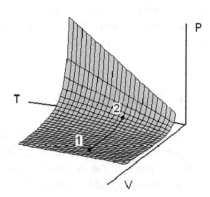

Fig. 2.1 The (P,V,T) surface. This is the graphical representation of the thermal equation of state. Figures 2.4, 2.6, 2.7, and 2.13 in reference [145] are beautiful representations of such surfaces

The relationship $P = P(V,T)$ is called the *thermal equation of state* for the substance and the surface in Fig. 2.1, is referred to as a (P,V,T) surface. A substance such as a gas, for which there are two independent thermodynamic properties, is called simple. Most substances of interest in applications are of this type. So we shall tentatively assume simple substances, or systems, in this text.

In general the thermal equation of state will not be known as an algebraic function. In these circumstances we must rely on tabulated data. The thermodynamic properties for many substances of interest have been tabulated (cf. [87] and [88]). The National Institute of Standards and Technology also has a database, *NIST Reference Fluid Thermodynamics and Transport Properties - REFPROP* [102], which can be purchased and used on a personal computer. *REFPROP Version 8* (April, 2007) has been used as a source of data extensively in this text.

2.2.5 Quasistatic Processes

The most efficient heat engines are those for which the changes in the thermodynamic state of the working substance are slow enough that the system may be considered to remain in an equilibrium state. In such processes there are very small differences between internal and external forces and an infinitesimal change in the external force will result in a reversal of the process. These processes are called quasistatic or reversible. An example is the process $1 \rightarrow 2$ in Fig. 2.1. In this process the state of the substance lies on the line drawn on the surface at each instant. Each of the states on this line are equilibrium states. The process then carries the substance (in the system) from one equilibrium state to another.

Real processes of interest are seldom if ever quasistatic. It is often possible, however, ignore the details and base the analysis on changes in thermodynamic properties.

For situations in which the variations in thermodynamic conditions are small over dimensions of interest or slow over measurement times, an assumption of local equilibrium may be made. That is, we may consider that thermodynamic equilibrium exists for small volumes and over short times resulting in an ability to apply the results of equilibrium thermodynamics to the situation. We may then speak meaningfully about thermodynamic variables as having values which depend on position and on the time within a system.

2.2.6 Properties and Processes

A thermodynamic property may be any general quantity that has a unique value for each thermodynamic state of a substance. That is a thermodynamic property a unique function of the thermodynamic state of the system. Two such functions are the internal energy and the entropy.

Fig. 2.2 A graphical
representation of a
thermodynamic property
$F(T,V)$. Separate processes
to change the thermodynamic
state from (1) to (2) are
shown

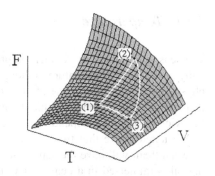

Changes in the value of a thermodynamic properties in a process are independent of the details of the process. The function $F(T,V)$ plotted in Fig. 2.2 is an example of a thermodynamic property. The value of the property F is the height of the surface above the (T,V) plane. We consider a quasistatic change in the thermodynamic state from (1) to (2). We may change the thermodynamic state of the system by holding the temperature constant and varying the volume. This is the direct path from (1) to (2) in Fig. 2.2. The change in F in this process is $\Delta F = F(2) - F(1)$ and is the difference in the heights of the surface $F(T,V)$ above the (T,V) plane at points (1) and (2). We may also hold the volume constant while we take the system first through a quasistatic change in the thermodynamic state from (1) to (3) and then follow this by a general quasistatic change in state from (3) to (2) during which both T and V vary. The resulting change in the value of F in the second process (1) → (3) → (2). is also $\Delta F = F(2) - F(1)$ and simply equal to the difference in the heights of the surface $F(T,V)$ above the (T,V) plane at points (1) and (2).

Calculations of the changes in values of thermodynamic properties may then be carried out choosing any process whatsoever carrying the system from the initial to the final thermodynamic state.

2.3 Properties and the Laws

Die Energie der Welt ist Konstant. Die Entropie der Welt strebt einem Maximum zu. (The energy of the universe is constant. The entropy of the universe tends to a maximum)

Rudolf Clausius, 1865.

Three thermodynamic properties are defined by the laws of thermodynamics. These are temperature, internal energy and entropy. In our historical outline of the preceding chapter we presented the background behind the first and second laws. These provide a basic understanding of heat and the motive power of heat. The first law defines the internal energy of a substance and the second law defines the entropy. Integral to the discussion of these laws was the concept of temperature, which we treated as an intuitive concept. It was not until the 20th century that we recognized that the definition of temperature requires another law. Because this should logically precede the first and second laws it is called the *zeroth law*.

2.3.1 Temperature

The sense of "hotter" or "colder" is physiological. Items in a room are all essentially at the same temperature, although a table may feel colder than a book. The physiological sensation is based on a rate of heat transfer upon touch rather than temperature. The thermodynamic definition of temperature is based on observable behavior of systems independent of human physiology.

We noted above that two systems in equilibrium with a diathermal boundary between them will have the same temperature. For example if a test tube containing oil is immersed in a beaker of water both the oil and the water will have the same temperature when a state of thermodynamic equilibrium is reached. To verify the equality in temperatures we must perform a measurement. And to measure the temperature of the water or of the oil we require a third system: a thermometer. Introducing this necessary third system brings us beyond the statement of equality of temperatures of the first two systems. We must logically make a statement about the third system.

In 1931, while reviewing a thermodynamics text by the great Indian astrophysicist, Megh Nad Saha,[1] Ralph H. Fowler[2] observed that for consistency the statement that [150]

> If systems A and B are both in equilibrium across a diathermal boundary with system C, then systems A and B will be in equilibrium with one another if separated by a diathermal wall.

should be included in the formulation of thermodynamics.

System C is our thermometer. We only need to arrange the system C in some contrived fashion so that we can easily determine its thermodynamic equilibrium state when it is in contact with the systems A and B. This statement then resolves our logical problem and serves to make the definition of an empirical (experimentally determined) temperature possible. This is the zeroth law.

We have been using T to designate temperatures. At this stage in our discussion T is an empirical temperature. With the zeroth law we have a more robust and definitive statement of the meaning of temperature than our previous intuitive concept. But this is not yet what we shall identify as a thermodynamic temperature. The thermodynamic or absolute temperature is a quantity, which will result from the second law.

An example of an empirical temperature scale is that determined by a mercury in glass thermometer. Our third system here is a glass capillary tube which is partially filled with mercury under vacuum so that only mercury vapor exists above the liquid mercury. If we neglect the vapor pressure of the mercury we have a system at a constant pressure (equal to zero).

---

[1] Meghnad Saha FRS (1893–1956) was an Indian astrophysicist. He is best known for the Saha equation, which gives the density of electrons in an ionized gas.
[2] Sir Ralph Howard Fowler OBE FRS (1889–1944) was a British physicist and astronomer.

The expansion of the liquid mercury with changes in temperature is much greater than that of the glass so we may safely assume that only the length of the mercury column varies with temperature. We have then contrived our system so that the length of the mercury column is the only property we need to measure in order to determine the temperature. The length is the thermometric property. The scale we choose as a measure of the length of the mercury column is arbitrary. For simplicity we may choose the scale to be linear. This requires two constants to fix the scale, or the specification of two standard temperatures. If these are taken to be separated by 100 degrees we have a centigrade scale. If the standard temperatures are also the ice and steam point of water the scale is the Celsius scale.

To specify our empirical temperature scale we must describe the system (mercury in a glass capillary tube filled under vacuum), our thermometric property (length of mercury column), our scale (linear), and the fixed points (ice and steam).

In the appendix we present the ideal gas temperature scale as an empirical scale which requires only a single fixed point and for which there is an absolute zero fixed by the universal properties of gases as their pressures approach zero. This scale is identical with the thermodynamic scale.

In the appendix we find that the ideal gas has a thermal equation of state

$$P(V,T) = nR\frac{T}{V}$$

in which n is the number of mols and R is a universal constant known as the (universal) gas constant. This defines a fictitious substance that will, nevertheless, prove very useful.

2.3.2 Internal Energy

A definition of internal energy as a thermodynamic property comes from the first law. Thomson pointed out that Joule's experiments identified this property and provided a method for its measurement. We have chosen to formulate the first law, following Clausius, as an energy conservation theorem. Nevertheless our formulation reduces to that of Joule and Thomson for the relationship between work and internal energy. We, therefore, still require a definition of thermodynamic work.

Thermodynamic Work

Joule's original experiment provides an unambiguous definition of thermodynamic work. The work done by or on a thermodynamic system may always be used to raise or lower a body of mass m a distance $\pm h$ in the earth's gravitational field. This work is $\pm mgh$, where g is the acceleration due to the gravitational field of the earth. Thermodynamic work is then defined in unambiguous mechanical terms as

Definition 2.3.1. *Thermodynamic work is defined as any interaction of the thermo-dynamic system with the surroundings that can be reduced to the raising or lowering of a material body in the gravitational field.*

The intended use to which we put the work done may not be to raise or lower a material body. The question of whether or not thermodynamic work has been done in a process is based, however, on whether or not we can imagine an arrangement of pulleys and gears that can reduce the final effect to raising or lowering of a material body.

This definition of thermodynamic work allows us to obtain an expression for the quasistatic (reversible) work done on or by a closed system. In a closed system it is only possible to perform quasistatic thermodynamic work by moving a boundary such as a piston. Figure 2.3 illustrates how this work can be reduced to the lowering of a material body in the gravitational field. The cylinder contains a gas and the force of the weight is transmitted to the piston by a cam arrangement. For practical purposes we may consider that the suspended mass is on a balance pan. If we remove a small amount of mass from the pan the piston and the suspended mass will both move upward. The motion of the body through the distance δh corresponds to a motion of the piston through a distance δs. In this process the internal force on the piston is a product of the system pressure, P, and the piston area, A. As the piston moves a distance δs the work done by the system on the surroundings is $PA\delta s = P\delta V$, where $dV \, (= A\delta s)$ is the differential change in the system volume. If there is no friction in the cam system or the piston this is equal to $mg\delta h$ in which m is the suspended mass remaining. Then, for a very small step in a quasistatic process occurring in a closed system, the work done by the system on the surroundings is

$$\delta W_{\mathrm{rev}} = P dV. \tag{2.1}$$

We have taken the limit $\delta V \to dV$ as the weight removed becomes infinitesimal and have written δW_{rev} for the reversible quasistatic work done.

The dependence of P on V is determined by the process involved. Therefore the infinitesimal quasistatic work done δW_{rev} is process dependent, as we pointed out in Chap. 1.

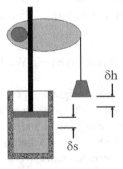

Fig. 2.3 Thermodynamic work is done by the closed system as the piston moves raising or lowering the mass

If $dV > 0$ work is done by the system on the surroundings and the mass is raised. If $dV < 0$ work is done by the surroundings on the system and the mass is lowered.

2.3.2.1   Closed Systems

For an infinitesimal, quasistatic (reversible) process occurring in a closed system we write the first law in (1.2) as

$$\delta Q_{rev} = dU + PdV. \tag{2.2}$$

The differential change in the internal energy is then given by

$$dU = \delta Q_{rev} - PdV. \tag{2.3}$$

Both terms on the right hand side of (2.3) are process dependent. But their difference is the differential of a thermodynamic property and is not process dependent.

The internal energy is defined by a differential expression provided by the first law. That is, only differences in internal energy are defined. The internal energy in integrated form is then only determined to within a constant, which may be chosen to be zero at an arbitrary point. This choice of a zero for the internal energy implies nothing about the actual value of the energy of a system of atoms or molecules, which never vanishes (see Sects. 10.3.3 and 10.4.2)

2.3.2.2   Open Systems

General. Matter may cross the boundary of a system as a result of thermodynamic forces on the boundary or, as in many engineering applications, because of entrance and exit ports in the system. In an open system there will be contributions to the energy associated with the transport of matter across the system boundary. The identity of these contributions depends on the system, the process taking place and the system design. We shall combine these additional contributions to the system energy in an infinitesimal process to form the term $d\Phi$. In engineering applications open systems are often used to produce work other than the PdV work associated with the motion of the system boundary. This work is traditionally called "shaft work" and is designated as δW_s. The first law for an open system is then

$$dU = \delta Q + d\Phi - \delta W_s - PdV. \tag{2.4}$$

In this equation δQ is a general heat transfer, which may include irreversible as well as reversible contributions.

Engineering Applications. In engineering applications matter enters and leaves the system through a number of ports such as steam lines. In this case the term $d\Phi$ contains contributions from energy transported by the matter and the work done on or by the system in transporting the matter. Work done in the transport of matter is considered to be an energy term. We shall designate the ports by superscripts i and the number of mols of substance λ entering or leaving the ith port in an infinitesimal process as $d_e n_\lambda^{(i)}$.

At the ith port the component λ has a specific (molar) energy determined by the conditions at the ith port,

$$e_\lambda^{(i)} = \mathfrak{e}_\lambda^{(i)} + u_\lambda^{(i)}. \tag{2.5}$$

Here $\mathfrak{e}_\lambda^{(i)}$ contains all forms of energy besides the (thermodynamic) internal energy, which we designate as $u_\lambda^{(i)}$.

The work required to transport matter into or out of port i is equal to the product of the pressure force at the port, P_i, and the volume occupied by the matter transported, $d_e n_\lambda^{(i)}$, which is $V_i = v_\lambda^{(i)} d_e n_\lambda^{(i)}$. This results in an increase of the energy of the system if the matter is transported into the system and a decrease if the matter is transported out of the system.

The term $d\Phi$ resulting from the transport of matter is then

$$d\Phi = \left(\mathfrak{e}_\lambda^{(i)} + u_\lambda^{(i)} + P_i v_\lambda^{(i)} \right) d_e n_\lambda^{(i)}$$
$$= \left(\mathfrak{e}_\lambda^{(i)} + h_\lambda^{(i)} \right) d_e n_\lambda^{(i)}. \tag{2.6}$$

Here

$$h_\lambda^{(i)} = u_\lambda^{(i)} + P_i v_\lambda^{(i)} \tag{2.7}$$

is the specific (molar) enthalpy[3] of the component λ at the port i.

In engineering applications the system of interest generally possesses energies in addition to the thermodynamic internal energy. We, therefore, write $d\mathscr{E}$ in place of dU. The general conservation of energy in (2.4) for open engineering systems then takes the form

$$d\mathscr{E} = \delta Q - \delta W_s + \sum_i \mathfrak{e}_i dn_i + \sum_i h_i dn_i. \tag{2.8}$$

In (2.8) any PdV work is involved in transport of matter.

Specific Heats. From (2.7) we have the enthalpy of a single component system as

$$H = nh = U + PV. \tag{2.9}$$

---

[3] The origin of the term enthalpy is with the Dutch physicist Heike Kamerlingh Onnes who derived its meaning from the Greek word "en-thal´-pos" ($\varepsilon\nu\theta\alpha\lambda\pi o\zeta$). Accent is on the second syllable.

The first law for a closed system can then be written as either (2.2) or as

$$\delta Q_{rev} = dH - V dP. \tag{2.10}$$

Since the heat transferred will produce a change in the system temperature at either constant volume or constant pressure we realize that dU in (2.2) and dH in (2.10) must both be proportional to dT and that U and H must be functions of the temperature T. Because (2.2) already contains dV we are led to assume that U will depend on the temperature T and the volume V. And because (2.10) already contains dP we similarly assume that H depends on T and P. We may then measure the temperature increase dT in the system resulting from an amount of heat transferred quasistatically δQ_{rev}. The quantity $\delta Q_{rev}/dT$ is called the *heat capacity*.

The concept of the heat capacity of a substance predates the work of Joule and, therefore, our present understanding of heat transfer as the transport of energy. The unit of the *calorie* was defined as the amount of heat required to raise the temperature of one gram of water by 1°C when the pressure is 1 atm (101.325 kPa). The properties of water are temperature dependent and the amount of heat required to raise the temperature of one gram of water by 1°C depends upon the temperature at which the measurement is made. The 20°C calorie, or 20° cal (4.1816 J) was the amount of heat required to raise one gram of water form 19.5 to 20.5°C at 1 atm pressure and the 15° cal (4.1855 J) was the amount of heat required to raise the temperature from 14.5 to 15.5°C at a pressure of 1 atm. These values defined the heat capacity of water as a function of temperature [50].

With our present understanding of thermodynamics we realize that the measurement required in this definition of the calorie is given in (2.10) with $dP = 0$. That is

$$\text{heat capacity} = \frac{\delta Q_{rev}]_P}{dT} = \frac{1}{n}\left(\frac{\partial H(T,P)}{\partial T}\right)_P, \tag{2.11}$$

in which n is the number of mols of the substance present in the system. Because $H = H(T,P)$ is a property of the substance, the quantity defined in (2.11) is also a property of the substance. The term defined in (2.11) is no longer called the heat capacity, but the (molar) *specific heat at constant pressure*, $C_P(T,P)$, defined by

$$C_P(T,P) \equiv \frac{1}{n}\left(\frac{\partial H(T,P)}{\partial T}\right)_P. \tag{2.12}$$

Similarly (2.2) provides a definition of a (molar) *specific heat at constant volume*, $C_V(T,V)$, defined as

$$C_V \equiv \frac{1}{n}\left(\frac{\partial U(T,V)}{\partial T}\right)_V. \tag{2.13}$$

Because $U = U(T,V)$ is a property of the substance (it is a thermodynamic potential) the specific heat at constant volume is a property of the substance.

Perpetual Motion. Let us consider an isolated, adiabatically enclosed system consisting of interconnected subsystems, which we designate with the subscript σ. This system can be made to go through a cyclic process by performing work on the system and allowing work to be done by the system. That is we may arrange a system of pulleys and gears so that weights are raised and lowered. If the system undergoes a cyclic process all subsystems must also undergo cyclic processes. The internal energy change of the system, and of each subsystem taken separately must be zero if the process is cyclic. The first law requires that the net work done by each subsystem is then equal to the net heat input to the subsystem in the cycle,

$$W_\sigma = Q_\sigma. \tag{2.14}$$

But the system is adiabatic. So there is no heat transferred to the system in any process. The net heat transferred to the system is the algebraic sum of the heats transferred to all the subsystems. Then

$$\sum W_\sigma = \sum Q_\sigma = 0, \tag{2.15}$$

Therefore no net work can be obtained from a cyclic process occurring in a closed adiabatic system. In order to obtain work from a cycle there must be a transfer of heat to the cycle.

A thermodynamic cycle that produces a net work without the input of heat is called a perpetual motion machine of the first kind. The status of the first law of thermodynamics is such that we can deny the possibility of a perpetual motion machine of the first kind with absolute confidence.

2.3.3 Entropy

The thermodynamic property introduced by the second law is the entropy. We already introduced the entropy in terms of reversible heat transfer and the (not yet defined) thermodynamic temperature in Sect. 1.3.2 (see (1.3)). There we also referred to the introduction of the entropy and its relation to irreversibility as one of the great steps in theoretical physics. We should not then expect that a simple mathematical definition and a single paragraph will provide all we require.

In this section we shall outline the steps that carry us from the original experimental evidence, showing that there are limitations on what can be done, to a succinct mathematical expression of those limitations. The difficulty in the enterprise is in the fact that we are dealing with limitations rather than positive statements. In our presentation here we shall lift out the principle steps in the development, leaving details to the appendix.

Because the limitations of the second law are formulated in terms of thermodynamic cycles our mathematical development must be first carried out in those terms. We will later be able to cast the entropy in statistical terms (see Sect. 9.7.1).

The Verbal Statements. There are two fundamental statements of the second law. These are due, respectively, to Clausius, and to Thomson (Lord Kelvin). They are

I. Clausius *No cyclic process exists which has as its sole effect the transference of heat from a body at a temperature Θ_2 to a body at a temperature Θ_1 if $\Theta_1 > \Theta_2$.*
II. Kelvin *No cyclic process exists which has as its sole effect the transference of heat from a single body and the complete conversion of that into an equivalent amount of thermodynamic work.*

We can show that these statements are logically equivalent by demonstrating that a violation of one implies a violation of the other. Traditionally the argument is based on Carnot's most efficient engine.

Carnot's Theorem. The ideas leading to the formulation of the second law are Carnot's. So we formally introduce the entropy S and the thermodynamic temperature T through what is now known as Carnot's theorem.

Theorem 2.3.1. *(Carnot's Theorem) There exist two functions of state, S and T, where T is a positive function of the empirical temperature alone, such that in any infinitesimal quasistatic change of state in a system the heat transfer is $\delta Q_{rev} = T dS$.*

To make this definition of the entropy complete we must add the statement that the entropy increases in any irreversible process. Proof of this is based on the Clausius inequality, which is the final goal of this section.

Carnot Cycle. Carnot realized that any difference in temperature can be used to obtain work, and that the most efficient cycle possible will operate quasistatically (see Sect. 1.2). Any transfer of heat must then be carried out isothermally and quasistatically.

To see how this can be done we consider the system in Fig. 2.3 but assume that the cylinder is not insulated. Removal of a very small mass from the balance pan will produce an expansion of the gas at constant temperature. This results in work done by the system and heat transfer to the system. Heat can be transferred from the system isothermally by adding weights.

In the cycle we must transfer heat to the system at a high temperature and from the system at a low temperature. Plotting pressure against volume we have the situation shown in Fig. 2.4. We consider that the cycle begins at point (1). Heat is taken in along the isotherm at temperature T_1 and the system volume expands until the system reaches the state (2). Heat is exhausted along the isotherm at T_2 from (3) to (4) with a decrease in volume.

Fig. 2.4 Reversible isothermal heat transfer to and from
a Cycle. Heat is transferred
to the system at T_1 and exhausted from the system at T_2

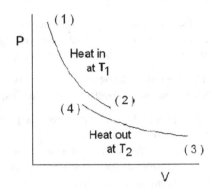

Because a cycle must take the system back to the starting point (1), the two
isotherms in Fig. 2.4 must be connected by two other processes. Since heat must
only be transferred isothermally, these connecting processes cannot transfer any
heat. They must be adiabatic. The completed cycle is that shown in Fig. 2.5.

Fig. 2.5 The Carnot cycle
composed of two isotherms
and two adiabats

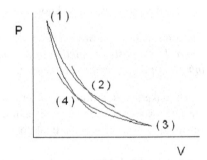

The cycle shown in Fig. 2.5 is then the most efficient that can possibly be constructed to operate between the two heat reservoirs at temperatures T_1 and T_2. In this
cycle

- 1–2 Heat is transferred into the system isothermally while work is done by the
  system.
- 2–3 Work is done by the system during adiabatic expansion.
- 3–4 Heat is transferred from the system isothermally while work is done on the
  system.
- 4–1 Work is done on the system by adiabatic compression.

This is the Carnot Cycle.

We may also represent the Carnot cycle symbolically as shown in Fig. 2.6. The
heat transferred to the cycle at T_1 is Q_1 and that transferred from the cycle at T_2 is
Q_2. The work done is W. From the first law, $W = Q_1 - Q_2$.

The Carnot cycle is an ideal cycle. Each of the isothermal and adiabatic steps
alone can be carried out using a cylinder with a piston. But the combination becomes

Fig. 2.6 The Carnot cycle represented symbolically. The *blocks* indicated by T_1 and T_2 are the high and low temperature reservoirs. Q_1 and Q_2 are heats transferred. W is the work done in the cycle

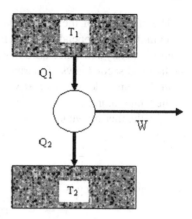

impractical. We would have to insulate the cylinder during parts of the cycle and then remove the insulation for others. The Carnot cycle is primarily a thinking piece for the proof of Carnot's theorem.

We can, however, approximate the Carnot cycle in practice. Water boils producing steam at a constant temperature and the steam condenses again to water at a single lower temperature. And because steam and water pass through them very rapidly, turbines and pumps are approximately adiabatic. A steam power plant then has the characteristics of a Carnot cycle.

Carnot Efficiency. From Fig. 2.6 the thermodynamic efficiency of the Carnot cycle η_C is

$$\eta_C = \frac{W}{Q_1} = 1 - \frac{Q_2}{Q_1}. \tag{2.16}$$

In the appendix we provide an outline of the classical development of the second law. There we show that the ratio of the heats transferred to and from a Carnot cycle is equal to a ratio of two functions each of which depends solely on the temperature of one of the reservoirs. That is

$$\frac{Q_2}{Q_1} = \frac{\tau \,(\text{temperature 2})}{\tau \,(\text{temperature 1})}, \tag{2.17}$$

where τ is a function of the (empirical) temperature alone. These functions of temperature can never be zero, since the rejected heat Q_2 can never vanish by the second law. Thomson (Lord Kelvin) suggested that we use the relationship in (2.17) to define a temperature, which is the *positive function of the empirical temperature alone* referred to in Carnot's theorem. That is $\tau = T$. This is the thermodynamic or absolute (Kelvin) temperature.

The efficiency of the Carnot cycle is independent of the working substance in the cylinder (see Appendix). Therefore the thermodynamic temperature requires no reference to any substance used in its measurement. Its definition depends solely on the first and second laws of thermodynamics.

In the remainder of this text we will always understand T to be the thermodynamic temperature.

With the thermodynamic temperatures T_1 and T_2, (2.17) becomes

$$\frac{Q_2}{Q_1} = \frac{T_2}{T_1},\qquad(2.18)$$

and Carnot efficiency is

$$\eta_C = \frac{T_1 - T_2}{T_1}.\qquad(2.19)$$

Clausius Inequality. We may use the fact that the entropy must increase in a spontaneous process occurring in an isolated system to obtain an integral inequality that elegantly expresses this property. We consider that an isolated system is in a thermodynamic state (1) and that a spontaneous process occurs within the system after which the system is in state (2). We then devise a set of reversible processes that will bring the system back to state (1). The change in the system entropy during the reversible processes bringing it from (2) back to (1) is

$$S_1 - S_2 = \int_2^1 \frac{\delta Q_{\text{rev}}}{T}.$$

Since the original spontaneous process produced a positive entropy change we know that $S_2 > S_1$. Therefore

$$\int_2^1 \frac{\delta Q_{\text{rev}}}{T} < 0.\qquad(2.20)$$

We now consider that the spontaneous process resulting in the initial state change from $(1) \rightarrow (2)$ becomes infinitesimal, but does not vanish. In the limit the states (1) and (2) will then become indistinguishable and the integral in (2.20) becomes an integral around a cycle. That is we have

$$\oint_{\text{cycle}} \frac{\delta Q}{T} < 0\qquad(2.21)$$

if an irreversible process occurs anywhere in the cycle. Because the integral around the cycle now incorporates an infinitesimal irreversible process we designate the infinitesimal heat transfer as δQ rather than δQ_{rev}. If no irreversible process occurs the integral vanishes because $\delta Q_{\text{rev}}/T$ is the differential of the property S. In general, then, the second law requires that

$$\oint \frac{\delta Q}{T} \leqslant 0. \tag{2.22}$$

This is the Clausius inequality. It is a complete mathematical statement of the second law, provide we append the statement that the equality only holds in the absence of any irreversible process in the cycle. In that case $\delta Q \rightarrow \delta Q_{rev}$ and the definition of the entropy as

$$dS = \delta Q_{rev}/T \tag{2.23}$$

emerges from the fact that the integral of dS around a cycle vanishes.

2.4 Combining the Laws

The first and second laws appear to express different aspects of what, with Carnot, we may call the motive power of heat. But there must be a unifying principle. We are speaking of the interaction of matter with energy. The fact that we are employing this in the production of useful work is incidental to the understanding we seek.

The unifying principle appears in (2.23). Although the term δQ_{rev} is path dependent and not itself a differential of a function, the entropy, whose differential is defined in (2.23), is a thermodynamic property. That is we convert the path dependent (problematic) term δQ_{rev} to the differential of a property if we multiply it by $1/T$. If we use the form of δQ_{rev} provided by the first law, we have

$$dS = \frac{1}{T}dU + \frac{P}{T}dV. \tag{2.24}$$

as a differential for the entropy of a closed system. This is equivalent to equating the two definitions of the quantity δQ_{rev} appearing in (2.3) and (2.23).

$$TdS = dU + PdV. \tag{2.25}$$

This was first done by Gibbs [57]. And (2.25) is known as the *Gibbs equation*. The Gibbs equation is the foundational mathematical statement of thermodynamics.

In an address entitled *The Problems of Mathematics* delivered to the *Second International Congress of Mathematicians* in Paris in 1900, David Hilbert[4] defined 23 unsolved problems confronting mathematicians at the beginning of the new (20th) century. Number six among these was to axiomatize all of physics. Because the physics that came out of the 20th century was not the physics Hilbert envisaged, physics will not be axiomatized. Nevertheless, in 1909 Constantin Carathéodory[5]

---

[4] David Hilbert (1862–1943) was a German mathematician. Hibert was professor of mathematics at Göttingen.

[5] Constantin Carathéodory (1873–1950) was born in Berlin of Greek parents, grew up in Brussels, and obtained his Ph.D. at Göttingen. The work referred to here was "Investigations on the Foundations of Thermodynamics" (Untersuchungen über die Grundlagen der Thermodynamik, Math. Ann., 67 p. 355–386 (1909))

was able to establish rigorously that dS in (2.25) is an exact differential. Herbert Callen[6] referred to the analysis of Carathéodory as a tour de force of delicate and formal logic [20]. The importance of this contribution cannot be overstated because of the centrality of thermodynamics as a science.[7]

The laws of thermodynamics are the three discussed in this section and the third law, or Nernst hypothesis, which we shall consider in a later chapter. The zeroth law is a logical statement which clarifies the concept of temperature, but contributes nothing to the structure of thermodynamics. The third law, as we shall see, provides understanding of the behavior of thermodynamic properties as the thermodynamic temperature approaches absolute zero. The third law opened an entire new area of investigation of the behavior of matter at low temperatures and provides absolute values for certain thermodynamic properties critical in physical chemistry. But the third law has no direct consequence for the motive power of heat or for the behavior of matter at moderate or high temperatures. The basis of much of our study and application of thermodynamics is then a combination of the first and second laws. And this combination is presented succinctly, and beautifully, in the Gibbs equation.

The Gibbs equation is a relationship that must hold among infinitesimal changes in entropy, internal energy, and volume of any physical substance. That is it defines a surface in a space in which the axes are U, V, and S. Thermodynamics requires that all possible states accessible to the system must lie on this surface. The geometrical form of this fundamental thermodynamic surface must then hold the key to our understanding of the behavior of matter in its interaction with energy. This is the key to an understanding of thermodynamics.

2.5 Summary

In this chapter we have presented the basic concepts required for a study of thermodynamics. These included systems, thermodynamic properties and states, and the principles or laws of thermodynamics. Each of these concepts is important for our understanding of the power, beauty, and the limitations of the science.

All of thermodynamics is based on studies of systems. When we speak of processes and cycles we must bear in mind that these occur in systems with well defined boundaries. The laws of thermodynamics will provide relations among thermodynamic properties. In any application of thermodynamics, however, it is critical that we understand the identity of the system involved.

The formulation of thermodynamics must include a discussion of the laws. We have, therefore, included a brief, but basically complete discussion of the first three laws of thermodynamics. In this we have chosen to make the laws understandable. We have, however, deviated from an approach based only on experimental evidence.

---

[6] Herbert B. Callen (1919–1993) was professor of physics at the University of Pennsylvania. He is author of *Thermodynamics and an Introduction to Thermostatistics*, one of the most widely cited books in physics.

[7] We have provided an outline of Carathéodory's principle in the appendix.

The claim that energy (or the total mass-energy of relativity) is conserved is presently accepted based on our experience of more than a century since the first law was formulated. But this was not the case for Clausius. We have followed Clausius' path to accepting conservation of energy as the fundamental principle. An alternative, philosophically more positivist approach would be based solely on experiment. The final mathematical statement of the first law is unchanged. In the positivist approach heat transfer is, however, a derived concept defined in terms of the measurable thermodynamic work.

The second law can also be treated in a more positivist framework. Although satisfying, the development in this framework is involved and is not transparent. We have, therefore, elected to present the concepts of the second law without proof based on either direct experimental evidence or on rigorous mathematics. The standard, classical development of the second law is presented in the appendix.

Exercises

2.1. You want to study heating of a moving, nontoxic liquid. You mount a section of pipe on the laboratory table, connect one end of the pipe to a cold source of the liquid, and the other end you exhaust into a large reservoir. Around the pipe you wrap heating wire and insulation. You can measure the power supplied to the heater.

What is your system if you are interested in the temperature rise of the liquid? Classify the system.

2.2. Fuel and oxidant are sprayed into a rocket combustion chamber and burning takes place. Hot gases, which may still be reacting, leave the chamber and enter the nozzle. Assume that the combustion chamber is approximately spherical.

(a) You are interested in the combustion process. Where will you draw your system? Define the boundaries. Classify the system.

(b) If you are interested in the mixing process for the fuel and oxidant, where is the system boundary? Classify the system.

2.3. You want to measure the thermal conductivity of a fluid. To do this you propose to place a platinum wire in the fluid and use the wire as a heat source. An electrical current in the wire will produce the heat and this will be transferred by conduction into the fluid. The wire will then itself be used as a thermometer so that the temperature field around the wire will not be affected by the introduction of a separate measuring device. In the experiment the temperature of the wire will be measured as a function of time.

It is known that the resistance of a platinum wire varies linearly with the temperature. Describe the empirical temperature scale you intend to set up for your experiment.

2.4. You are designing a steam power plant and have decided that it helps your thought pattern to return to Clausius' way of picturing heat transfer and work done. Use arrows for heat and work. In your plant you will have a boiler operating at a temperature of T_1. You may assume that the steam coming out of the boiler has that temperature. You have cooling coils which you can locate either in cooling towers or in the local river. Call the low temperature of the cooling coils T_2.

A turbine does the work, which is then used to drive a generator. The best you can hope for is an efficiency $\eta = W/Q_{in} = (T_1 - T_2)/T_1$.

    (a) Redraw the Clausius cycle indicating clearly the location of the boiler, cooling coils, heat engine.

    (b) The temperature of the river is ΔT (degrees) lower than the cooling coils. You want to increase efficiency. You can either raise the boiler temperature by ΔT or opt out for the river over the towers. Which do you choose to obtain the greatest increase in efficiency?

2.5. The ideal gas is a simple model to use in calculations. We obtained equations for δQ_{rev} in terms of both dU and dH for the ideal gas we also will find that the specific heats are constants. Defining $\gamma = C_P/C_V$, and using the equations for dU and dH in terms of δQ_{rev} show that for the ideal gas $PV^\gamma = $ constant in an adiabatic reversible (isentropic) process.

2.6. Show that the empirical ideal gas temperature T_g, which we can take to be defined by the ideal gas equation of state obtained from a gas thermometer (see Appendix) as $P = RT_g/V$, is identical to the thermodynamic temperature, which is defined in terms of the heats transferred to and from a Carnot engine $Q_{in}/Q_{out} = T_{high}/T_{low}$.

2.7. Assume that the atmosphere is isothermal. If ρ is the air density the mass of a volume $dV = Adh$, where $A = $ area and $dh = $ height, of air is $dm = \rho Adh$. The weight of this mass of air is supported by the difference in pressures $dP = P_2 - P_1$ between the bottom of dV and the top (a distance dh). Find the variation in pressure with altitude of an isothermal atmosphere if air is an ideal gas.

2.8. A helium balloon, which is a sphere of radius r, rises in an isothermal atmosphere. Assume also that the difference in pressure of the helium inside the balloon and that of the atmosphere outside is given by $P_{He} - P_a = \alpha/r$. This is known as the Laplace equation for the excess pressure inside a bubble, where the constant α is proportional to the surface tension. The temperature of the helium will be that of the atmosphere. Assume helium is an ideal gas.

What is the relationship between the balloon radius and the height to which the balloon rises?

2.9. A rubber bag contains m kilograms of sand at a temperature T_0. The sand has a specific heat of
$$C_V = (\alpha T + \beta) \, kJ \, kg^{-1},$$

where α and β are constants. Assume that the rubber bag is adiabatic to a reasonable approximation. The bag of sand is dropped from a height of h m onto the laboratory floor, which is hard epoxy. Assume no energy in any form is transferred to this hard floor. The bag also does not bounce.

What is the rise in temperature of the sand?

2.10. Consider a volume contained in an adiabatic boundary. The volume consists of two separate, but identical parts with an adiabatic boundary between them. One compartment contains n_A mols of gas A at temperature T_A and pressure P_A. The other contains n_B mols of gas B at temperature $T_B > T_A$ and pressure P_B. Let the specific heat of the gas A be $(3/2)R$ and of the gas B be $(5/2)R$. The two compartments are connected through a valve (a stopcock). You can open the valve and the gases will flow freely between compartments. Assume the gases to be ideal so that the molecules behave completely independently of one another.

What is the entropy produced in the process?

2.11. Consider a gas contained in a cylinder with a movable piston. The gas is taken sequentially through constant pressure and constant volume processes. The gas is first in state we shall call A. By placing a particular mass on the top of the piston the pressure is maintained at a pressure P_2 while the volume is increased from $V_1 \rightarrow V_2$ to point B. Then the piston is clamped in place and the pressure is decreased to a value of P_1 to point C. The next step is like the first in that a mass on the piston keeps the pressure constant as the volume is decreased to V_1 at point D. The final step goes back to A with the piston clamped in place.

Plot the cycle on the (P,V) – plane.

(a) Which steps require doing work on the system? In which steps is work done by the system? In which steps is no work done at all?

(b) Is there net work done on or by the system? What is the net work?

(c) In which processes is heat flow into the system? In which is heat flow out of the system?

(d) What is the total change in internal energy in the cycle?

(e) Recalling Carnot's idea that heat engines should work on a cycle, what you have is a cycle that could be used as a heat engine. The engine may be difficult to build in practice. But the cycle is there. Do you need additional information to compute the efficiency of this engine?

2.12. Find the work done by an ideal gas in (a) an isothermal process and (b) an isentropic process. Which is greater between the same initial states and the same final volumes? Indicate this graphically by drawing both processes on a (P,V) plane and showing the work done in each process.

2.13. In the system shown here a n mols of a gas are contained in an adiabatic vessel. The vessel has an internal heating coil which supplies an amount of heat, Q, to the system. The gas temperature increases and the spring is compressed. You make three measurements on the system: the power supplied to the coil

P and the time taken Δt (i.e. you measure the total energy supplied by the battery to the coil) and the compression, *x*, of the spring. Find the increase in temperature of the gas. Assume the gas to be ideal.

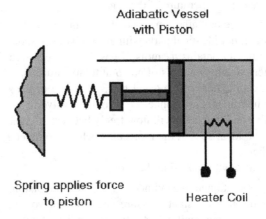

Adiabatic Vessel
with Piston

Spring applies force
to piston

Heater Coil

2.14. Consider the isothermal heat transfer steps in a Carnot engine. The engine is in contact with large isothermal reservoirs during these steps. The working substance you may assume is an ideal gas.

(a) What is the change in entropy for each reservoir?
(b) what is the change in entropy for the gas during a cycle in terms of any change in thermodynamic state variables? In terms of the heat transferred?
(c) What is the total change in entropy of the gas system and the surroundings?

2.15. You have a thermodynamic cycle which, when plotted in the (T, S) plane is a rectangle. The cycle is from *A* (in the upper right hand corner) $\rightarrow B \rightarrow C \rightarrow D \rightarrow A$.

(a) Show that the work done in this cycle is equal to the area enclosed in the (T, S) diagram.
(b) What is the heat transferred to or from the system in each step of the cycle?
(c) What is the entropy change in the cycle?
(d) What is the efficiency of this cycle?

Do you recognize the cycle?

2.16. A nozzle converts thermal energy into kinetic energy. The combustion chamber is to the left of the nozzle shown here. Designate the conditions in the combustion chamber by subscript 1. The thermodynamic state of the gas is known and the velocity is essentially zero. Designate the conditions of the gas at the exit by the subscript 2. The velocity at the exit is $\mathcal{V}$.

(a) Find $\mathscr{V}$ in terms of the temperature difference and the mass flow rate.
(b) What has the nozzle done in energy terms.
(c) explain the concept of thrust of the rocket engine.

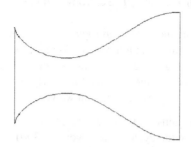

2.17. (a) A $100\,\Omega$ resistor is immersed in an oil bath. The oil bath remains at a constant temperature of $27\,^\circ$C. A current of $100\,$mA flows in the resistor for $50\,$S. Find the entropy change of the resistor and of the oil bath. At the beginning at at the end of the experiment the temperature of the resistor is that of the oil bath.

(b) Instead of placing the resistor in the above problem in an oil bath assume it is thermally insulated. That is you enclose it in an adiabatic boundary. The resistor has a mass of $5\,$g and the material from which it is made has a specific heat at constant pressure of $850\,$J$\,$kg$^{-1}\,$K^{-1}. The other parameters for the experiment are the same. What is the change in entropy of the resistor? You may assume that the experiment is conducted under constant pressure. Assume also that the resistor volume change is negligible during the experiment.

2.18. The vessel shown here is insulated (adiabatic) and the two compartments are separated by a frictionless adiabatic piston. In compartment A is a heater with electrical connections to the outside. The total volume of the vessel is V. Initially the two compartments A and B are filled with equal amounts of the same monatomic ideal gas at the same temperature (T_0) and pressure (P_0). The heater is turned on and the piston moves to the right very slowly until the volume of the right compartment is $V_B < V/2$. What is the heat supplied?

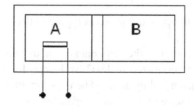

2.19. A metal cylinder is divided into two compartments by a metal disk welded to the walls. One compartment contains a gas under pressure. The other is empty. The empty compartment is 9 tenths of the total volume. Initially the gas in the compartment behind the disk has temperature T_1, pressure P_1, and volume V_1. the disk is puncture remotely and the gas expands to fill the vessel. The final and initial temperatures are the same (that of the room).

(a) The process is irreversible. The entropy will increase. What is the increase in entropy?

(b) We realize that an increase in entropy means we have lost the ability to do work. We decide to put in a movable (frictionless) piston with a rod passing through the evacuated part of the vessel to the outside where we can use the work done. We move the piston slowly so the process is isothermal and quasistatic. How much work can we get?

2.20. A reversible engine gains heat from a single reservoir at 400 K and exchanges heat, at two points in the cycle, to reservoirs at 300 K and at 200 K. During a number of cycles the engine absorbs 1200 J of heat from the reservoir at 400 K and performs 200 J of thermodynamic work.

(a) find the quantities of heat exchanged with the reservoirs at 300 K and at 200 K and decide whether the engine absorbs or exhausts heat at each reservoir.

(b) Find the change in entropy of each reservoir.

(c) What is the change in entropy of the engine plus surroundings?

2.21. You have two identical metal blocks at different temperatures T_1 and T_2 with $T_1 > T_2$. The block mass of each block is m and the specific heat of the metal at constant pressure is C_P. You may assume that there is no change in volume of the blocks. The temperature in the laboratory is $T_0 = (T_1 - T_2)/2$. The temperatures of the two blocks may be made equal in a number of ways. You consider three ways.

(a) Bring the blocks into contact with one another.

(b) Place the blocks in a large constant temperature oil bath (T_0). The oil bath has mass M_{bath} and specific heat $C_{P,bath}$.

(c) Use the two blocks as reservoirs for a Carnot cycle and run the cycle until the temperatures are the same.

What is the final temperature and the entropy produced in each method? You should find a greater increase in entropy when the bath is used as an intermediary. Think about this.

How does the final temperature in (c) compare to the laboratory temperature? Comment on this.

2.22. An experiment to measure the dependence of the internal energy of a gas on volume was first conducted in 1807 by Joseph-Louis Gay-Lussac (1778–1850) and later reproduced by Joule. The experimental apparatus is shown in Fig. 2.7. A glass vessel with two parts separated by a stopcock was immersed in a water bath. One part of the vessel contains the gas of interest and the other is initially evacuated. In the experiment the stopcock was opened and the gas flowed freely into the evacuated part of the vessel. A thermometer in the water bath measures the change in the water temperature resulting from the free expansion of the gas.

Fig. 2.7 Gay-Lussac apparatus

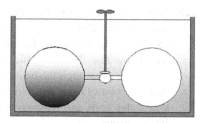

Analyze the experiment based on the first law and relate the temperature measurement to $(\partial U/\partial V)_T$. Do you make any assumptions about the apparatus?

2.23. In a classic experiment designed by Joule and Thomson to measure the dependence of enthalpy on pressure a gas was forced through a porous plug of cotton wool [84, 85]. The core of the experimental apparatus of Joule and Thomson is shown in Fig. 2.8. The pipe was made of beechwood and insulated. Thermometers were located on both sides of the cotton wool. Pressure on each side of the wool plug were maintained by cylindrical reservoirs with weighted covers [[4], p. 138]. The gas flowed very slowly through the cotton wool plug under pressure. The gas velocities on either side of the plug could then be neglected. The dashed line is a system boundary enclosing a constant amount of air.

Using the first law for open systems, show that in this experiment the thermal enthalpy is a constant.

In the experiment pressures and temperatures were measured on both sides of the wool plug. So the experiment measured $(\partial T/\partial P)_H$, which has become known as the Joule-Thomson, or simply Joule coefficient $\mu_J = (\partial T/\partial P)_H$. How to extract $(\partial H/\partial P)_T$ from μ_J is a mathematical question that will be investigated in Chap. 3.

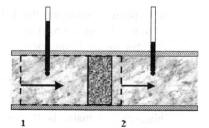

Fig. 2.8 The Joule-Thomson apparatus

1   2

2.24. *Throttling.* Rather than the wool plug of the Joule-Thomson experiment consider a pipe with a constriction, such as a partially opened valve or a long and narrow capillary. The process is known as throttling and has industrial applications [[89], p. 248]. Analyze flow through a throttling valve using the first law. Assume that the incoming fluid is moving slowly. If you neglect the kinetic energy compared to the thermal enthalpy change, you should have the result that throttling is an isenthalpic process.

2.25. You are impressed that the most efficient cycle must transfer heat isothermally and that the other legs of the cycle must be isentropic. The cycle plotted on (S,T) coordinates is a rectangle.

You know that the isentropic assumption may be made when a fluid flows rapidly, because then the heat transfer per mol is small. Turbines do this very well. So you can use a turbine to obtain work isentropically. You also know that water boils and condenses at constant temperatures if the pressures are held constant.

In Fig. 2.9 we have plotted the idea in (S,T) coordinates for water. The saturation line and the phases of water are indicated.

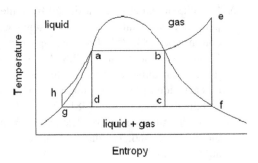

Fig. 2.9 Carnot and Rankine cycles plotted in the (S,T) plane of water. The saturation line is shown

The Carnot cycle is a→b→c→d. The leg b→c is the turbine, and you propose to pump the partially condensed liquid from d→a to enter the boiler. The difficulty becomes clear after some thought. So you elect to heat the steam to a higher temperature before entrance to the turbine and you allow the condensation to run to completion. The result is the cycle a→b→e→f→c→d→g→h. The small vertical leg g→h is the liquid pump, which is also isentropic.

(a) Explain the problems that led to super heating and full condensation.
(b) In which legs are work done and is this by or on the system?
(c) Draw the cycle on a sheet of paper and indicate heat into the cycle and heat exhausted.
(d) What represents the work done?
(e) Where are boiling and condensation?

This cycle is actually the Rankine cycle and forms the basis of power production.

2.26. Conduct a first law analysis of the boiler and the turbine in for the cycle in Exercise 2.25. Use the Eq. (2.8) written for a time dependent process, i.e.

$$d\mathcal{E}/dt = \dot{Q} - \dot{W}_s + \sum_i e_i dn_i/dt + \sum_i h_i dn_i/dt.$$

Here all terms are rate terms. $\dot{Q}$ and $\dot{W}_s$ are rate of heat transfer and rate at which work is done. For steady state processes $d\mathscr{E}/dt = 0$ because the total energy of the system does not change in time. The terms dn_i/dt are mass flow rates at the ports with flow in positive.

2.27. Consider running the cycle in Exercise 2.25 in reverse. That is we replace the turbine with a compressor, condense the fluid at the high temperature and boil it at the low temperature. We no longer use water for this. Describe what you have.

Chapter 3
Mathematical Background

> The miracle of the appropriateness of the
> language of mathematics for the
> formulation of the laws of physics is a
> wonderful gift which we neither
> understand nor deserve.
>
> Eugene Wigner [161]

3.1 Introduction

We concluded Chap. 2 with a combination of the two foundational laws of thermodynamics to produce a single differential equation known as the Gibbs equation. In principle the Gibbs equation may be integrated to obtain a fundamental surface $U = U(S,V)$ for a substance. The natural, or characteristic dependence of the internal energy U on the variables S and V appearing here is a result of the physics of the work/energy relationship of the first law and the heat/entropy relationship of the second law. We have then a single geometrical surface that contains all the information we require, or can possess, about the substance we are studying. And thermodynamics becomes the study of the geometry of this surface. We must only make this study practical in terms of what can be discovered in the laboratory and what is required in application. In this we identify two problems.

1. The function $U = U(S,V)$ is an abstract relationship. We can measure neither energy nor entropy directly in the laboratory. Energy and entropy must be calculated in terms of laboratory variables, which are temperature, pressure, and volume or density. We must then find ways to relate the properties of this fundamental surface to laboratory variables.
2. Although the fundamental surface must exist for each substance, we can only obtain this as a function in closed form for the simplest of substances. Our practical work will involve small variations in one property or another. In essentially all cases we must make approximations and we must deal with tabulated data.

To deal with these problems we must have a basic understanding of the structure of general surfaces. This is a mathematical problem. In this chapter we shall discuss

C.S. Helrich, *Modern Thermodynamics with Statistical Mechanics*,
DOI 10.1007/978-3-540-85418-0_3, © Springer-Verlag Berlin Heidelberg 2009

the basis of the mathematics required. Specifically we shall introduce the concept of
the partial derivative, the integration of a general differential form, and the Legendre
transformation, which will allow us to obtain functions equivalent to $U = U(S,V)$
dependent on more accessible laboratory coordinates.

3.2 Exact Differentials

We begin by considering the variable z as an arbitrary function of x and y, that is
$z = z(x,y)$. This is a surface in (x,y,z) – space such as that illustrated in Fig. 3.1.

Let us consider a differential vector displacement d**C** between the points (0) and
(1) on the surface $z = z(x,y)$. This is illustrated in Fig. 3.2, where dx, dy, and dz
are drawn larger than infinitesimal and d**C** is drawn as a curve for illustration. The
components of the vector d**C** are the displacements dx, dy, and dz along the axes
(x,y,z) as shown in Fig. 3.2. The vector d**C** must remain on the surface $z = z(x,y)$.
Therefore there must be a relationship among the displacements dx, dy, and dz.
For example if we hold y constant and change x by an infinitesimal amount $[dx]_y$ the
result will be a definite infinitesimal change $[dz]_y$ in z, in order to stay on the surface.
We use subscripts to indicate that the variable y is being held constant. In the same
way we obtain a definite change $[dz]_x$ from a change $[dy]_x$ in y holding x constant.
Because $[dx]_y$ and $[dy]_x$ are infinitesimals we may write the general change in z
resulting from changes in both x and y as a linear sum of the two separate changes.

$$dz = [dz]_x + [dz]_y$$

$$= \frac{[dz]_y}{[dx]_y}[dx]_y + \frac{[dz]_x}{[dy]_x}[dy]_x. \tag{3.1}$$

The quantity $[dz]_y / [dx]_y$ is the partial derivative $(\partial z/\partial x)_y$ of z with respect to
x holding y constant. And $[dz]_x / [dy]_x$ is the partial derivative $(\partial z/\partial y)_x$ of z with
respect to y holding x constant. The partial derivatives are the slopes of the two

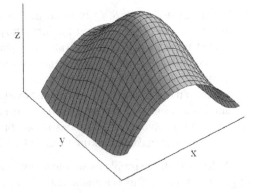

Fig. 3.1 A surface $z = z(x,y)$.
The coordinate z is the vertical
distance above the (x,y)
plane. The surface is defined
by the values of z specified by
$z = z(x,y)$

Fig. 3.2 Graphical illustration of a differential vector dC from (0) to (1) on the surface $z = z(x,y)$

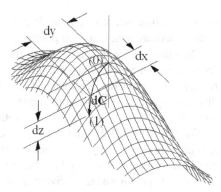

tangent lines to the surface shown in Fig. 3.3. The slopes of these lines depend on the point on the surface at which they are evaluated. Partial derivatives of z are then generally functions of (x,y).

Fig. 3.3 Tangent lines. The *two lines* shown lie in planes perpendicular to the x- and the y-axes. These lines define a plane tangent to the surface $z = z(x,y)$

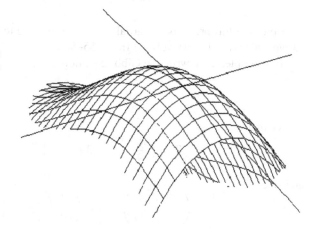

The linear sum in (3.1) is

$$dz = \left(\frac{\partial z}{\partial x}\right)_y dx + \left(\frac{\partial z}{\partial y}\right)_x dy, \tag{3.2}$$

where we drop the subscripts on dx and dy as superfluous. This is called Pfaff's[1] differential form, or a Pfaffian.[2] The Pfaffian is a linear differential form relating

---

[1] Johann Friedrich Pfaff (1765–1825) was one of Germany's most eminent mathematicians during the 19th century. He is noted for his work on partial differential equations of the first order, which became part of the theory of differential forms. He was also Carl Friedrich Gauss's formal research supervisor.

[2] Pfaff's differential form for the function $\Psi(\xi_1,\ldots,\xi_n)$ is defined as

$$d\Psi = \sum_{j}^{n} \left(\frac{\partial \Psi}{\partial \xi_j}\right) d\xi_j.$$

arbitrary differential changes in dx and dy to the resultant differential change dz with the requirement that each point (x, y, z) lies on the surface $z = z(x, y)$.

The subscripts on the partial derivatives in (3.2) would be superfluous if the variables on which z depended were fixed and did not change. This is not true in thermodynamics. For example, the change in internal energy of a system with pressure differs if a process is carried out at constant temperature or at constant volume, and $(\partial U / \partial P)_T \neq (\partial U / \partial P)_V$. So we shall retain the subscripts.

The Pfaffian in (3.2) is an exact differential. That is (3.2) is the differential of a specific function $z = z(x, y)$. A linear differential form that is an exact differential can, in principle, be integrated to obtain the original function.

We may obtain partial derivatives of a function to any order. Second partial derivatives are of particular interest to us because the order of partial differentiation makes no difference. That is for the function $z = z(x, y)$ we have

$$\left(\frac{\partial}{\partial y} \left(\frac{\partial z}{\partial x} \right)_y \right)_x = \left(\frac{\partial}{\partial x} \left(\frac{\partial z}{\partial y} \right)_x \right)_y . \tag{3.3}$$

A proof of this property of partial derivatives may be found in any text on multivariant calculus (cf. [34], volume II, pp. 55–56).

For example, if we write the Gibbs equation (2.25) as

$$dU = T dS - P dV$$

we see that

$$T = \left(\frac{\partial U}{\partial S} \right)_V \text{ and } P = - \left(\frac{\partial U}{\partial V} \right)_S ,$$

and since

$$\left(\frac{\partial}{\partial V} \left(\frac{\partial U}{\partial S} \right)_V \right)_S = \left(\frac{\partial}{\partial S} \left(\frac{\partial U}{\partial V} \right)_S \right)_V ,$$

we have

$$\left(\frac{\partial T}{\partial V} \right)_S = - \left(\frac{\partial P}{\partial S} \right)_V .$$

3.3 Integration

Let us assume that we only have the Pfaffian for z given in (3.2) and we require the function (surface) $z = z(x, y)$. To find $z = z(x, y)$ we must integrate the Pfaffian and this must be done along a contour.

Let us assume that we know, or specify the value of the function $z = z(x, y)$ at some point (x_0, y_0). This point is indicated in the horizontal (x, y) plane below the

Fig. 3.4 Surface $z = z(x,y)$
with contours for integration.
To determine the value of
z at any arbitrary point on
the surface we must move
along the surface. This is the
concept of contour integration

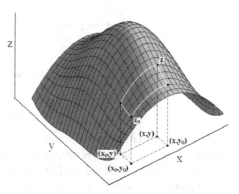

surface $z = z(x,y)$ in Fig. 3.4. The value of z at this point, which we shall call z_0, is
indicated as the point at which the vertical line from (x_0, y_0) intersects the surface.
If we knew the form of the surface $z = (x,y)$ we could move our point on the surface
from z_0 following either of the white contours shown in Fig. 3.4.

This is actually what is done in contour integration. We do not know the surface.
But we do know the tangents to the surface as shown in Fig. 3.3. We can use these
tangents to make certain we stay on the surface as we move from z_0 to z.

If we choose to move along the white contour from (x_0, y_0) to the point (x, y_0)
we hold y constant at the value y_0 and move a distance $dz = (\partial z/\partial x)_y dx$ for each
dx. At the end of the contour we have then increased the elevation in z from z_0 to

$$z_0 + \int_{x=x_0,\ y \text{ held constant at } y_0}^{x} \left(\frac{\partial z}{\partial x}\right)_y dx.$$

Now we do the same thing following the second leg of the contour from the intermediate point above (x, y_0) to the final point above (x, y). We now hold x constant
at its final value and increase our value of z in accordance with $dz = (\partial z/\partial y)_x dy$ as
we move each distance dy. When we get to the final point above (x, y) we have then
a final value of z given by

$$z = z_0 + \int_{x=x_0,\ y \text{ held constant at } y_0}^{x} \left(\frac{\partial z}{\partial x}\right)_y dx$$

$$+ \int_{y=y_0,\ x \text{ held constant}}^{y} \left(\frac{\partial z}{\partial y}\right)_x dy. \tag{3.4}$$

When we recall that

$$\int_{z_0}^{z} dz = z(x,y) - z_0,$$

we realize that what we have is

$$\int_{z_0}^{z} dz = z(x,y) - z_0$$

$$= \int_{x=x_0,\ y\ \text{held constant at}\ y_0}^{x} \left(\frac{\partial z}{\partial x}\right)_y dx$$

$$+ \int_{y=y_0,\ x\ \text{held constant at}\ x}^{y} \left(\frac{\partial z}{\partial y}\right)_x dy,$$

which is the mathematical definition of a contour integral.

Our choice of contours is arbitrary as long as we have a Pfaffian, i.e. an exact differential. For example we could have gone in the y- direction first and taken the second contour. It is usually helpful, however, to draw the contours in the plane of the independent variables as a reminder of the variable to be held constant along each contour.

3.4   Differential Relationships

Much of our reasoning in thermodynamics is done with differential equations. In this section we shall derive certain relationships among partial derivatives that will be indispensable in dealing with differentials.

Reciprocals. The Pfaffian for the function $f = f(x,y)$ is

$$df = \left(\frac{\partial f}{\partial x}\right)_y dx + \left(\frac{\partial f}{\partial y}\right)_x dy, \tag{3.5}$$

If we hold one of the independent variables constant, for example the variable x, then (3.5) becomes

$$[df]_x = \left(\frac{\partial f}{\partial y}\right)_x [dy]_x. \tag{3.6}$$

We introduced the notation $[df]_x$ and $[dy]_x$ in Sect. 3.2. Using the definition of the partial derivative, (3.6) becomes

$$\left(\frac{\partial f}{\partial y}\right)_x = \frac{1}{\left(\frac{\partial y}{\partial f}\right)_x}. \tag{3.7}$$

This is a general relationship for the partial derivatives of functions of two independent variables. It can be extended to functions of an arbitrary number of variables (see Exercises). For a function of the variables (x,y,z) for example

$$\left(\frac{\partial f}{\partial y}\right)_{x,z} = \frac{1}{\left(\frac{\partial y}{\partial f}\right)_{x,z}}. \tag{3.8}$$

Cyclic Permutation. For constant f the Pfaffian in (3.5) has the form

$$df = 0 = \left(\frac{\partial f}{\partial x}\right)_y [dx]_f + \left(\frac{\partial f}{\partial y}\right)_x [dy]_f. \tag{3.9}$$

Using the definition of the partial derivative, (3.9) becomes

$$-\left(\frac{\partial f}{\partial x}\right)_y = \left(\frac{\partial f}{\partial y}\right)_x \left(\frac{\partial y}{\partial x}\right)_f, \tag{3.10}$$

With the help of (3.7) this can be written as

$$\left(\frac{\partial f}{\partial y}\right)_x \left(\frac{\partial x}{\partial f}\right)_y \left(\frac{\partial y}{\partial x}\right)_f = -1. \tag{3.11}$$

This is the *cyclic permutation relationship* among derivatives for functions of two independent variables.

3.5   Variable Transformations

In actual practice the fundamental surface $U = U(S,V)$ is not particularly convenient because the entropy is not a directly measurable property. Representations of thermodynamic properties in terms of (T,V) or (T,P) are more suitable. It is possible to transform $U = U(S,V)$ into equivalent fundamental surfaces, which have characteristic dependencies on other sets of variables. As fundamental surfaces these will each contain the same information as $U = U(S,V)$. This is accomplished by the Legendre transformation.

Because of the central importance of the Legendre transformation to thermodynamics we shall discuss the formulation in some detail. In Sects. 3.5.1 and 3.5.2 below we outline the idea of the Lengendre transformation applied to functions of one variable, which are curves, and functions of two variables, which are surfaces.

3.5.1   From Points to Tangent Lines

A single-valued curve in two dimensions, $y = y(x)$, represents a point relationship between the coordinates x and y. To each point x there corresponds a definite point y. This curve may also be constructed as an envelop of straight lines each of which is tangent to the curve. This is illustrated, geometrically for a limited number of such tangent lines, in Fig. 3.5.

Each straight tangent line can be defined by its slope α and intercept β (with the y-axis). There is then a continuous functional relationship $\beta = \beta(\alpha)$ defined by

Fig. 3.5 *Point and tangent line* representation of a curve. The original *curve* can be represented completely by *tangent lines*

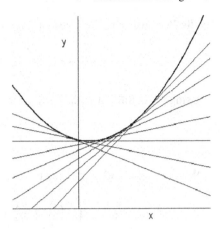

the continuum of points along the curve in Fig. 3.5. The function $\beta = \beta(\alpha)$ is a different, but completely equivalent representation of the relationship $y = y(x)$.

We now make this more definite. We begin with the general equation for the tangent lines, which is

$$y = \alpha x + \beta. \tag{3.12}$$

The slope and intercept are functions of the point through which the tangent line passes, which is $(x, y(x))$ on the original point curve. So we may write (3.12) as

$$\beta = y(x) - \alpha(x)x. \tag{3.13}$$

The differential of (3.13) is

$$d\beta = \frac{dy}{dx}dx - \frac{dy}{dx}dx - xd\alpha = -xd\alpha, \tag{3.14}$$

or

$$\frac{d\beta}{d\alpha} = x(\alpha), \tag{3.15}$$

since $\alpha(x) = dy/dx$ may be inverted to obtain $x = x(\alpha)$. That is β is actually a function only of α. That is the (3.13) is the transformation we sought.

The function β defined by (3.13) is certainly a function only of α. So we must also replace x with $x(\alpha)$ as we did in (3.15). The transform (Legendre transformation) we sought is then

$$\beta = y(x(\alpha)) - \alpha x(\alpha). \tag{3.16}$$

As an example of the Legendre transformation we consider the function

$$y(x) = 3x^2 - x + 5, \tag{3.17}$$

which is that in Fig. 3.5.

Example 3.5.1. Taking the derivative of $y(x)$ in (3.17) we have

$$\frac{dy}{dx} = \alpha(x) = 6x - 1. \tag{3.18}$$

Then

$$x(\alpha) = \frac{\alpha + 1}{6}$$

and Eq. (3.16) for this example is

$$\beta = 3\left(\frac{\alpha + 1}{6}\right)^2 - \frac{\alpha + 1}{6} + 5 - \left(\frac{\alpha + 1}{6}\right)\left[6\left(\frac{\alpha + 1}{6}\right) - 1\right]$$
$$= -\frac{1}{12}(\alpha + 1)^2 + 5. \tag{3.19}$$

We have plotted $\beta = \beta(\alpha)$ from (3.19) in Fig. 3.6. The form of the function $\beta = \beta(\alpha)$ in Fig. 3.6 differs from the form of $y = y(x)$ plotted in Fig. 3.5.

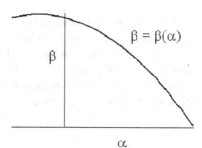

Fig. 3.6 Plot of $\beta = \beta(\alpha)$ from Eq. (3.19)

3.5.2 Surfaces and Contours

For functions of two independent variables the Legendre transformation is similar to that for one. We consider, for example, the arbitrary surface $g = g(x, y)$ in Fig. 3.7.

In Fig. 3.7 the dotted curve on the surface $g = g(x, y)$ is the intersection of the surface with the plane perpendicular to the y-axis and passing through a point $y = y'$. This curve, $g(x, y')$ with y' fixed, can be represented by an infinity of tangent lines as in the preceding section. The slopes of these lines $\alpha(x, y')$ are the partial derivatives $(\partial g/\partial x)_y$ evaluated at $y = y'$. At each value of y the expression corresponding to (3.13) is

$$\beta = g(x, y) - \alpha(x, y)x. \tag{3.20}$$

In (3.20 we have dropped the designation y', since this expression is valid for all points y.

Fig. 3.7 *Slope/intercept representation of a contour on a surface. This contour is similar to the curve in Fig. 3.5. It can be represented by *tangent lines*

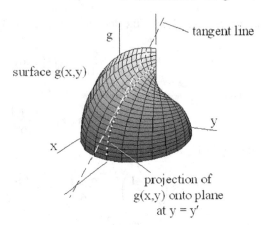

The differential of β in (3.20) is

$$d\beta = (\partial g/\partial y)_x \, dy - x d\alpha. \tag{3.21}$$

Therefore, in this instance of two independent variables, β is a function of y and α. The transformation in (3.20) has then resulted in a function $\beta(\alpha, y)$. Just as in the case of the single independent variable, we must now insert $x = x(\alpha, y)$, obtained from $\alpha(x, y) = (\partial g/\partial x)_y$, into the final expression for β.

As an example of the Legendre transform of a function of two independent variables, we transform the function

$$z(x, y) = x^2 \cos(y) \tag{3.22}$$

removing the dependence on x and replacing that with a dependence on $(\partial z/\partial x)_y$, which we shall continue to designate as α.

Example 3.5.2. The partial derivative of $z = z(x, y)$ in Eq. (3.22) with respect to x is

$$\left(\frac{\partial z}{\partial x}\right)_y = 2x\cos(y) = \alpha(x, y).$$

Then

$$x(\alpha, y) = \frac{\alpha}{2\cos(y)}.$$

Then (3.20) is

$$\beta = z(x, y) - \alpha(x, y)x$$

$$= -\frac{\alpha^2}{4\cos(y)}.$$

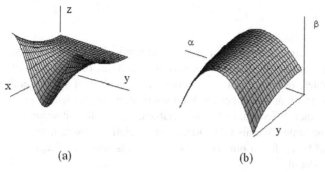

Fig. 3.8 The function $z = x^2 \cos(y)$ **(a)** and its Legendre transform **(b)**

We have plotted the original function and the transformed function of the preceding example in Fig. 3.8 (a) and (b). These plots illustrate the fact that the form of the transformed surface is in general very different from that of the original surface. Nevertheless, the surfaces are completely equivalent.

3.5.3 General Legendre Transformation

We may generalize the Legendre transformation to any number of variables, since the transformation only deals with a single variable and the partial derivative of the original function with respect to that variable. For example we may replace the dependence of a general function, $\Phi(x_1, x_2, \ldots, x_N)$, on the variable, x_j, in favor of a dependence on the partial derivative, $(\partial\Phi/\partial x_j)$. The transformation produces a completely equivalent function $\Psi\left(x_1, \ldots, x_{j-1}, (\partial\Phi/\partial x_j), x_{j+1}, \ldots, x_N\right)$ defined by

$$\Psi = \Phi - (\partial\Phi/\partial x_j)\, x_j. \tag{3.23}$$

We must also eliminate all dependence of Ψ on x_j using

$$x_j = x_j\left(x_1, \ldots, x_{j-1}, (\partial\Phi/\partial x_j), x_{j+1}, \ldots, x_N\right) \tag{3.24}$$

obtained from

$$(\partial\Phi/\partial x_j) = \text{function of } (x_1, x_2, \ldots, x_N) \tag{3.25}$$

The Legendre transform to eliminate the dependence of Φ on x_j in favor of a dependence on $(\partial\Phi/\partial x_j)$ then has two distinct steps.

1. Define the new function Ψ using (3.23).
2. Using (3.25) solve for the variable x_j in terms of $\left(x_1, \ldots, x_{j-1}, (\partial\Phi/\partial x_j), x_{j+1}, \ldots, x_N\right)$ as in (3.24) and substitute this result into (3.23) to obtain Ψ as a function of only $\left(x_1, \ldots, x_{j-1}, (\partial\Phi/\partial x_j), x_{j+1}, \ldots, x_N\right)$.

3.6 Summary

In this chapter we have outlined the most important mathematical concepts required for a study of thermodynamics. We began with partial derivatives of a function of two independent variables, relating these to differential paths along a surface and the exact differential or Pfaffian of a function. We then treated the problem of contour integration of an exact differential of two variables. This is the situation we will most frequently encounter. We also developed some relationships among partial derivatives that will be useful in our work. Finally we discussed the Legendre transformation in some detail.

The space devoted to this transform is indicative of its importance. Although the Legendre transform is only used infrequently, the power and scope of thermodynamics is based on the fact that functions related by a Legendre transform are completely equivalent to one another. The Legendre transformation will form the basis of Chap. 4.

Exercises

3.1. Which of the following are exact differentials?

(a)
$$df = \left(2xy^3 - y^4\right) dx + \left(3x^2y^2 - 4xy^3\right) dy$$

(b)
$$dg = \left(\frac{2y - x\ln y}{x}\right) dx + \left(\frac{y\ln x^2 - x}{y}\right) dy$$

(c)
$$d\phi = \left(\frac{y^2 - x\ln y}{x}\right) dx + \left(\frac{y\ln x - x^2}{y}\right) dy$$

3.2. Integrate the exact differentials in problem 3.1 along contours from $(1,1)$ to an arbitrary point (x,y). Check your result(s) by taking the differential of the final function(s).

3.3. Show that the equality of the second partial derivative of the entropy with respect to U and V to that with respect to V and U yields

$$\left(\frac{\partial}{\partial V}\frac{1}{T}\right)_U = \left(\frac{\partial}{\partial U}\frac{P}{T}\right)_V.$$

This differential relationship must then hold among the variables $P, T, U,$ and V.

(b) Show that this relationship applied to the ideal gas implies that the internal energy of the ideal gas must depend only on temperature.

(c) You are reviewing an article in which the author claims to have discovered the following relationships for a substance

$$U = aPV$$

and

$$PV^2 = bT$$

in which a and b are constants. Do you recommend publication?

3.4. For the function

$$f(x,y) = x^2 y + y^2 x$$

Legendre transform away the y-dependence to obtain $g = g(x,Z)$ where $Z = (\partial f / \partial y)_x$.

If you have access to appropriate software, compare the plots of the transformed and original functions.

3.5. If we are investigating the properties of a system experimentally two parameters are of great importance. These are the isothermal compressibility

$$\kappa_T = -\frac{1}{V}\left(\frac{\partial V}{\partial P}\right)_T$$

and coefficient of volume expansion

$$\beta = \frac{1}{V}\left(\frac{\partial V}{\partial T}\right)_P.$$

The differential volume is then

$$dV = \beta V dT - \kappa_T V dP.$$

(a) For the ideal gas,

$$\beta = \frac{1}{T}$$

and

$$\kappa_T = \frac{1}{P}.$$

Integrate the differential of the volume along an appropriate path in the (T,P) plane to obtain the equation of state for the ideal gas.

(b) Assume a more general case in which you have an empirical expression for β and κ.

$$\beta = \beta_0 + A_\beta T + B_\beta T^2 + L_\beta P$$

and

$$\kappa_T = \kappa_0 + A_\kappa P + B_\kappa P^2 + L_\kappa T.$$

What is the equation of state for this substance?

3.6. Show that the cyclic permutation relationship is valid for the functions

$$f = 5x^3 y^2,$$

$$g = x \ln y,$$

and

$$h = 3y \exp\left(-2x^2\right) + 4xy.$$

3.7. A particle in one dimension in a region with potential energy $V(x)$ has a velocity $\dot{x}$. The Lagrangian function for the particle is

$$L(x, \dot{x}) = \frac{1}{2} m\dot{x}^2 - V(x)$$

Note that the momentum is

$$p = \left(\frac{\partial L}{\partial \dot{x}}\right)_x.$$

Obtain the function $H = H(p, x)$ as the *negative* of the Legendre transform of $L(x, \dot{x})$. This is the Hamiltonian.

3.8. In Exercise 2.20 we encountered the Joule-Thomson experiment and showed that the thermal enthalpy was constant in the experiment. Joule and Thomson measured $\mu_J = (\partial T / \partial P)_H$. How can we extract $(\partial H / \partial P)_T$ from these data? [Hint: use the cyclic permutation relation (3.11).]

Chapter 4
Thermodynamic Potentials

One of the principal objects of
theoretical research in any department of
knowledge is to find the point of view
from which the subject appears in its
greatest simplicity.

Josiah Willard Gibbs

4.1 Introduction

The simplicity Gibbs is calling for in the quotation above he discovered in the mathematical combination of the laws of thermodynamics, which we carried out in Chap. 2. His great contribution to thermodynamics was to show that the consequences of the laws of thermodynamics were then contained in the geometry of a single fundamental surface obtained from this combination. This fundamental surface of thermodynamics is the representation of the internal energy of a system as a function of the entropy and the volume, $U = U(S,V)$. In this chapter we begin our exploration of this discovery and its consequences. We shall use the Legendre transformation to obtain three equivalent surfaces which are functions of (S,P), (T,V) and (T,P). Collectively these are called the thermodynamic potentials.

4.2 The Fundamental Surface

We shall limit our initial treatment to homogeneous systems containing pure substances. Inhomogeneities can only exist under conditions of nonequilibrium, and equilibrium is required to define thermodynamic variables such as temperature and pressure.

We begin our formulation of the fundamental surface with the Gibbs equation (2.24), which we repeat here in the form

C.S. Helrich, *Modern Thermodynamics with Statistical Mechanics*,
DOI 10.1007/978-3-540-85418-0_4, © Springer-Verlag Berlin Heidelberg 2009

$$dU = TdS - PdV \qquad (4.1)$$

for the sake of continuity of the discussion.

As we noted in Sect. 2.4 the Gibbs equation is an exact differential relationship among dU, dS and dV of a universal character. The surface $U = U(S,V)$, which can be obtained from (4.1) (see Sect. 3.3) is then a universal characteristic relationship [cf. [5], p. 98] among U, S and V. The relationships $T = T(S,V)$ and $P = P(S,V)$ required in the integration are unique to each substance. So the surface $U = U(S,V)$ is unique to each substance. This is the fundamental surface for the substance.

Since (4.1) is an exact differential, it is a Pfaffian (see Sect. 3.2). That is

$$dU = \left(\frac{\partial U}{\partial S}\right)_V dS + \left(\frac{\partial U}{\partial V}\right)_S dV. \qquad (4.2)$$

Comparing (4.1) and (4.2) we can identify the thermodynamic temperature and thermodynamic pressure as partial derivatives of the fundamental surface or potential $U = U(S,V)$ with respect to its characteristic variables. Specifically we have

$$T(S,V) = \left(\frac{\partial U}{\partial S}\right)_V \qquad (4.3)$$

and

$$P(S,V) = -\left(\frac{\partial U}{\partial V}\right)_S. \qquad (4.4)$$

Equations (4.3) and (4.4) are considered definitions of the thermodynamic temperature and thermodynamic pressure and are universally valid for any and all substances under conditions of thermodynamic equilibrium regardless of the form of the thermal equation of state for the substance (see Sect. 2.2.4).

We cannot overemphasize the importance of this identification. Max Planck considered (4.3) a crucial foundation of his investigations of the blackbody radiation spectrum that led to the first identification of the quantum of action in 1900 and this relationship is again central in Einstein's paper proposing the photon as a quantum of the electromagnetic wave. Neither Planck nor Einstein would relinquish the foundational importance of this definition [47, 131], and [M. Klein in [160]].

4.3 The Four Potentials

4.3.1 Internal Energy

The designation of the internal energy U as a thermodynamic potential has its origin in the science of mechanics where we speak of the kinetic and potential energies of a moving body. We consider a body moving in the (x,y) plane where there is a conservative force with components $F_x = -(\partial \Phi / \partial x)$ and $F_y = -(\partial \Phi / \partial y)$ obtained from the potential energy $\Phi = \Phi(x,y)$. In mechanics we refer to the forces F_x and F_y as conjugate to the coordinates x and y respectively.

The work done on the body in moving a differential distance with components dx and dy is

$$F_x dx + F_y dy = -\left(\frac{\partial \Phi}{\partial x}\right) dx - \left(\frac{\partial \Phi}{\partial y}\right) dy.$$

The right hand side of this equation is the Pfaffian for $-d\Phi$. That is the work done on the body in the differential distance (dx, dy) is $-d\Phi$. This work increases the kinetic energy of the body at the expense of a loss in potential energy.

We consider the internal energy $U = U(S, V)$ to be the thermodynamic analog of the mechanical potential energy $\Phi = \Phi(x, y)$. In this sense the partial derivatives of the thermodynamic internal energy are termed generalized forces conjugate to the generalized coordinates volume and entropy.

This analogy is intuitive for processes taking place at constant entropy. At constant entropy we have work done on the surroundings at the expense of a loss in internal energy, which from (4.2) and (4.4) is

$$-[dU]_S = -\left(\frac{\partial U}{\partial V}\right)_S dV = PdV. \tag{4.5}$$

The thermodynamic pressure, which is a physical force per unit area, is then the (generalized) force conjugate to the (generalized) coordinate V, since the product of P and dV is the work done on the surroundings during an infinitesimal change in the volume [[89], p. 138].

The reason for using the adjective "generalized" becomes more evident when we consider changes in the entropy. From (4.2) and (4.3) a differential change in entropy at constant volume contributes to a differential change in the internal energy given by

$$-[dU]_V = -\left(\frac{\partial U}{\partial S}\right)_V dS = -TdS. \tag{4.6}$$

In keeping with our analog of dU with $d\Phi$, the terms in (4.5) and (4.6) are both products of generalized forces and generalized displacements. In this sense the thermodynamic temperature is called the generalized force conjugate to the generalized coordinate entropy.

We have then two pairs of conjugate variables (T, S) and (P, V) appearing in the thermodynamic potential $U = U(S, V)$.

As an example we carry out the calculation of the potential $U = U(S, V)$ for the ideal gas. For this case we can do all calculations in closed form.

Example 4.3.1. The ideal gas has the thermal equation of state

$$PV = RT,$$

and constant specific heats. In Chap. 2 we found the molar entropy as a function of (T, V) from a contour integration as

$$S - S_0 = C_V \ln \frac{T}{T_0} + R \ln \frac{V}{V_0},$$

or, choosing $S_0(T_0, V_0) = 0$ and using non-dimensionalized $T = T/T_0$, $V = V/V_0$, and $S = S/R$, we have

$$\exp(S) = (T)^{\frac{1}{(\gamma-1)}} V, \qquad (4.7)$$

with $R - C_P - C_V$ for the ideal gas, and $\gamma = C_P/C_V$. From (4.7) the temperature $T(S,V)$ is

$$T(S,V) = \exp[(\gamma-1)S] V^{(1-\gamma)}, \qquad (4.8)$$

and $P(S,V)$ is

$$P(S,V) = \frac{T}{V} = \exp[(\gamma-1)S] V^{-\gamma}. \qquad (4.9)$$

The Gibbs equation is then

$$dU = \exp[(\gamma-1)S] V^{(1-\gamma)} dS - \exp[(\gamma-1)S] V^{-\gamma} dV.$$

Integrating this along a contour from an initial point (S_1, V_1) to a final point (S,V) we have

$$U(S,V) - U_1 = V_1^{(1-\gamma)} \int_{S_1, V=V_1}^{S} e^{(\gamma-1)S} dS - e^{(\gamma-1)S} \int_{V_1, S}^{V} V^{-\gamma} dV$$

$$= \frac{1}{(\gamma-1)} \exp[(\gamma-1)S] V^{(1-\gamma)} - \frac{1}{(\gamma-1)} V_1^{(1-\gamma)} \exp[(\gamma-1)S_1]$$

The second term on the right hand side depends only on the reference point (S_1, V_1) and may be identified as the term $-U_1$ on the left. Then we have for the fundamental surface

$$U(S,V) = \frac{1}{(\gamma-1)} e^{(\gamma-1)S} V^{(1-\gamma)} \qquad (4.10)$$

in non-dimensional form. From this we may obtain the thermodynamic temperature and pressure as functions of (S,V) by partial differentiation. The results are those we originally used for the contour integration of the Gibbs equation. For completion we carry out this step here.

$$\left(\frac{\partial U(S,V)}{\partial S}\right)_V = e^{(\gamma-1)S} V^{(1-\gamma)} = T(S,V) \qquad (4.11)$$

$$\left(\frac{\partial U(S,V)}{\partial V}\right)_S = -e^{(\gamma-1)S} V^{-\gamma} = -P(S,V). \qquad (4.12)$$

In Fig. 4.1 we have plotted the fundamental surface $U = U(S,V)$ for an ideal gas, which we found in (4.10),

Fig. 4.1 The fundamental surface $U = U(S, V)$ for an ideal gas

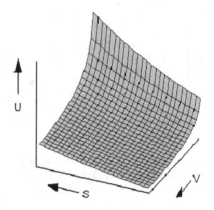

All possible equilibrium states for the substance are points on this fundamental surface. Therefore, any quasistatic reversible process that is physically possible must be representable as a line on this surface. The differential distance between points on the surface is given by the differential form in (4.1). We have illustrated this in Fig. 4.2.

In Fig. 4.2(a) a line of constant temperature (isotherm) is drawn on the fundamental surface. All equilibrium thermodynamic states, points on the fundamental surface, lying on this line have the same temperature.

In Fig. 4.2(b) this isotherm is connected to another isotherm, at a higher temperature, by two lines of constant entropy (isentropes). Figure 4.2(b) is then the cycle proposed by Carnot as the most efficient, which we encountered in Sect. 2.3.3.

The Carnot cycle of Fig. 4.2(b) can be projected either onto the (S, U), the (V, S) or the (V, U) plane to obtain two-dimensional representations of the cycle. This is done in Fig. 4.3. The plots in Fig. 4.3(a), (b) and (c) are not drawn to the same scale

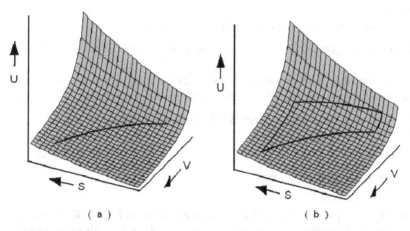

(a)   (b)

Fig. 4.2 The fundamental surface $U = U(S, V)$ for an ideal gas with an isotherm **(a)** and a Carnot cycle **(b)** depicted on the surface

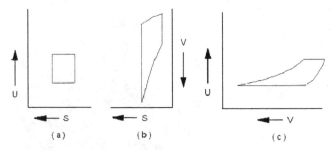

Fig. 4.3 Projections of the Carnot cycle for an ideal gas as working substance. Projection on the (S,U) plane (a), on the (V,S) plane (b) and on the (V,U) plane (c)

as the plot in Fig. 4.2(b). The orientation of the axes is, however, preserved for the sake of consistency. For the ideal gas U is proportional to T.

4.3.2 Transformations

The representation of the fundamental surface as $U = U(S,V)$ is awkward for many applications because neither the entropy nor the internal energy are laboratory variables. Both of these must be calculated. But we know that the function $U = U(S,V)$ can be transformed into a completely equivalent function using a Legendre transformation (see Sect. 3.5.3).

Using the Legendre transformation we may replace the entropy dependence in $U = U(S,V)$ by a dependence on $T = (\partial U/\partial S)_V$ (see (4.3)) or the volume dependence by a dependence on $P = -(\partial U/\partial V)_S$ (see (4.4)). Because of the properties of the Legendre transformation both of the functions obtained will be completely equivalent to $U = U(S,V)$.

The Legendre transformation involves two distinct steps.

1. first obtain the algebraic form of the new function in terms of the original function.
2. then algebraically eliminate the transformed variable from the equation obtained in the first step using the partial derivative defining the new variable.

These steps were carried out in Sect. 3.5.3, (3.23) and (3.24).

4.3.3 The Helmholtz Energy

To obtain the *Helmholtz energy*, which we shall designate as $F = F(T,V)$, we replace the entropy dependence in the internal energy $U = U(S,V)$ with a dependence on the thermodynamic temperature $T = (\partial U/\partial S)_V$. The first step in the Legendre transformation results in the algebraic form of the Helmholtz energy as

$$F = U - (\partial U / \partial S)_V S \qquad (4.13)$$

The variables in (4.13) are still (S,V).

We must now use the definition $T(S,V) = (\partial U / \partial S)_V$ to solve for $S = S(T,V)$ and eliminate S algebraically from the right hand side of (4.13). After this step we will have

$$F(T,V) = U(T,V) - TS(T,V) \qquad (4.14)$$

Equation (4.14) is the Helmholtz energy in terms of its characteristic variables (T,V).

Because of the equivalence of $U = U(S,V)$ and $F = F(T,V)$, the differential of the Helmholtz energy

$$dF = dU - TdS - SdT \qquad (4.15)$$

is equivalent to the Gibbs equation (4.1). Therefore the combination of differential terms $dU - TdS$ in (4.15 is equal to $-PdV$ (see (4.1)). The differential form in (4.15) therefore becomes

$$dF = -PdV - SdT, \qquad (4.16)$$

which is the Pfaffian of the fundamental surface in the coordinates T and V,

$$dF = \left(\frac{\partial F}{\partial V}\right)_T dV + \left(\frac{\partial F}{\partial T}\right)_V dT. \qquad (4.17)$$

The coefficients of the differentials dV and dT in (4.16) and (4.17) are then identical. That is

$$P(T,V) = -\left(\frac{\partial F}{\partial V}\right)_T \qquad (4.18)$$

and

$$S(T,V) = -\left(\frac{\partial F}{\partial T}\right)_V \qquad (4.19)$$

We recognize $P = P(V,T)$ as the thermal equation of state for the substance (see Sect. 2.2.4). Therefore, if we have the Helmholtz energy for a substance in terms of its characteristic variables (T,V) we have the thermal equation of state for that substance. But we cannot obtain the Helmholtz energy from the thermal equation of state. To obtain the Helmholtz energy we must integrate the differential in (4.17) along a contour in the (T,V) plane, which requires knowledge of both partial derivatives in (4.18) and (4.19). That is we need both the thermal equation of state and the entropy as a function of (T,V) for the substance in order to obtain the Helmholtz energy.

As an example we obtain the Helmholtz energy from the fundamental surface $U = U(S,V)$ for an ideal gas. We again use the non-dimensionalized forms introduced in the Example 4.3.1.

Example 4.3.2. From the first step in the Legendre transformation of the internal energy, the Helmholtz energy is

$$F(T,V) = U(T,V) - TS(T,V)$$

With (4.11) the internal energy in (4.10) as a function of (T,V) is

$$U(T,V) = \frac{1}{(\gamma-1)} T,$$

and from (4.11) the entropy as a function of (T,V) is

$$S(T,V) = \frac{1}{(\gamma-1)} \ln TV^{(\gamma-1)}.$$

Then

$$F(T,V) = \frac{1}{\gamma-1}(T - T\ln T) - T\ln V. \tag{4.20}$$

The partial derivatives of $F(T,V)$ are

$$S(T,V) = -\left(\frac{\partial F(T,V)}{\partial T}\right)_V = \frac{1}{\gamma-1}\ln T + \ln V \tag{4.21}$$

and

$$P(T,V) = -\left(\frac{\partial F(T,V)}{\partial V}\right)_T = \frac{T}{V}. \tag{4.22}$$

We see that the second of these is the thermal equation of state.

In Fig. 4.4 we have plotted the Helmholtz energy $F(T,V)$ for the ideal gas, which we found in (4.20). The difference in the geometrical form of the internal energy and the Helmholtz energy are apparent if we compare Fig. 4.1 and 4.4.

The Helmholtz energy is logically equivalent to the internal energy, but is not an analog of the mechanical energy. It is no longer appropriate to use the terminology of generalized forces conjugate to generalized coordinates here, even though the expression PdV still appears in the Pfaffian of $F = F(T,V)$. The terminology

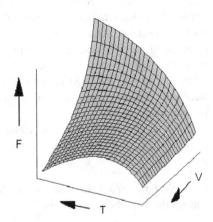

Fig. 4.4 Helmholtz energy
$F = F(T,V)$ for an ideal gas

generalized force is reserved for the terms resulting from the partial derivatives of the internal energy.

Using (4.16), we may write the reversible work done by a closed system as

$$\delta W_{\mathrm{rev}} = -\mathrm{d}F - S\mathrm{d}T, \tag{4.23}$$

For isothermal processes in closed systems we then have

$$\delta W_{\mathrm{rev}} = -[\mathrm{d}F]_T, \tag{4.24}$$

and the decrease in the Helmholtz energy at constant temperature is the reversible work done by a closed system. This is the maximum work that can be obtained from a system at constant temperature. For this reason the the original German designation of the Helmholtz energy was $A(T,V)$, with the A designating *Arbeit*, which is the German word for work.

In the present literature the Helmholtz energy is designated by either $F(T,V)$ or by $A(T,V)$. Aside from the fact that $F(T,V)$ is not used in the German literature, there is no complete uniformity. The reader should simply be aware of the two designations.

Another source of some confusion is the occasional identification of the Helmholtz energy as the Helmholtz *free* energy. The concept of free energy originated with von Helmholtz in the 19th century [68]. This terminology carries with it the burden of the intellectual climate of the 19th century when the concept of energy was not as clear as it is in the 21st century.

In its 1988 meeting the *International Union of Pure and Applied Chemistry* (IUPAC, referred to as the "Eye-You-Pack") supposedly banished the use of the terminology "free energy" replacing this simply with energy. And references to free energy are slowly vanishing from text books and articles. The *IUPAC Compendium of Chemical Terminology, second edition (1997)* carries the definition "Helmholtz energy (function), internal energy minus the product of thermodynamic temperature and entropy. It was formerly called free energy."[112]

The *IUPAC* has also suggested that we revert to the original designation of the Helmholtz energy as A rather than F. We shall use the designation F in this text as the most familiar in the physics literature at this time.

4.3.4 The Enthalpy

A Legendre transformation of $U(S,V)$ to replace the dependence on V by a dependence on $P = -(\partial U/\partial V)_S$ results in the thermodynamic potential *enthalpy* $H = H(S,P)$. We previously encountered the enthalpy in the application of the first law to open systems as a heat flow term resulting from the transport of mass (see Sect. 2.3.2).

The enthalpy has sometimes, perhaps unfortunately, been called the heat function or heat content, which arises from the number of points at which the enthalpy appears to take on the role of heat. For example the heat transferred to a closed system at constant pressure is equal to the change in the enthalpy of the system. Also in thermal chemistry the so-called heat of reaction is identified as the change in enthalpy during the reaction. And, as we discovered in Sect. 2.3.2 the enthalpy appears as a heat carried by the flow of matter.

The Dutch physicist and discoverer of superconductivity, Heike Kamerlingh Onnes,[1] initially introduced the term enthalpy at the first meeting of the Institute of Refrigeration in Paris in 1908. He had derived the name enthalpy from the Greek word "enthalpos" ($\varepsilon\nu\theta\alpha\lambda\pi o\zeta$) [63]. In Dutch the word enthalpy is pronounced with accent on the second syllable (en-THAL-py). This pronunciation easily distinguishes the enthalpy from the entropy in any discourse on the subject. The chemist J.P. Dalton wrote, "... I strongly advocate this pronunciation (en-THAL-py) in English, so as to distinguish it clearly from EN-tro-py. Unfortunately, most chemists accent the first syllable ..." [cited by [76]].

The first step in the Legendre transformation results in the enthalpy as

$$H(S,P) = U - (\partial U/\partial V)_S V. \tag{4.25}$$

Using the definition of thermodynamic pressure in (4.4) to eliminate V this becomes

$$H(S,P) = U(S,P) + PV(S,P) \tag{4.26}$$

Because the differential of the enthalpy

$$dH = dU + PdV + VdP$$

is equivalent to the Gibbs equation (4.1), we may then replace $dU + PdV$ with its equivalent TdS obtaining

$$dH = TdS + VdP \tag{4.27}$$

as the differential of the enthalpy. Since this is a Pfaffian, (4.27) is

$$dH = \left(\frac{\partial H}{\partial S}\right)_P dS + \left(\frac{\partial H}{\partial P}\right)_S dP. \tag{4.28}$$

We may then identify the partial derivatives as the thermodynamic temperature

$$T = \left(\frac{\partial H}{\partial S}\right)_P \tag{4.29}$$

and the volume

$$V = \left(\frac{\partial H}{\partial P}\right)_S. \tag{4.30}$$

---

[1] Heike Kamerlingh Onnes (1853–1926) was professor of physics at the University of Leiden and founder of the cryogenics laboratory there. Onnes discovered superconductivity in 1911 and was awarded the Nobel Prize in Physics in 1913 for his work in cryogenics.

As an example we obtain the enthalpy of an ideal gas from the fundamental surface $U = U(S,V)$.

Example 4.3.3. From the Legendre transformation of the internal energy, the enthalpy is

$$H(S,P) = U(S,P) + PV(S,P).$$

To get $U(S,P)$ we begin with the fundamental surface defined by (4.10), and the volume as a function of entropy and pressure, which from (4.12) is

$$V(S,P) = P^{-\frac{1}{\gamma}} \exp\left[\frac{(\gamma-1)}{\gamma}S\right] \tag{4.31}$$

Then from Eqs. (4.10) and (4.31) we have the internal energy $U = U(S,P)$ as

$$U(S,P) = \frac{1}{(\gamma-1)} \exp\left[\frac{(\gamma-1)}{\gamma}S\right] P^{\frac{(\gamma-1)}{\gamma}}. \tag{4.32}$$

The enthalpy is then

$$H(S,P) = \frac{\gamma}{(\gamma-1)} \exp\left[\frac{(\gamma-1)}{\gamma}S\right] P^{\frac{(\gamma-1)}{\gamma}}. \tag{4.33}$$

The partial derivatives are

$$T(S,P) = \left(\frac{\partial H(S,P)}{\partial S}\right)_P = \exp\left[\frac{(\gamma-1)}{\gamma}S\right] P^{\frac{(\gamma-1)}{\gamma}} \tag{4.34}$$

and

$$V(S,P) = \left(\frac{\partial H(S,P)}{\partial P}\right)_S = \exp\left[\frac{(\gamma-1)}{\gamma}S\right] P^{-\frac{1}{\gamma}} \tag{4.35}$$

In Fig. 4.5 we have plotted the enthalpy $H(S,P)$ for an ideal gas, which we found in (4.33), This can be compared with the plots in Figs. 4.1 and 4.4.

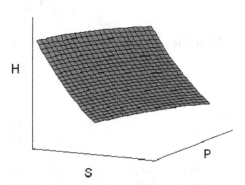

Fig. 4.5 Enthalpy $H = H(S,P)$ for an ideal gas

4.3.5 Gibbs Energy

The *Gibbs energy*, which is a function of (T,P), is the final potential to be obtained by Legendre transformations originating with the fundamental surface $U = U(S,V)$. The Gibbs energy results from either a Legendre transformation replacing the volume in the Helmholtz energy $F = F(T,V)$ by the thermodynamic pressure $P = -(\partial F/\partial V)_T$ or by replacing the entropy in the enthalpy $H = H(S,P)$ by the thermodynamic temperature $T = (\partial H/\partial S)_P$. According to (3.23), the Gibbs energy is then either

$$G(T,P) = F - \left(\frac{\partial F}{\partial V}\right)_T V$$
$$= F + PV. \tag{4.36}$$

or

$$G(T,P) = H - \left(\frac{\partial H}{\partial S}\right)_P S$$
$$= H - TS. \tag{4.37}$$

From (4.13) and (4.26) we see that (4.36) and (4.37) are identical and that the Gibbs energy may be written as

$$G(T,P) = U - TS + PV. \tag{4.38}$$

This form of the Gibbs energy indicates that it is obtained by a second Legendre transformation from the original $U = U(S,V)$.

Using either (4.16) or (4.27) the differential of the Gibbs energy in either (4.37) or (4.38) is

$$dG = -SdT + VdP, \tag{4.39}$$

which is the Pfaffian

$$dG = \left(\frac{\partial G}{\partial T}\right)_P dT + \left(\frac{\partial G}{\partial P}\right)_T dP. \tag{4.40}$$

Identifying the terms in (4.39) and (4.40) we have

$$S = -\left(\frac{\partial G}{\partial T}\right)_P \tag{4.41}$$

and

$$V = \left(\frac{\partial G}{\partial P}\right)_T. \tag{4.42}$$

Because $G = G(T,P)$, (4.42) is the thermal equation of state.

As an example we obtain the Gibbs energy of an ideal gas from a transformation of the enthalpy $H = H(S,P)$.

Example 4.3.4. From a Legendre transformation of the enthalpy, the Gibbs energy is

$$G(T,P) = H(T,P) - TS(T,P).$$

Using (4.34) we have entropy as a function of (T,P) as

$$S(T,P) = \frac{\gamma}{(\gamma-1)} \ln T - \ln P. \tag{4.43}$$

And from (4.33) and (4.43) we have enthalpy as a function of (T,P)

$$\begin{aligned} H(T,P) &= \frac{\gamma}{(\gamma-1)} \exp\left[\frac{(\gamma-1)}{\gamma}S\right] P^{\frac{(\gamma-1)}{\gamma}} \\ &= \frac{\gamma}{(\gamma-1)} T. \end{aligned} \tag{4.44}$$

Then using (4.43) and (4.44) the Gibbs energy is

$$\begin{aligned} G(T,P) &= H(T,P) - TS(T,P) \\ &= \frac{\gamma}{(\gamma-1)} (T - T\ln T) + T\ln P \end{aligned} \tag{4.45}$$

The partial derivatives of $G(T,P)$ then are

$$S(T,P) = -\left(\frac{\partial G(T,P)}{\partial T}\right)_P = \frac{\gamma}{(\gamma-1)} \ln T - \ln P \tag{4.46}$$

and

$$V(T,P) = \left(\frac{\partial G(T,P)}{\partial P}\right)_T = \frac{T}{P} \tag{4.47}$$

In Fig. 4.6 we have plotted the Gibbs energy $G(T,P)$ for an ideal gas, which we found in (4.45), From a comparison of the forms of the surfaces in Figs. 4.1, 4.4, 4.5, and 4.6 we can see how the geometrical form of a simple surface can change in a set of Legendre transformations.

We may relate the Gibbs energy to the possible reversible work that can be done in a process. With (2.1), (2.25) becomes

$$\delta W_{\text{rev}} = T\,dS - dU \tag{4.48}$$

Written in terms of the Gibbs energy, (4.48) is

$$\delta W_{\text{rev}} = -dG - S\,dT + P\,dV + V\,dP. \tag{4.49}$$

Under conditions of constant temperature and constant pressure (4.49) becomes

$$\delta W_{\text{rev}} - P\,dV = -[dG]_{T,P}. \tag{4.50}$$

Fig. 4.6 Gibbs energy
$G = G(T,P)$ for an ideal
gas

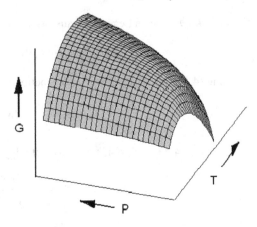

That is the negative change in the Gibbs energy, under conditions of constant thermodynamic temperature and pressure, is the difference between the maximum (reversible) work that can be done and the PdV work performed in the process. This is a similar result to that of (4.24) and is the source of the earlier identification of the Gibbs energy as a *free* energy.

The *IUPAC* has the same comment regarding the terminology for the Gibbs energy as for the Helmholtz energy, "It was formerly called free energy or free enthalpy."

4.4 Relations Among Potentials

In each of the Legendre transformations of the thermodynamic potentials the general term $(\partial \Phi / \partial x_j) x_j$ in (3.23) produced products of the conjugate variables (T,S) and (P,V) respectively (see (4.13), (4.26), (4.37) and (4.36)). This is because the conjugate variables (T,S) and (P,V) are linked as TdS and PdV in the differential of the internal energy and the equivalence of the differential of each potential to the Gibbs equation (4.1) results in a preservation of terms TdS, SdT, PdV or VdP in the forms of the differentials dU, dF, dH, and dG.

Each Legendre transformation of a thermodynamic potential replaces one of the conjugate variables in either of the sets (T,S) or (P,V) with the other variable in that set. We have illustrated the set of possible transformations in Fig. 4.7. Each potential must depend upon one variable from each of the two sets of conjugate variables. And there are only four possible combinations of these variables: (S,V), (S,P), (T,V) and (T,P) as indicated in Fig. 4.7.

We have excluded external fields from consideration. Including an external magnetic field and considering a magnetic material, for example, introduces an additional term $\mathbf{H} \cdot d\mathbf{B}$ into the differential form for the internal energy, where $\mathbf{H}$ and $\mathbf{B}$ are the magnetic field intensity and magnetic flux density (or magnetic induction)

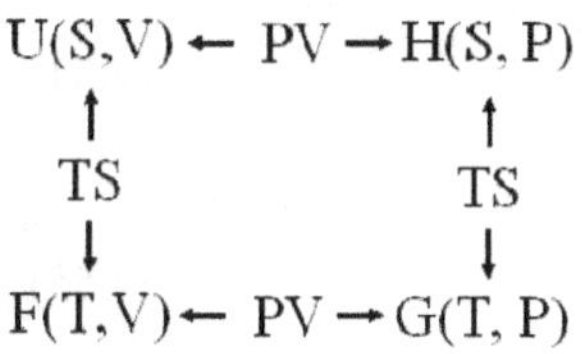

$$U(S,V) \leftarrow PV \rightarrow H(S, P)$$
$$\uparrow \qquad\qquad \uparrow$$
$$TS \qquad\qquad TS$$
$$\downarrow \qquad\qquad \downarrow$$
$$F(T,V) \leftarrow PV \rightarrow G(T, P)$$

Fig. 4.7 Legendre transformations among the thermodynamic potentials. The *arrows* indicate possible transormations of each potential. The products of conjugate variables PV and TS used in each transformation are also indicated

respectively [[100], pp. 73–77]. In a magnetic material the relationship between $\mathbf{H}$ and $\boldsymbol{B}$ may be nonlinear and temperature dependent. Similar considerations result from external electric fields and polarizable materials. These concepts are developed in any text on electromagnetic field theory [cf. [140]].

4.5   Maxwell Relations

The Legendre transformations produce definitions of the thermodynamic variables T, S, P, and V in terms of partial derivatives of the thermodynamic potentials. These are (4.3), (4.4), (4.18), (4.19), (4.30), (4.29), (4.41), and (4.42). Because the order of partial differentiation is immaterial these definitions of thermodynamic variables in terms of partial derivatives of the potentials result in new relationships among the variables.

For example for $U = U(S, V)$,

$$\left(\frac{\partial}{\partial V} \left(\frac{\partial U}{\partial S} \right)_V \right)_S = \left(\frac{\partial}{\partial S} \left(\frac{\partial U}{\partial V} \right)_S \right)_V . \qquad (4.51)$$

But from (4.3) and (4.4) we have

$$\left(\frac{\partial U}{\partial S} \right)_V = T \text{ and } \left(\frac{\partial U}{\partial V} \right)_S = -P.$$

Equation (4.51) then becomes

$$\left(\frac{\partial T}{\partial V} \right)_S = - \left(\frac{\partial P}{\partial S} \right)_V . \qquad (4.52)$$

This is one of what are called the *Maxwell Relations* of thermodynamics.

We obtain the remaining Maxwell Relations in the same way by equating second partial derivatives of the enthalpy, Helmholtz energy and Gibbs energy. These are

$$\left(\frac{\partial T}{\partial P} \right)_S = \left(\frac{\partial V}{\partial S} \right)_P , \qquad (4.53)$$

$$\left(\frac{\partial P}{\partial T}\right)_V = \left(\frac{\partial S}{\partial V}\right)_T, \tag{4.54}$$

and

$$\left(\frac{\partial V}{\partial T}\right)_P = -\left(\frac{\partial S}{\partial P}\right)_T. \tag{4.55}$$

Each of the Maxwell Relations expresses a relationship between the two pairs of conjugate variables (T, S) and (P, V) specified by each of the thermodynamic potentials. We identify the potential which is the source for each Maxwell Relation from the subscripts on the partial derivatives, which are the characteristic variables for that potential. For example, (4.55) comes from the Gibbs energy.

Because $(\partial f/\partial y)_x = [(\partial y/\partial f)_x]^{-1}$ (see Sect. 3.4, (3.7)), we can write (4.52) and (4.55) as

$$\left(\frac{\partial T}{\partial V}\right)_S = -\frac{1}{\left(\frac{\partial S}{\partial P}\right)_V} \text{ and } \left(\frac{\partial T}{\partial V}\right)_P = -\frac{1}{\left(\frac{\partial S}{\partial P}\right)_T} \tag{4.56}$$

and (4.53) and (4.54) as

$$\left(\frac{\partial T}{\partial P}\right)_S = -\frac{1}{\left(\frac{\partial S}{\partial V}\right)_P} \text{ and } \left(\frac{\partial T}{\partial P}\right)_V = -\frac{1}{\left(\frac{\partial S}{\partial V}\right)_T}. \tag{4.57}$$

When we write the Maxwell Relations in the form (4.56) and (4.57) we see that they are sets relations between $(\partial T/\partial V)$ and $(\partial S/\partial P)$ and between $(\partial T/\partial P)$ and $(\partial S/\partial V)$ under various conditions expressed by the subscripts on the partial derivatives. That is the dependencies of (T, S) on (P, V) are such that all four variables appear in each Maxwell relation.

The primary practical use to which we will put the Maxwell relations is to convert some expressions into more understandable forms. We cannot measure the entropy in the laboratory and, therefore, we normally will want to convert partial derivatives involving the entropy into forms involving measurable quantities. The Maxwell relations are often useful in accomplishing this, as the example here shows.

Example 4.5.1. We wish to find the dependence of the enthalpy of an incompressible substance (a liquid or a solid) on thermodynamic pressure while thermodynamic temperature is held constant. We begin with the Pfaffian for the enthalpy $H = H(S, P)$ given by (4.27). In a process occurring at constant T we have

$$\frac{[dH]_T}{[dP]_T} = T\frac{[dS]_T}{[dP]_T} + V,$$

which is, using the definition of partial derivative,

$$\left(\frac{\partial H}{\partial P}\right)_T = T\left(\frac{\partial S}{\partial P}\right)_T + V.$$

Using the Maxwell relation in (4.55), this may be written as

$$\left(\frac{\partial H}{\partial P}\right)_T = -T\left(\frac{\partial V}{\partial T}\right)_P + V.$$

The dependence of the enthalpy of a substance on pressure at constant temperature can then be obtained from mechanical measurements of the volume as a function of temperature at constant pressure.

If the substance is incompressible $(\partial V/\partial T)_P = 0$ and

$$\left(\frac{\partial H}{\partial P}\right)_T = V.$$

Since V is a constant for an incompressible substance, the change in enthalpy with pressure for an incompressible substance is a constant. The enthalpy of a liquid or solid is then a linear function of pressure.

4.6   Intensive and Extensive Variables

The original form of the fundamental surface $U = U(S,V)$ was obtained from a combination of the first and second laws written for a single system containing a specific number of mols of a pure homogeneous substance. Because the substance is pure and homogeneous the relationships obtained will apply to any large or small amount of the substance. The only limitation is that any subdivision we choose will still have a sufficient number of molecules to be treated as a continuum. We may then define a specific molar internal energy, entropy, or volume as as

$$u = \frac{1}{n}U, s = \frac{1}{n}S, \text{ and } v = \frac{1}{n}V$$

where n is the number of mols in the original system. Quantities for which the value depends on the amount of substance present, are termed *extensive* variables or properties.

Each of the three other thermodynamic potentials $H(S,P)$, $F(T,V)$, and $G(T,P)$ is equivalent to $U(S,V)$ and is, therefore, also an extensive property. We may, then, define specific molar enthalpy, Helmholtz energy and Gibbs energy as

$$h = \frac{H}{n}, f = \frac{F}{n}, \text{ and } g = \frac{G}{n}.$$

Partial differentiation of one of the potentials U, H, F, or G with respect to an extensive variable results in a variable that is independent of the number of mols present. That is from (4.3), (4.4), (4.29), and (4.18) we have

$$T = \left(\frac{\partial U}{\partial S}\right)_V = \frac{n}{n}\left(\frac{\partial u}{\partial s}\right)_v = \left(\frac{\partial u}{\partial s}\right)_v$$
$$= \left(\frac{\partial H}{\partial S}\right)_P = \left(\frac{\partial h}{\partial s}\right)_P,$$

and

$$P = -\left(\frac{\partial U}{\partial V}\right)_S = -\left(\frac{\partial u}{\partial v}\right)_S$$
$$= -\left(\frac{\partial F}{\partial V}\right)_T = -\left(\frac{\partial f}{\partial v}\right)_T.$$

The thermodynamic temperature and pressure are uniform throughout an equilibrium system and independent of the amount of substance present in the system. Here we neglect any variation of thermodynamic pressure with elevation, which will be present in very large liquid systems. Variables which are independent of the amount of the substance in the system, are called *intensive* variables.

We note that each of the pairs of conjugate variables (T,S) and (P,V) that link the thermodynamic potentials consists of one extensive and one intensive variable. This is because the product of the variables in each set must be extensive, as are the potentials.

We obtain the intensive variable in each pair by partial differentiation of a thermodynamic potential with respect to the extensive variable of the pair. We see this in (4.3), (4.4), (4.18), and (4.29). Correspondingly we obtain the extensive variable in each pair by partial differentiation of a thermodynamic potential with respect to the intensive variable of the pair. We see this in (4.19), (4.30), (4.41), and (4.42).

4.7 Variable Composition

We have thus far considered only closed systems. This limitation is too severe for many practical considerations. Most engineering systems are open. And all biological systems are open. So almost all applications of thermodynamics involve open systems.

In this section we shall investigate the dependence of the thermodynamic potentials on a variation in the number of mols present in the system. We shall first consider only single substances and then the effects of mixing of chemically distinct substances.

4.7.1 Single Components

If the number of mols in the system is variable the Pfaffian of each potential must include one of the terms

$$\left(\frac{\partial U}{\partial n}\right)_{S,V} dn, \quad \left(\frac{\partial H}{\partial n}\right)_{S,P} dn, \quad \left(\frac{\partial F}{\partial n}\right)_{T,V} dn, \quad \text{and} \quad \left(\frac{\partial G}{\partial n}\right)_{T,P} dn. \tag{4.58}$$

The first three of the partial derivatives appearing in (4.58) present difficulties because in each of these the characteristic dependence of the potential is on one of the extensive quantities S and V. These are proportional to the number of mols present, but are held constant in the differentiation. The term involving the Gibbs energy, $(\partial G/\partial n)_{T,P} dn$, however, presents no such difficulty because the Gibbs energy depends only on intensive properties. For this reason the Gibbs energy is the potential on which the formulation for systems of variable composition is based.

Once we have the Pfaffian for $G(T,P,n)$ the Pfaffians for the other potentials can be obtained directly by writing them in terms of the Gibbs energy and obtaining the differential of the result.

Gibbs Energy. For a single component system with a variable number of mols, n, the Gibbs energy is

$$G(T,P,n) = ng(T,P), \tag{4.59}$$

in which $g(T,P)$ is the specific (molar) Gibbs energy. We define the *chemical potential* for a single substance $\mu = \mu(T,P)$ as

$$\mu(T,P) = \left(\frac{\partial G}{\partial n}\right)_{T,P} \tag{4.60}$$

Because T and P are intensive variables, from (4.59) we have

$$\left(\frac{\partial G}{\partial n}\right)_{T,P} = g(T,P). \tag{4.61}$$

That is the chemical potential for a single substance is the Gibbs energy per mol for the substance.

The Pfaffian for the specific (molar) Gibbs energy is, from (4.39),

$$dg = -sdT + vdP. \tag{4.62}$$

Therefore, from (4.59), (4.60), and (4.62) we have the Pfaffian for $G(T,P,n)$ as

$$\begin{aligned} dG(T,P,n) &= ndg(T,P) + g(T,P)dn \\ &= n(-sdT + vdP) + \mu dn \\ &= -SdT + VdP + \mu dn. \end{aligned} \tag{4.63}$$

Then

$$S = -\left(\frac{\partial G}{\partial T}\right)_{P,n} \tag{4.64}$$

and

$$V = \left(\frac{\partial G}{\partial P}\right)_{T,n} \tag{4.65}$$

for single component systems of variable composition.

Helmholtz Energy. Equation (4.36) provides the Helmholtz energy in terms of the Gibbs energy. Then using (4.63) we have

$$dF(T,V,n) = d[G(T,P,n) - PV]$$
$$= -SdT - PdV + \mu dn. \tag{4.66}$$

By identifying terms in this Pfaffian, we have

$$S = \left(\frac{\partial F}{\partial T}\right)_{V,n},$$

$$P = -\left(\frac{\partial F}{\partial V}\right)_{T,n},$$

and

$$\mu(T,V) = \left(\frac{\partial F}{\partial n}\right)_{T,V} \tag{4.67}$$

Here $\mu(T,V)$, the chemical potential, is still the molar Gibbs energy, but written as a function of (T,V). It is not the specific (molar) Helmholtz energy, which is defined as $f(T,V) = F(T,V,n)/n$.

Internal Energy and Enthalpy. In the same way the reader can show that

$$dU = TdS - PdV + \mu dn \tag{4.68}$$

and

$$dH = TdS + VdP + \mu dn, \tag{4.69}$$

so that

$$T = \left(\frac{\partial U}{\partial S}\right)_{V,n} = \left(\frac{\partial H}{\partial S}\right)_{P,n},$$

$$P = \left(\frac{\partial U}{\partial V}\right)_{S,n},$$

and

$$V = \left(\frac{\partial H}{\partial P}\right)_{S,n}.$$

The chemical potential appears in (4.68) as a function of (S,V),

$$\mu(S,V) = \left(\frac{\partial U}{\partial n}\right)_{S,V} \tag{4.70}$$

and in (4.69) as a function of (S, P),

$$\mu(S,P) = \left(\frac{\partial H}{\partial n}\right)_{S,P} \tag{4.71}$$

Again, these are not the specific molar potentials, u or h.

4.7.2 Mixing

General. Mixing of distinct gases or liquids is an irreversible process. A fluid mixture does not spontaneously unmix. The entropy of a collection of fluids separated from one another by partitions is then different from the entropy of the same collection of fluids if the partitions are removed.

The chemical potentials of each of the components of a system will depend on the exact composition of the system [cf. [123], Chap. 5]. Specifically the interactions among the molecules differ depending on the identity of the interacting partners. This will generally be a function of the composition, which will determine the environment of each molecule. If we neglect these complications due to interactions and consider the system to be made up of ideal gases we can, however, gain insight into an important aspect of the entropy increase in a mixture fluids.

Ideal Gas Mixing Entropy. Because the entropy is a thermodynamic function of state the contribution to the entropy due to mixing cannot depend on the way in which the gases are mixed. For the purposes of the present discussion we shall, however, contrive a method of mixing the gases that will involve only the physical mixing. We will then be able to calculate the mixing contribution to the entropy in a straightforward fashion.

Let us consider a diathermal vessel with r compartments. We assume that the partitions separating the compartments can be removed from outside the vessel without performing any thermodyanmic work or transferring any heat. We fill each of the compartments with a different ideal gas. The amount of each of the gases is determined by the requirement that the pressures in the compartments are all equal. Because the system is diathermal the initial temperature of the gas in each compartment and the final temperature of the mixture are equal to the laboratory temperature.

Because the gases are ideal the molecules of the gases do not interact initially or in the final mixture. After the partitions are removed each gas contributes a separate partial pressure force to the vessel walls

$$P_\lambda = \frac{n_\lambda}{V}RT \tag{4.72}$$

and the total pressure is the sum of these contributions. The total pressure after mixing is then

$$P_{\text{total}} = \sum_\lambda^r \frac{n_\lambda}{V}RT,$$

and is equal to the initial pressure, which is the atmospheric pressure of the laboratory.

Because there is no change in the temperature the difference in entropy due to mixing of the ideal gases is

$$\Delta S_{\text{mix}}]_{\text{ideal gas}} = S_{\text{final}} - S_{\text{initial}}$$

$$= -nR \sum_{\lambda}^{r} \frac{n_\lambda}{n} \ln \left(\frac{\frac{n_\lambda}{V} RT}{P_{\text{total}}} \right)$$

$$= -nR \sum_{\lambda}^{r} \chi_\lambda \ln (\chi_\lambda), \tag{4.73}$$

where

$$\chi_\lambda = \frac{n_\lambda}{n}$$

is the mol fraction of the component λ and n is the total number of mols present.

The final entropy of the mixture is the sum of the individual entropies and the entropy increase due to mixing. This is

$$S = \sum_{\lambda}^{r} n_\lambda s_\lambda + \Delta S_{\text{mix}}]_{\text{ideal gas}}, \tag{4.74}$$

Potentials with Mixing. We may calculate the Gibbs energy for a mixture of ideal gases most easily from the integrated equation

$$G = H - TS. \tag{4.75}$$

Where the entropy now includes the mixing entropy. That is

$$S(T,P,\{n_\lambda\}) = \sum_{\lambda} n_\lambda \left(C_{P,\lambda} \ln \frac{T}{T_{\text{ref}}} - R \ln \frac{\chi_\lambda P}{P_{\text{ref}}} + s_{\text{ref},\lambda} \right) \tag{4.76}$$

The second term on the right hand side of (4.76) is the mixing term at the pressure $P = \sum_\lambda P_\lambda$ from (4.74). The reference state for the entropy is $(T_{\text{ref}}, P_{\text{ref}})$ and the reference molar entropy for the component λ in this state is $s_{\text{ref},\lambda}$.[2] We choose the same reference point for all components of the mixture.

The molar enthalpy of the ideal gas with constant specific heats is

$$h_\lambda (T) = C_{P,\lambda} T.$$

---

[2] The ideal gas does not obey the third law and the ideal gas entropy does not vanish at $T = 0$. We, therefore, choose an arbitrary reference point for entropy.

The Gibbs energy for a mixture of ideal gases is then

$$G(T,P,\{n_\lambda\}) = \sum_\lambda n_\lambda \left[C_{P,\lambda} \left(T - T \ln \frac{T}{T_{\text{ref}}} \right) + RT \ln \frac{P}{P_{\text{ref}}} \right.$$
$$\left. - T s_{\text{ref},\lambda} + RT \ln \frac{n_\lambda}{n} \right]. \qquad (4.77)$$

The Gibbs energy for n mols of a single ideal gas is found by setting $n_\lambda = n$ in (4.77).

Straightforward partial differentiation of (4.77) results in

$$\left(\frac{\partial}{\partial T} G(T,P,\{n_\lambda\}) \right)_{P,\{n_\lambda\}} = -S(T,P,\{n_\lambda\}), \qquad (4.78)$$

$$\left(\frac{\partial}{\partial P} G(T,P,\{n_\lambda\}) \right)_{P,\{n_\lambda\}} = -V(T,P,\{n_\lambda\}) \qquad (4.79)$$

and

$$\mu_\lambda(T,P,\{n_\lambda\}) = \left(\frac{\partial G}{\partial n_\lambda} \right)_{T,P,\,n_\rho \neq n_\lambda}$$
$$= C_{P,\lambda} \left[T - T \ln \frac{T}{T_{\text{ref}}} \right] + RT \ln \frac{P}{P_{\text{ref}}} - T s_{\text{ref},\lambda} + RT \ln \frac{n_\lambda}{n}. \quad (4.80)$$

Equation (4.80) is the chemical potential for the component λ in a mixture of ideal gases. The subscript $n_\rho \neq n_\lambda$ on the partial derivative with respect to n_λ indicates that all mol numbers other than n_λ are held constant.

To obtain (4.80) some care must be taken in differentiating the pressure dependent term in (4.76) because $n = \sum n_\lambda$. We note that

$$\frac{\partial}{\partial n_\lambda} \sum n_\lambda \ln \frac{n_\lambda P}{n P_{\text{ref}}} = \ln \frac{n_\lambda P}{n P_{\text{ref}}} + \frac{n - n_\lambda}{n} - \sum_{\beta \neq \lambda} \frac{n_\beta}{n}$$
$$= \ln \frac{n_\lambda P}{n P_{\text{ref}}}.$$

Using equations (4.78), (4.79), and (4.80) we have

$$dG(T,P,\{n_\lambda\}) = -S(T,P,\{n_\lambda\})\,dT - V(T,P,\{n_\lambda\})\,dP$$
$$+ \sum_\lambda \mu_\lambda(T,P,\{n_\lambda\})\,dn_\lambda \qquad (4.81)$$

Because (4.81) differs from (4.66) only in the replacement of $\mu\,dn$ by $\sum_\lambda \mu_\lambda\,dn_\lambda$, we may immediately write

$$dF(T,V,\{n_\lambda\}) = -S(T,V,\{n_\lambda\})\,dT - P(T,V,\{n_\lambda\})\,dV$$
$$+ \sum_\lambda \mu_\lambda(T,V,\{n_\lambda\})\,dn_\lambda, \qquad (4.82)$$

$$dU\left(S,V,\{n_\lambda\}\right) = T\left(S,V,\{n_\lambda\}\right)dS - P\left(S,V,\{n_\lambda\}\right)dV$$
$$+ \sum_\lambda \mu_\lambda\left(S,V,\{n_\lambda\}\right)dn_\lambda \qquad (4.83)$$

and

$$dH\left(S,P,\{n_\lambda\}\right) = T\left(S,P,\{n_\lambda\}\right)dS + V\left(S,P,\{n_\lambda\}\right)dP$$
$$+ \sum_\lambda \mu_\lambda\left(S,P,\{n_\lambda\}\right)dn_\lambda \qquad (4.84)$$

for the differentials of the potentials for mixtures of ideal gases. In each case μ_λ is the molar chemical potential for the mixture expressed in the characteristic variables for the respective potential.

Gibbs Paradox. Let us consider the application of (4.73) to a particularly simple situation. We partition the diathermal vessel into two volumes containing equal numbers of mols of gas. Then (4.73) yields

$$\Delta S_{\text{mix}}]_{\text{ideal gas}} = nR\ln\left(2\right). \qquad (4.85)$$

This result is completely independent of the manner in which the gases may differ from one another. It is only important that we are able to distinguish the gases from one another and, subsequently, the mixture from the pure gases.

Gibbs was struck by the fact that this numerical result is independent of the individual identities of the gases. If the gases contained in the two compartments are indistinguishable from one another there is no entropy change. This has come to be known as the Gibbs paradox.

There was, for Gibbs, however, no paradox. Gibbs understood the situation quite well as early as 1874. He discussed this in the first paper on heterogeneous equilibrium [80], [57]. To resolve the apparent paradox we need only be clear about the meaning of thermodynamic state in these two separate situations, i.e. for distinguishable and indistinguishable gases.

If we are able to distinguish between the two gases we will be able to separate them again into the two volumes. The process of separation will require a change in the surroundings, such as a lowering of a mass (thermodynamic work) and the change in temperature of a bath (heat transfer). But the separation can be done if we claim to be able to distinguish the gases. And we need only to restore the initial state. Restoration of the initial state means that in each compartment we have the correct initial temperature, pressure, and the correct mol numbers of the distinguishable gases. We can make no pretense at being able to posses more information about the system than this. The change in entropy of the surroundings required to accomplish this separation is equal to the mixing entropy because the original state has been restored.

If we are unable to distinguish between the two gases based on any measurements we are able to make, then the gases in each of the two compartments are identical. There will then be no entropy change on mixing because to return the system to the original state we need only push the partition back into place, which requires no work and no heat transfer.

As Jaynes points out, the resolution of what has been considered by many authors to be a paradox is then based simply on the appropriate understanding of the meaning of thermodynamic state in terms of measurable quantities and what must be done to restore initial states. The reader should be aware that this situation was resolved by Gibbs. But the terminology "Gibbs paradox" is in common usage.

4.7.3 Gibbs-Duhem Equation

The internal energy is the potential which is characteristically dependent on two extensive variables. Therefore if we change the numbers of mols of each component in the system by an infinitesimal amount $dn_\lambda = cn_\lambda$, where c is an infinitesimal number, there will be infinitesimal changes in the total number of mols, internal energy, volume, and entropy of the system given by

$$dn = cn, \tag{4.86}$$
$$dU = cU, \tag{4.87}$$
$$dV = cV, \tag{4.88}$$
$$dS = cS. \tag{4.89}$$

The amount of material cn is of the same composition as that already present. The Pfaffian for the internal energy (4.83) is then

$$cU = \left(\frac{\partial U}{\partial S}\right)_{V,n_\lambda} cS + \left(\frac{\partial U}{\partial V}\right)_{S,n_\lambda} cV + \sum_\lambda \left(\frac{\partial U}{\partial n_\lambda}\right)_{S,V,n_{\beta\neq\lambda}} cn_\lambda. \tag{4.90}$$

Because $c \neq 0$ we may divide by the common factor c, which results in

$$U = \left(\frac{\partial U}{\partial S}\right)_{V,n_\lambda} S + \left(\frac{\partial U}{\partial V}\right)_{S,n_\lambda} V + \sum_\lambda \left(\frac{\partial U}{\partial n_\lambda}\right)_{S,V,n_{\beta\neq\lambda}} n_\lambda. \tag{4.91}$$

Since $T = (\partial U/\partial S)_{V,n_\lambda}$, $P = -(\partial U/\partial V)_{S,n_\lambda}$, and $(\partial U/\partial n_\lambda)_{S,V,n_{\beta\neq\lambda}} = \mu_\lambda$, (4.91) becomes

$$U = TS - PV + \sum_\lambda n_\lambda \mu_\lambda. \tag{4.92}$$

Equation (4.92), which is an integrated form of the Gibbs equation, is referred to as the *Euler equation* because (4.91) is of the form of *Euler's relation* [[34], p. 109]. A *function* $f(x,y,z,\ldots)$ *is homogeneous of degree* h *if it satisfies*

$$f(cx, cy, \ldots) = c^h f(x,y,\ldots). \tag{4.93}$$

If $f(x,y,z,\ldots)$ is differentiable then, differentiating (4.93) with respect to $\mathfrak{c}$, we have

$$\left(\frac{\partial f(\mathfrak{c}x,\mathfrak{c}y,\ldots)}{\partial \mathfrak{c}x}\right)\frac{\mathrm{d}(\mathfrak{c}x)}{\mathrm{d}\mathfrak{c}}+\cdots = h\mathfrak{c}^{h-1}f(x,y,\ldots).$$

If we set $\mathfrak{c}=1$ this is

$$x\left(\frac{\partial f}{\partial x}\right)_{y,z,\ldots}+y\left(\frac{\partial f}{\partial y}\right)_{x,z,\ldots}+\cdots = hf(x,y,\ldots), \qquad (4.94)$$

which is Euler's relation. Comparing (4.94) and (4.91) we see that the internal energy satisfies Euler's relation for a homogeneous function of degree 1.

If we take the differential of (4.92) and use (4.83) we have,

$$V\mathrm{d}P = S\mathrm{d}T + \sum_{\lambda} n_{\lambda}\mathrm{d}\mu_{\lambda}. \qquad (4.95)$$

This is the *Gibbs-Duhem equation*,[3] which is an equation involving differentials of the chemical potentials rather than differentials of the numbers of mols.

4.8   The Gibbs Formulation

The principles outlined in this chapter are the basis for the formulation of thermodynamics first introduced by Gibbs. Geometrical representations did not originate with Gibbs. Pressure, volume, and temperature diagrams were common. But the surface $U(S,V)$ was first introduced by Gibbs in the paper "A Method of Geometrical Representation of the thermodynamic Properties of Substances by Means of Surfaces" read before the Connecticut Academy of Sciences on October 22, 1873. This paper and an earlier one "Graphical Methods in the Thermodynamics of Fluids" were published in the *Transactions of the Connecticut Academy of Sciences, vol ii,* p. 309–342 and 382–404 respectively. Although the *Transactions of the Connecticut Academy of Sciences* was not a widely read journal, these papers came to the attention of Maxwell who was, at that time, director of the Cavendish Laboratory in Cambridge, England. Maxwell immediately recognized the significance of Gibbs' work and added a new chapter into the next edition of his *Theory of Heat* [108] entitled, "Prof. Gibbs' Thermodynamic Model." There he wrote: "Prof. J. Willard Gibbs of Yale College, U.S., to whom we are indebted for a careful examination of the different methods of representing thermodynamic relations by plane diagrams, has introduced an exceedingly valuable method of studying the properties of a substance by means of a surface." Gibbs himself termed the equation $U = U(S,V,\{n\})$, in which $\{n\}$ is a set of mol fractions of components with distinct chemical identities, the *fundamental equation* [[57], p. 86].

---

[3] Some authors reserve this name for the constant temperature and pressure form of this equation, i.e. $\sum_{\lambda} n_{\lambda} d\mu_{\lambda} = 0$.

Gibbs' classic work, *On the Equilibrium of Heterogeneous Substances* was an extension of the method of geometric surfaces to heterogeneous substances. This lengthy and very important paper was published in 1875 and 1878 also in the *Transactions of the Connecticut Academy of Sciences*. It became widely known only after translation into German by Wilhelm Ostwald[4] in 1891 and into French by Henri-Louis Le Chatelier[5] in 1899. In this work Gibbs discussed the transformation the fundamental equation into equivalent forms with functional dependence on (T, V), (T, P), and (S, P). These he designated as ψ, ζ, and χ respectively. He also referred to these functions as *thermodynamic potentials*, a designation which has roots in analytical mechanics, as we have indicated in Sect. 4.3.1. These were obtained using the Legendre transformation, although Gibbs did not discuss the Legendre transformation directly in the paper. Rather he referred to a French publication by M. Massieu for the details. The point is that the graphical-visual approach we have introduced above is actually central to the methods introduced by Gibbs. He did not introduce them as a visual aid, but as the basis of a formulation.

The fact that this 19th century formulation by Gibbs is still, in the 21st century, the mathematical basis of thermodynamics is a tribute to the genius of Gibbs.

4.9 Summary

In this chapter we have obtained the four potentials of thermodynamics. We have shown that each of these is a completely equivalent alternative representation of the fundamental surface of thermodynamics. These potentials, considered as geometrical surfaces, therefore, contain all possible equilibrium states available to the substance. In this text we will only consider transitions from one identifiable equilibrium state to another, whether the transition itself is reversible or irreversible. Our studies of all processes in thermodynamics will then be between states represented on this fundamental surface. If the process we are considering is quasistatic it will be a contour or path on this fundamental surface and we shall use whichever representation is the most convenient for the application.

We have, therefore, obtained the principal object of theoretical research to which Gibbs referred in the quotation at the beginning of this chapter. Throughout the years since Gibbs cast thermodynamics in the form of a study of differentials on surfaces no simpler approach has ever been found.

We concluded the chapter with a development of the Maxwell relationships of thermodynamics and considerations of variable composition and of mixtures.

This chapter is then an almost self-contained discussion of the thermodynamic potentials.

---

[4] Friedrich Wilhelm Ostwald (1853–1932) was a German chemist. He received the Nobel Prize in Chemistry in 1909 for his work on catalysis, chemical equilibria and reaction velocities.

[5] Henry Louis Le Chatelier (1850–1936) was a French chemist.

Exercises

4.1. Consider a closed cavity in a block of metal. Assume the block of metal to be maintained at an arbitrary temperature. This cavity will be filled with electro-magnetic radiation emitted by the electrons in the metal. This was the situation studied Max Planck and Albert Einstein, who proposed the photon based on the entropy of the radiation field.

The radiation inside the cavity has an internal energy given by the Stefan-Boltzmann Law $U = bVT^4$, the pressure of the radiation is $P = \frac{U}{3V}$, and the Euler equation $U = TS - PV$ is a relationship between U and S.

(a) Find the Helmholtz energy for the radiation field.
(b) Find the relationship between temperature and volume in the radiation field if the entropy is constant.
(c) Obtain the isentropic decrease in temperature with volume.
(d) Assuming an isentropic expansion of the universe, obtain a formula for the universe volume in terms of the present volume and the background radiation temperatures now (2.7 K) and later.

Comment on the assumption of an isentropic expansion of the universe.

4.2. Consider a closed system with n compartments. The λth compartment contains n_λ mols of the gas λ at the pressure P_λ. The partitions are removed remotely doing no work. The gases mix isothermally and the final pressure of the system is $P = \sum_\lambda P_\lambda$. How much heat is transferred into or out of the system?

4.3. A substance has an enthalpy $H = C_1 S^2 \ln (P/P_0)$ where S is the entropy and P_0 is a reference pressure.

Find C_V for this substance.

4.4. The Helmholtz energy for a particular substance was found to be $F = AT \exp (\alpha T) - nRT [\ln (V - B) + D/V]$ where A, B, D, and α are constants and n is the number of mols present.

(a) Find the thermal equation of state and comment on any relationship to the ideal gas.
(b) Find the internal energy.

4.5. Show that you can obtain the Gibbs energy directly from the internal energy as $G = U - S(\partial U/\partial S)_V - V(\partial U/\partial V)_S$.

4.6. A certain substance has a fundamental surface $(S - S_0)^4 = AVU^2$, where $A > 0$ is a constant.

(a) Find the Gibbs energy as a function of characteristic variables.
(b) Find the thermal equation of state.
(c) Find C_P.

4.7. A substance has a Gibbs energy $G = nRT \ln(P/P_0) - A(T)P$, where $A(T)$ is a positive valued function of the temperature.

(a) Find the equation of state.
(b) Find $C_P(T,P)$.
(c) Find the entropy.
(d) Find β.
(e) Find κ_T.

Chapter 5
Structure of the Potentials

Our job in physics is to see things simply,
to understand a great many complicated
phenomena in a unified way, in terms of
a few simple principles.

Steven Weinberg

5.1 Introduction

In the preceding chapter we established the four thermodynamic potentials as the
basis of thermodynamics. Using the ideal gas as an example we showed that the
geometric form of the potentials varies considerably. Specifically the curvatures of
the potentials differ one from the other.

In this chapter we shall relate the curvatures to one another and to actual laboratory quantities. Our final goal is to show specifically how the potentials can be
obtained directly from laboratory measurements.

5.1.1 General Curvature Relationships

There are general relationships among the curvatures of surfaces related by Legendre transformation. To establish these we consider a function $f(x,y)$, which we
wish to transform into a function $g(x,z)$ by replacing the variable y with

$$z = (\partial f/\partial y)_x. \tag{5.1}$$

Equation (5.1) defines a new function $z = z(x,y)$, which can be inverted to obtain
$y = y(x,z)$. The Legendre transform is then

$$g(x,z) = f(x,y(x,z)) - y(x,z)z. \tag{5.2}$$

If we take the partial derivative of both sides of (5.2) with respect to the
untransformed variable we have

C.S. Helrich, *Modern Thermodynamics with Statistical Mechanics*,
DOI 10.1007/978-3-540-85418-0_5, © Springer-Verlag Berlin Heidelberg 2009

$$\left(\frac{\partial g}{\partial x}\right)_z = \left(\frac{\partial f}{\partial x}\right)_y . \tag{5.3}$$

Partially differentiating (5.3) with respect to the untransformed variable results in

$$\left(\frac{\partial^2 g}{\partial x^2}\right)_z = \left(\frac{\partial^2 f}{\partial x^2}\right)_y + \left(\frac{\partial^2 f}{\partial y \partial x}\right)\left(\frac{\partial y}{\partial x}\right)_z . \tag{5.4}$$

If we take the partial derivative of both sides of (5.2) with respect to the transformed variable we have

$$\left(\frac{\partial g}{\partial z}\right)_x = -y \tag{5.5}$$

Then

$$\left(\frac{\partial^2 g}{\partial z^2}\right)_x = -\left(\frac{\partial y}{\partial z}\right)_x$$
$$= -\frac{1}{\left(\frac{\partial^2 f}{\partial y^2}\right)_x} . \tag{5.6}$$

Equations (5.4) and (5.6) relate the curvatures of surfaces connected by a Legendre transformation.

5.2 Gibbs-Helmholtz Equations

In this section we shall obtain explicit expressions for the curvatures of the Helmholtz and Gibbs energies along the temperature axis. The path will, lead us through the Gibbs-Helmholtz equations and a general thermodynamic relationship for the difference between the specific heats. We could write down the final relationships almost immediately. Taking the path we have chosen will provide a deeper insight into the potentials and introduce the Gibbs-Helmholtz equations in a readily understandable context.

From Fig. 4.7 we realize that we can obtain U and H from F and G by Legendre transform. That is

$$U(T,V) = F(T,V) - T\left(\frac{\partial F(T,V)}{\partial T}\right)_V , \tag{5.7}$$

and

$$H(T,P) = G(T,P) - T\left(\frac{\partial G(T,P)}{\partial T}\right)_P , \tag{5.8}$$

In (5.7) and (5.8) we have chosen not to introduce the entropy for the (negative) partial derivatives. We realize that (5.7) and (5.8) define the internal energy U and the enthalpy H. However, in choosing not to introduce the entropy in either expression we are forcing the internal energy to remain a function of (T,V), which are the characteristic variables of the Helmholtz energy and the enthalpy to remain a function of (T,P), which are the characteristic variables of the Gibbs energy.

Equations (5.7) and (5.8) may be written as

$$\frac{U(T,V)}{T^2} = -\left(\frac{\partial}{\partial T}\frac{F(T,V)}{T}\right)_V,$$ (5.9)

or

$$\frac{H(T,P)}{T^2} = -\left(\frac{\partial}{\partial T}\frac{G(T,P)}{T}\right)_P.$$ (5.10)

Equations (5.9) and (5.10) are known as the *Gibbs-Helmholtz equations*. The Gibbs-Helmholtz equations are particularly useful because the entropy, which is not measurable, is missing from them. They provide the internal energy and the enthalpy in terms of very practical laboratory coordinates (T,V) and (T,P) respectively.

We argued previously, based on the first law written in terms of dH and dU, that it was natural to consider $U = U(T,V)$ and $H = H(T,P)$ (see Sect. 2.3.2). Our argument there was heuristic and based on some physical intuition. The Gibbs-Helmholtz equations are based solidly on what we know of the potentials and the Legendre transformation. With the Gibbs-Helmholtz equations we have a more logical basis for the specific choices of the laboratory variables in which to express U and H.

The Pfaffians for $U(T,V)$ and $H(T,P)$ are

$$dU = nC_V dT + \left(\frac{\partial U}{\partial V}\right)_T dV,$$ (5.11)

and

$$dH = nC_P dT + \left(\frac{\partial H}{\partial P}\right)_T dP.$$ (5.12)

where we have used (2.12) and (2.13). With (5.11) and (5.12) (4.1) and (4.27) become

$$TdS = nC_V dT + \left[\left(\frac{\partial U}{\partial V}\right)_T + P\right] dV,$$ (5.13)

and

$$TdS = nC_P dT + \left[\left(\frac{\partial H}{\partial P}\right)_T - V\right] dP.$$ (5.14)

These are general expressions from which we may obtain the entropy in terms of (T,V) or (T,P) provided we have determined the specific heats and the dependence of U on V and H on P for the substance of interest.

We may obtain general expressions for the dependence of U on V and H on P using the Maxwell Relations. Under conditions of constant temperature the Pfaffians for dU and dH, (4.1) and (4.27), are

$$[dU]_T = T[dS]_T - P[dV]_T,$$ (5.15)

and

$$[dH]_T = T[dS]_T + V[dP]_T. \tag{5.16}$$

Then, using the Maxwell Relation (4.54), (5.15) becomes

$$\left(\frac{\partial U}{\partial V}\right)_T = T\left(\frac{\partial P}{\partial T}\right)_V - P. \tag{5.17}$$

Similarly, using the Maxwell Relation (4.55), (5.16) becomes

$$\left(\frac{\partial H}{\partial P}\right)_T = -T\left(\frac{\partial V}{\partial T}\right)_P + V. \tag{5.18}$$

Equations (5.17) and (5.18) are completely general and show that the volume dependence of the internal energy and the pressure dependence of the enthalpy can be determined if the thermal equation of state is known.

With (5.17) and (5.18) we may combine (5.13) and (5.14) to produce a universal equation relating dT, dV, and dP,

$$n(C_P - C_V)dT = T\left(\frac{\partial P}{\partial T}\right)_V dV + T\left(\frac{\partial V}{\partial T}\right)_P dP. \tag{5.19}$$

If P is held constant (5.19) becomes

$$n(C_P - C_V) = T\left(\frac{\partial P}{\partial T}\right)_V \left(\frac{\partial V}{\partial T}\right)_P. \tag{5.20}$$

Using the cyclic permutation relation (3.11),

$$\left(\frac{\partial P}{\partial T}\right)_V = -\left(\frac{\partial P}{\partial V}\right)_T \left(\frac{\partial V}{\partial T}\right)_P.$$

Then

$$n(C_P - C_V) = -T\left(\frac{\partial P}{\partial V}\right)_T \left(\frac{\partial V}{\partial T}\right)_P^2. \tag{5.21}$$

For any real substance an isothermal increase in pressure results in a decrease in volume. That is

$$\left(\frac{\partial P}{\partial V}\right)_T < 0. \tag{5.22}$$

Therefore for any real substance $C_P > C_V$ and the difference can be obtained from the thermal equation of state. Or, even if the thermal equation of state is unknown, the partial derivatives in (5.21) are quantities readily measured in the laboratory. The difference in specific heats can then always be found and it is only necessary to measure one of the specific heats in the laboratory.

The specific heats may be obtained directly from the Helmholtz and Gibbs energies. From (5.7) and (5.8) we have

$$C_V(T,V) = -\frac{1}{n}T\left(\frac{\partial^2 F(T,V)}{\partial T^2}\right)_V,$$

(5.23)

and

$$C_P(T,P) = -\frac{1}{n}T\left(\frac{\partial^2 G(T,P)}{\partial T^2}\right)_P.$$

(5.24)

Even though the specific heats are laboratory quantities, (5.23) and (5.24) show that they are defined thermodynamically in terms of the structure (curvatures) of the fundamental surface.

Finally we observe that if the Helmholtz and Gibbs energies are known as functions of their characteristic variables, then the internal energy and enthalpy are known as functions of laboratory coordinates. But the Helmholtz and Gibbs energies cannot be obtained from integration of the Gibbs-Helmholtz equations because these equations contain only single partial derivatives of the Helmholtz and Gibbs energies. the partial integration of the Gibbs-Helmholtz equations can only provide $F(T,V)$ to within an arbitrary function of the volume and $G(T,P)$ to within an arbitrary function of the pressure.

5.3 Curvatures of the Fundamental Surface

5.3.1 Caloric Properties and Curvatures

Because there is no known substance for which the input of heat results in a decrease in temperature, we have $C_V > 0$ and $C_P > 0$. Therefore from (5.23) and (5.24) we find that

$$\left(\frac{\partial^2 F(T,V)}{\partial T^2}\right)_V < 0$$

(5.25)

and

$$\left(\frac{\partial^2 G(T,P)}{\partial T^2}\right)_P < 0.$$

(5.26)

The curvatures along the temperature axes of both the Helmholtz and Gibbs energies are then negative for any real substance. This is illustrated in Figs. 4.4 and 4.6 for the ideal gas.

The specific heats are a measure of the response of a substance to transfer of heat. They are, therefore, referred to as "caloric" quantities. Since we require knowledge of only one specific heat (see (5.21)) and $C_V(T,V)$ can be found from $U = U(T,V)$, traditionally the function $U = U(T,V)$ is referred to as the caloric equation of state [40].

Since the entropy as a function of (T,V) or of (T,P) is obtained from the (negative) partial derivatives of the Helmholtz and Gibbs energies (see (4.19) and (4.41)), some authors also refer to the entropy as a function of (T,V) or of (T,P) as caloric equations of state. The specific heats, as functions of the sets of variables (T,V) and (T,P) are sometimes alone referred to as caloric equations of state. We shall adhere to the *IUPAC* recommendation and refer to $U(T,V)$ as the caloric equation of state and acknowledge $C_V(T,V)$ and $C_P(T,P)$ as caloric quantities.

5.3.2 Mechanical Properties and Curvatures

There are two experimental properties that are strictly mechanical. Their definitions have no reference to heat transfer. They are the *coefficient of volume expansion* β defined as

$$\beta \equiv \frac{1}{V}\left(\frac{\partial V}{\partial T}\right)_P, \tag{5.27}$$

and the *isothermal coefficient of compressibility*, or simply *isothermal compressibility*, κ_T defined as

$$\kappa_T \equiv -\frac{1}{V}\left(\frac{\partial V}{\partial P}\right)_T. \tag{5.28}$$

The coefficient of volume expansion is sometimes referred to as the *volume expansivity* and designated as α. We note that these mechanical properties are completely known if the thermal equation of state is known. For any real substance $(\partial V/\partial P)_T < 0$ (see Sect. 5.2). Therefore $\kappa_T > 0$ for all real substances.

Using (4.42) in (5.27) we have

$$\beta \equiv \frac{1}{V}\left(\frac{\partial^2 G}{\partial T \partial P}\right). \tag{5.29}$$

So the coefficient of volume expansion is known if the Gibbs energy is known as a function of characteristic variables.

If we conduct compressibility studies adiabatically rather than isothermally we step beyond the bounds of the strictly mechanical. We then obtain what is termed the *adiabatic* or *isentropic coefficient of compressibility*, or simply *isentropic compressibility*, κ_S defined as

$$\kappa_S \equiv -\frac{1}{V}\left(\frac{\partial V}{\partial P}\right)_S. \tag{5.30}$$

Then, since

$$\frac{C_P}{C_V} = \frac{\kappa_T}{\kappa_S}, \tag{5.31}$$

(see Exercises) we also know that $\kappa_S > 0$. Because $C_P > C_V$ (see Sect. 5.2) we see from (5.31) that $\kappa_T > \kappa_S$.

With (5.27) and (5.28) we can simplify (5.21) to obtain

$$C_P - C_V = T \frac{\beta^2 v}{\kappa_T},$$ (5.32)

which is a general thermodynamic relationship involving only measurable laboratory properties. Combining (5.31) and (5.32) we have

$$\kappa_T - \kappa_S = T \frac{\beta^2 v}{C_P},$$ (5.33)

which is also a general thermodynamic relationship involving measurable laboratory properties.

Using again (4.42) for the volume as a function of thermodynamic temperature and pressure, we have

$$\kappa_T(T,P) = -\frac{1}{V} \left(\frac{\partial^2 G}{\partial P^2} \right)_T.$$ (5.34)

The curvature of the Gibbs energy along the P axis is then determined by the mechanical property κ_T and is negative. This may be seen for the ideal gas in Fig. 4.6.

5.3.3 Curvatures of the Potentials

The connections among the potentials are provided in Fig. 4.7. There we see that the Helmholtz energy and the enthalpy may each be obtained from the Gibbs energy. From the equations in Sect. 5.1.1 we may then find the curvatures of the Helmholtz energy and the enthalpy from the curvatures of the Gibbs energy.

For example from (5.4) we have, using (5.24) and (5.29),

$$\left(\frac{\partial^2 F}{\partial T^2} \right)_V = \left(\frac{\partial^2 G}{\partial T^2} \right)_P + \left(\frac{\partial^2 G}{\partial P \partial T} \right) \left(\frac{\partial P}{\partial T} \right)_V$$
$$= -\frac{nC_P}{T} + \beta V \left(\frac{\partial P}{\partial T} \right)_V.$$ (5.35)

The partial derivative $(\partial P/\partial T)_V$ can, with the help of (3.11), (5.27), and (5.28), be shown to be

$$\left(\frac{\partial P}{\partial T} \right)_V = \frac{\beta}{\kappa_T}.$$ (5.36)

Then using (5.32) we see that (5.35) becomes (5.23). And from (5.6) we have

$$\left(\frac{\partial^2 F}{\partial V^2} \right)_T = \frac{1}{\left(\frac{\partial^2 G}{\partial P^2} \right)_T} = \frac{1}{V \kappa_T}.$$ (5.37)

We may continue in this fashion to establish all of the second partial derivatives (curvatures) for all of the potentials. The results are

$$\left(\frac{\partial^2 U}{\partial S^2}\right)_V = \frac{T}{nC_V} \quad \text{and} \quad \left(\frac{\partial^2 U}{\partial V^2}\right)_S = \frac{1}{V\kappa_S}, \tag{5.38}$$

$$\left(\frac{\partial^2 H}{\partial S^2}\right)_P = \frac{T}{nC_P} \quad \text{and} \quad \left(\frac{\partial^2 H}{\partial P^2}\right)_S = -V\kappa_S, \tag{5.39}$$

$$\left(\frac{\partial^2 G}{\partial T^2}\right)_P = -\frac{nC_P}{T} \quad \text{and} \quad \left(\frac{\partial^2 G}{\partial P^2}\right)_T = -V\kappa_T. \tag{5.40}$$

and

$$\left(\frac{\partial^2 F}{\partial T^2}\right)_V = -\frac{nC_V}{T} \quad \text{and} \quad \left(\frac{\partial^2 F}{\partial V^2}\right)_T = \frac{1}{V\kappa_T}, \tag{5.41}$$

We have arranged these to correspond to the order shown in Fig. 4.7.

We see that the curvatures are all determined from quantities readily obtained in the laboratory. The curvatures along the T and S axes are all determined by caloric properties and the curvatures along the V and P axes are determined by the compressibilities. Of these κ_S is not strictly mechanical, since its measurement specifies no heat transfer. So there is only a separation into caloric and mechanical curvatures in the case of the Helmholtz and Gibbs energies.

All of these properties of the potentials are consequences of the fact that they are connected by the Legendre transformation. As such they are mathematical results and the reader should not necessarily attach any deeper significance to them at this point in the discussion. The deeper significance will emerge when we discuss thermodynamic stability in later chapter. At this stage we have only the separate requirements of caloric stability that the specific heats are positive, and the requirements of mechanical stability that the compressibilities are positive. We will later link these as general requirements of real substances (see Sect. 12.4).

5.3.4   From Curvatures to Potentials

The curvatures of the potentials are all laboratory measurable quantities. But we cannot obtain functions from knowledge of their second derivatives alone. We must perform a contour integral of a Pfaffian. That is we must have first partial derivatives of the function.

The first partial derivatives of the potentials are equal to (positive or negative values of) T, P, V, or S. We know $T(P,V)$, $P(T,V)$, and $V(T,P)$ if we know the thermal equation of state of the substance. And then we must calculate the entropy S. That is we must have knowledge of the thermal equation of state and of the entropy in order to obtain the potentials.

We may obtain the thermal equation of state from the Pfaffian for V, which is

$$\frac{dV(T,P)}{V(T,P)} = \beta(T,P)\,dT - \kappa_T(T,P)\,dP. \tag{5.42}$$

This results from either identifying the partial derivatives of V as second partial derivatives of the Gibbs energy and using (5.29) and (5.40), or by using the definitions of β and κ_T in (5.27) and (5.28) directly to form dV.

The Pfaffians for $S(T,P)$ and $S(T,V)$ can be obtained from the curvatures of the Gibbs and Helmholtz energies. These are

$$dS = \left(\frac{\partial S}{\partial T}\right)_P dT + \left(\frac{\partial S}{\partial P}\right)_T dP$$

$$= -\left(\frac{\partial^2 G}{\partial T^2}\right)_P dT - \left(\frac{\partial^2 G}{\partial P \partial T}\right) dP \tag{5.43}$$

using (4.41), and

$$dS = \left(\frac{\partial S}{\partial T}\right)_V dT + \left(\frac{\partial S}{\partial V}\right)_T dV$$

$$= -\left(\frac{\partial^2 F}{\partial T^2}\right)_V dT - \left(\frac{\partial^2 F}{\partial V \partial T}\right) dV. \tag{5.44}$$

using (4.19). Then with (5.29) and the first of the equations in (5.40) the Pfaffian in (5.43) becomes

$$dS = nC_P(T,P)\frac{dT}{T} - \beta(T,P)V(T,P)dP. \tag{5.45}$$

With the help of the cyclic permutation relation (3.11), and (5.27) and (5.28), we have

$$\frac{\partial^2 F}{\partial V \partial T} = -\frac{\beta(T,V)}{\kappa_T(T,V)}$$

Then using the first equation in (5.41) the Pfaffian in (5.44) becomes

$$dS = \frac{nC_V(T,V)}{T}dT + \frac{\beta(T,V)}{\kappa_T(T,V)}dV. \tag{5.46}$$

The contour integration of (5.45) and (5.46) will then yield the functions $S(T,P)$ and $S(T,V)$.

We have, therefore, obtained the functions $V(T,P)$ and $S(T,P)$ required for the contour integration of (4.39) and the functions $P(T,V)$ and $S(T,V)$ required for the contour integration of (4.16). To carry out this program we have required only the laboratory quantities C_V, C_P, β, and κ_T. The difficulty in this program is, however, that the reference value for the entropy S_0 contributes a temperature dependence of $T\,S_0$ to the final forms of the Gibbs and Helmholtz energies. That is the part of the temperature dependence of the Gibbs and Helmholtz energies is proportional to the reference value of the entropy. The reference value of the entropy is, therefore, not arbitrary in the same sense that the reference value for the energy was arbitrary. We must, however, specify some value for S_0.

The resolution of this difficulty will be with the third law of thermodynamics. A consequence of that law, although not the law itself, will be that the entropy asymptotically approaches a constant value as the thermodynamic temperature approaches absolute zero. At the suggestion of Max Planck, this constant is chosen is to be zero [132]. We mention this resolution for completeness here.

5.4 Mechanical and Caloric Consistency

The two types of laboratory quantities we have identified, i.e. the caloric (C_P, C_V, and κ_S) and the mechanical (β and κ_T) are not separate. They are linked by the laws of thermodynamics. This is linkage is indicated by (5.31), (5.32) and (5.33). We may also obtain self-consistency relations between mechanical and caloric quantities from the Pfaffians for the entropy in (5.45) and (5.46). Because the order of partial differentiation is immaterial we have $(\partial/\partial P (\partial S/\partial T)_P)_T = (\partial/\partial T (\partial S/\partial P)_T)_P$, which results in

$$\frac{n}{T}\left(\frac{\partial}{\partial P}C_P(T,P)\right)_T = -\left(\frac{\partial}{\partial T}\beta(T,P)V(T,P)\right)_P, \tag{5.47}$$

and $(\partial/\partial V (\partial S/\partial T)_V)_T = (\partial/\partial T (\partial S/\partial V)_T)_V$, which results in

$$\frac{n}{T}\left(\frac{\partial}{\partial V}C_V(T,V)\right)_T = \left(\frac{\partial}{\partial T}\frac{\beta(T,V)}{\kappa_T(T,V)}\right)_V. \tag{5.48}$$

With the help of (5.27) and (5.42) we can partially integrate (5.47) and (5.48) to obtain $C_P(T,P)$ and $C_V(T,V)$, provided we have the values of the specific heats as functions of temperature at reference pressure and volume, which we designate here as P^0 and V^0.

$$C_P(T,P) = C_P\left(T,P^0\right) - \frac{T}{n}\int_{P^0}^{P}\left(\frac{\partial^2 V}{\partial T^2}\right)_P dP. \tag{5.49}$$

$$C_V(T,V) = C_V\left(T,V^0\right) + \frac{T}{n}\int_{V^0}^{P}\left(\frac{\partial^2 P}{\partial T^2}\right)_V dV. \tag{5.50}$$

Because of the universal relationship in (5.32), (5.48) and (5.47) are not independent. There is then only one self consistency relationship for the mechanical and caloric properties of any substance. This is either (5.48) or (5.47). Since β and κ_T together determine the thermal equation of state, this is fundamentally a relationship between the thermal equation of state and the caloric behavior of any substance that must hold if the substance is real.

5.5 Summary

In this chapter we have shown that the curvatures of the potentials are proportional to specific heats and compressibilities, which are quantities that can be measured in the laboratory. We have also been able to find general relationships among specific heats, compressibilities, and the expansivity, which must be valid for the properties of any real substance. This indicates an interdependence of material properties related to the transfer of heat (caloric properties) and those independent of any heat transfer considerations (mechanical properties). This interdependence is reflected in the fact that to determine one of the thermodynamic potentials, such as the Gibbs energy, requires knowledge of the thermal equation of state and of the entropy for the substance.

We have ended the chapter by outlining a path that leads, in principle, directly from laboratory measurable quantities to the potentials. And we have obtained equations that must hold if the laboratory quantities are actually self-consistent. That is we have laid the groundwork for Chap. 6 on laboratory measurements.

Exercises

5.1. Using (5.45) and (5.46) show that $\kappa_T/\kappa_S = C_P/C_V$.

5.2. Consider a solid material for which $1/\kappa_T \approx [\varepsilon/(2V_0)]\left[(2\Gamma C_V T/\varepsilon)(V_0/V) - 3(V_0/V)^3\right]$ and $\beta \approx (1/T)\left(1 + 3(\varepsilon/(2\Gamma C_V T))(V_0/V)^2\right)$, where ε is a constant with the units of energy, Γ is a dimensionless constant and V_0 is a reference volume less than V [103]. The temperature range is such that we may assume that the specific heat at constant volume C_V is independent of temperature.

Find the thermal equation of state.

5.3. Find $U(T,V)$ for the solid in Exercise 5.2.

5.4. For the solid in (5.2)–(5.3) we have the approximate relationship $\beta/\kappa_T = \Gamma C_V/V + O\left((V_0/V)^5\right) \approx \Gamma C_V/V$. Use this to obtain the entropy and the Helmholtz energy for the solid.

5.5. From $S(T,V)$ found in Exercise 5.4 obtain $T(S,V)$ and then U in characteristic variables (S,V).

5.6. Use (5.32) and the expressions for β and $1/\kappa_T$ in Exercise 5.2 to obtain C_P.

5.7. Is the approximation $\beta/\kappa_T = \Gamma C_V/V + O\left((V_0/V)^5\right) \approx \Gamma C_V/V$ in Exercise 5.4 self consistent?

5.8. The speed of sound in a medium is $u^2 = (\partial P/\partial\rho)_S$. Using the thermal equation of state in the form $\rho = \rho(T,P)$, where $\rho = 1/v$, obtain the speed of sound as $u^2 = \left[(\partial\rho/\partial P)_T - (\kappa_S/\kappa_T)(1/\rho^2)(T/C_V)(\partial\rho/\partial T)_P^2\right]^{-1}$

5.9. The speed of sound in a medium is $u^2 = (\partial P/\partial\rho)_S$. Using the thermal equation of state in the form $P = P(T,\rho)$, where $\rho = 1/v$, obtain the speed of sound as $u^2 = (\partial P/\partial\rho)_T + (n/\rho^2)[T/(nC_V)](\partial P/\partial T)_\rho^2$.

5.10. Using the result of either (5.9) or (5.10), find the speed of sound in an ideal gas.

5.11. Prove that any substance for which $U = U(T)$ alone and $H = H(T)$ alone must have the thermal equation of state $PV = $ constant $\times T$.

5.12. We obtained mechanical and caloric consistency expressions (5.47) and (5.48). We may also consider self consistency relations for a hypothetical substance based on the Gibbs equation.

    (a) Show that equality of cross partial derivatives in the Gibbs equation demands that $[\partial (1/T)/\partial V]_U = [\partial (P/T)/\partial U]_V$ for any substance.

    (b) Is a substance satisfying $U = C_1 PV$ and $PV^2 = C_2 T$ possible?

    (c) Is the substance satisfying $U = C_1 PV^2$ and $PV^2 = C_2 T$ possible?

5.13. For the possible substance in Exercise 5.12

    (a) find the entropy

    (b) find the Helmholtz energy

    (c) find the thermal equation of state from the Helmholtz energy.
       [Hint: consider (5.17) and (5.18)]

5.14. (a) Find β for an ideal gas.

    (b) Find κ_T for an ideal gas.

5.15. For a particular substance $\kappa_T = aT^3/P^2$ and $\beta = bT^2/P$ where a and b are constants.

    (a) What is the relationship between a and b?

    (b) Find the thermal equation of state.

Chapter 6
Laboratory Measurements

Everything should be made as simple as possible, but not simpler.

Albert Einstein

6.1 Introduction

In the preceding chapter we identified the thermodynamic variables that must be determined experimentally in order to obtain the thermodynamic potentials for matter in bulk. These are the specific heats, the expansivity, and the compressibilities. We have also explicitly assumed that measurements of temperature, volume (or density), and pressure are easily accomplished. In this chapter we shall discuss some of the actual laboratory measurements that form the basis of our present practical knowledge of the thermodynamic behavior of matter.

The types of measurements that must be made, as well as the instruments designed for these measurements, are constantly evolving. For basic reasons in our investigation of the behavior of matter and because of our technological developments we now require knowledge of these variables under extreme conditions that were beyond our capabilities short decades ago. Our requirements result in new developments in technique, which often indicate the necessity of further investigation of fundamental properties. This requires creativity and ingenuity and a depth of understanding of the physics.

Because of the breadth of the problems encountered in obtaining accurate measurements of thermodynamic variables, we shall only discuss some of the basic issues and techniques. Much of our treatment discussion is based on the reviews in the book *Measurement of the Thermodynamic Properties of Single Phases, Experimental Thermodynamics Volume VI*, edited by A.R.H. Goodwin, K.N. Marsh, and W.A. Wakeham [60]. This is volume VI of the Experimental Thermodynamics Series of the International Union of Pure and Applied Chemistry (IUPAC), Physical Chemistry Division, Commission on Thermodynamics. We shall limit our discussion primarily to gases and liquids.

C.S. Helrich, *Modern Thermodynamics with Statistical Mechanics*,
DOI 10.1007/978-3-540-85418-0_6, © Springer-Verlag Berlin Heidelberg 2009

Our treatment is organized into sections on (1) temperature, (2) pressure, (3) density, (4) speed of sound, and (5) calorimetry. Measurements of the speed of sound are of great importance for determining the thermodynamic properties of substances. We, therefore, present a separate section dealing with this alone.

In no way is this treatment intended to be exhaustive. It is included in the text in order to familiarize the reader with some of the modern techniques being used at the time of this writing.

6.2   Temperature

6.2.1   International Scale

The *International Temperature Scale*, ITS, is defined by the *Comité International des Poids et Measure* (CIPM). This is regularly revised and updated in a cycle of approximate 20 years. At the time of this writing the most recent revision is known as ITS-90. These were previously known as *International Practical Temperature Scales* (IPTS). The previous version was IPTS-68 of 1968. There have been noted deviations between ITS-90 and IPTS-68 in the m K regime.

It is the intention of ITS-90 and all predecessors to define measurement techniques over ranges of the thermodynamic temperature scale that may be used in the laboratory to perform measurements of thermodynamic temperature. Precision is the goal of ITS-90. Accuracy, however, cannot be guaranteed.

The thermodynamic scale is defined by Carnot's Theorem 2.3.1 and direct measurement in terms of the definition is impractical. Recent measurements indicate that near 300 K ITS-90 differs from the thermodynamic scale by about 5 mK.

The fact that the ITS is updated implies that the accuracy with which temperatures are known at specific points, not defined by international agreement, may vary. For example IPTS-68 determined the normal boiling point of water to be 373.15 K (at $P = 0.101325$ MPa) while ITS-90 has found this to be 373.124 K. The difference is 26 mK. The fixed point is the triple point of water and is defined as 273.16 K.

ITS-90 covers a temperature range of between 0.65 K up to the highest temperatures that can be determined practically by pyrometric measurements. Laser heating (Nd:YAG laser) may result in temperatures as high as 5,000–10,000 K [[60], p. 498]. The thermometers specified by ITS-90 for specific temperature ranges are indicated in Table 6.1.

6.2.2   Fixed Points

The fixed points for ITS-90 are provided by Goodwin et al. [[60], p. 11]. These are phase transformation points (liquid to gas (l+g) and solid to liquid (s+l)) and triple

Table 6.1 ITS-90 temperature ranges and thermometers specifies for those ranges. Interpolation formulae are defined within ranges. Reprinted from Experimental Thermodynamics, Vol 6/J.V. Nicholas and D.R. White, Temperature, Fig. 2.1, 2003, with permission from Elsevier

Temperature range (K)	Thermometer
0.65–5	He Vapor Pressure
3–25	Gas
14–1234.93	Pt Resistance
1234.93–	Radiation

points (s+l+g). More detailed and practical information on the use of ITS-90 can be found in the English version of the official text *Supplementary Information for the International Temperature Scale of 1990* [19].

The primary use for the He vapor pressure and the radiation thermometers is to transfer the scale to more convenient measurement instruments. Standard platinum resistance thermometers (SPRTs) can be used directly or for the calibration of more readily available instruments. Thermometers constructed and calibrated to ITS-90 specifications should be traceable.

The triple point of water requires Standard Mean Ocean water (SMOW) with composition given in Table 6.2. Any variation from SMOW will result in a different value for the temperature. For example light water ($^1H_2$ ^{16}O) has a triple point temperature of 273.1587 K, which is approximately 1.3 mK below that of SMOW [[60], p. 15]. Triple point cells are available commercially and manufacturer's instructions should be carefully followed. For example, because of the need to relax the strain in the ice after freezing, the cell must be maintained for at least 14 d for equilibration to assure the maximum accuracy. After equilibration corrections must then be made for hydrostatic pressure variation vertically in the cell.

For simpler calibration with accuracies of about 5 mK an ice bath (distilled water) can be used. The ice should be shaved into small chips and the flask packed with ice.

Many fixed points on ITS-90 are melting points of metals. These are more difficult to use than the points for water. But they are reliably reproducible. Most of these are available in cells with furnaces and temperature control systems. A major

Table 6.2 Isotopic mol fraction χ of SMOW. Reprinted from Experimental Thermodynamics, Vol 6/J.V. Nicholas and D.R. White, Temperature, Table 2.2, 2003, with permission from Elsevier

Component	Mol fraction (χ)
^1H	0.999842
^2H (deuterium)	0.000158
^3H (tritium)	0
^{16}O	0.997640
^{17}O	0.000371
^{18}O	0.001989

expense is the pure metal for the fixed point. SPRTs are used up to the melting point
of silver. Melting points of gold and copper are in the range covered by radiation
thermometers.

6.2.3   Thermometers

Cryogenic Range. The vapor pressure and gas thermometers specified in ITS-90
for the very low temperature range are difficult to use in practice. They are normally
used only to calibrate electrical thermometers that are then employed in laboratory
work.

Vapor Pressure Thermometers. There are no fixed points specified for the extremely
low temperature regime. Instead ITS-90 specifies a numerical relationship for the
calculation of temperatures from vapor pressures for the range $0.65\,\mathrm{K} \leq T \leq 3.2\,\mathrm{K}$
(based on ^{3}He) and for ranges $1.25\,\mathrm{K} \leq T \leq 2.1768\,\mathrm{K}$ and $2.1768\,\mathrm{K} \leq T \leq 5.0\,\mathrm{K}$
(based on ^{4}He) [[60], p. 35]. With care the uncertainties can be limited to 0.5 mK.

Gas Thermometers. We discussed the ideal gas thermometer in our derivation of the
ideal gas thermal equation of state (see Sect. 17.1). In principle the ideal gas thermometer requires only one fixed point. In the interest of accuracy, however, ITS-90
specifies three fixed points over the range 3 K to the neon triple point at 24.5561 K.
The gas is either ^{3}He or ^{4}He over the wider range. It is possible to achieve accuracies
of 0.1 mK.

6.3   Pressure

6.3.1   Electronic Pressure Measurement

An electronic pressure gauge is a device which converts the mechanical strain resulting from a pressure into an electrical signal. We provide examples of some of
these here.

Piezoresistive Silicon (Strain) Gauges (PSG). Silicon has excellent elastic properties, silicon diaphragms can be easily constructed by photolithography, and the
resistivity of silicon is a function of the strain.

The change in resistance R of a material under a strain ε is $\Delta R = KeR$ where K
is the gauge factor

$$K = 1 + 2\sigma + R_{\mathrm{p}}E.$$

Here σ is Poisson's ratio, E is Young's modulus, and R_{p} is the piezoresistive coefficient. Resistive strain gauges may be fabricated from silicon, poly-silicon, or metal
foil.

Fig. 6.1 Rectangular pressure transducer membrane etched into silicon wafer. *Top* is section through transducer showing piezoresistors. *Bottom* is top view of transducer showing four piezoresistors with connections. Reprinted from Experimental Thermodynamics, Vol 6/J.V. Nicholas and D.R. White, Pressure, Fig. 3.2, 2003, with permission from Elsevier

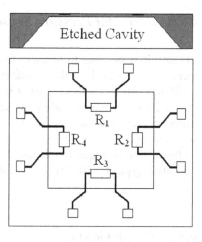

Piezoresistive strain gauges are mounted on a membrane resulting from photoetching of a silicon wafer. Deflection of the membrane then produces a resistance change that can be measured in a bridge. This arrangement is shown schematically for four piezoresistors mounted on the membrane in Fig. 6.1.

Capacitive Diaphragm Gauges (CDG). In a CDG the pressure is applied to a diaphragm and the motion of the diaphragm is determined by measuring the change in capacitance. The capacitance of a parallel plate capacitor is

$$C = \frac{\varepsilon A}{d}, \tag{6.1}$$

in which ε is the permittivity of the dielectric material between the plates, A is the plate area, and d is the plate spacing. The plate facing the fluid medium is the sensor and d is the measured quantity. Changes in the area are neglected, as are any effects of the plate deflection on (6.1).

At the applied pressure P one plate deforms, the spacing changes to $d + \Delta d$, and the capacitance becomes $C(P)$. For $\Delta d/d \ll 1$ the ratio of the new capacitance to that with no deflection is

$$\frac{C(P)}{C} = 1 - \frac{\Delta d}{d}.$$

The capacitive changes can be determined from an AC capacitance bridge, which has a relative uncertainty of 10^{-8}. Corrections can be obtained using finite element analysis.

The insensitivity of the CDG to stresses in the silicon membrane makes it attractive for micro electromechanical systems (MEMS). A CDG diaphragm may be fabricated in a silicon wafer of the order of 1 mm^2. The fixed plate may then be a metal vapor deposited on glass and mounted above the wafer.

6.3.2 Resonant Pressure Sensors

Because it is possible to measure frequencies very accurately, pressure measurements can be based on the measurement of resonant frequencies of mechanical oscillators. Monocrystalline quartz is stable upon cycling of both temperature and pressure and, therefore, makes an excellent material for use as a resonator. Resonance frequency of the shear modes are determined for both temperature and pressure sensing.

Resolution of a quartz pressure transducer is better than 10^{-6}. The long-term stability is typically 10^{-4} times the full scale pressure [[60], p. 50].

6.3.3 Piston Gauges

A piston gauge consists of a loaded piston and a matching coaxial cylinder. A very thin film of the gas flows in the space between the piston and the cylinder preventing contact and centering the piston. The piston freely rotates and does not touch the cylinder. Sometimes the order is reversed with the cylinder moving and the piston fixed. A schematic of a piston gauge is shown in Fig. 6.2.

The capabilities of piston pressure gauges have been improved through advanced electronics, advances in materials science, and simpler design. The range of application has also been extended to 0.5 GPa. Piston-cylinder geometry has had a greater effect than previously anticipated. The annular gap between piston and cylinder affects the time over which a measurement can be made and hence is a critical component of pressure gauge design.

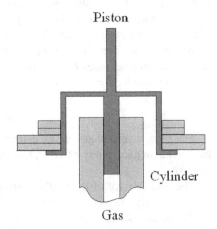

Fig. 6.2 Piston gauge schematic with cylinder stationary. Reprinted from Experimental Thermodynamics, Vol 6/J.V. Nicholas and D.R. White, Pressure, Fig. 3.18, 2003, with permission from Elsevier

6.3.4 Low Pressure

Low pressures are considered to be in the range 0.1–1 000 Pa.

Liquid Column Manometers. Manometry remains the most accurate method for the measurement of low pressures. Relative uncertainties are $\approx 3 \times 10^{-6}$. Mercury is the liquid of choice in most measurements. The high density of Hg makes manometer heights reasonable. There are, however, several problems associated with Hg manometry.

Because of the high surface tension of Hg the manometer tube must have a large diameter ($\approx 45\,\text{mm}$) in order to keep the capillary depression low and the meniscus flat enough for interferometric measurements. There may also be ripples on the meniscus surface arising from external vibrations and variations in the contact between the mercury and the tube. These ripples can destroy interference of a laser beam used to measure the meniscus location. The coefficient of thermal expansion of Hg is also large. There may then be sufficient variations in density with temperature to affect the measurements. Finally, the vapor pressure of Hg (171 mPa at $T = 293.15\,\text{K}$) makes determination of low pressures difficult and the contamination of the sample by mercury affects the sample pressure in low pressure regimes.

The location of the surface of the Hg meniscus may be determined with a Michelson interferometer. To eliminate the effect of surface disturbances a cat's-eye float reflector is used. This is illustrated in Fig. 6.3. With the cat's-eye float reflector uncertainties in the meniscus location of the order of 0.1 mm can be attained [[60], p. 76].

The height of the Hg column may also be measured by ultrasonic waves. To do this a transducer serving as both transmitter and detector is placed at the bottom of the mercury column. The ultrasonic wave is reflected by the meniscus and returns to the transducer. The wavelength of the ultrasonic wave is sufficiently large ($\sim 150\,\mu\text{m}$) that the wave interference is unaffected by disturbances on the mercury surface.

A disadvantage is the effect of temperature variations on the measurement. The temperature variation of the speed of sound in mercury with temperature (see Sect. 6.5) is high. To control temperature variations ultrasonic interferometric

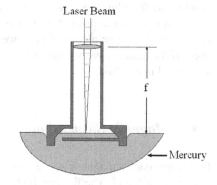

Fig. 6.3 Cat's-eye float reflector. The laser beam is focused at a point on the mercury surface by the lens. Reprinted from Experimental Thermodynamics, Vol 6/J.V. Nicholas and D.R. White, Pressure, Fig. 3.14, 2003, with permission from Elsevier

manometers are surrounded by aluminum (12 mm) and thermal foam (50 mm). Generally, however, ultrasonic interferometric measurements are the most accurate possible for low pressure manometry.

Oil Manometers. Oil has specific advantages over mercury as a manometric fluid. The vapor pressure, density, and surface tension are low. The column heights may then be more accurately measured with a much lower capillarity and sensitivity to vibrations. The pressure uncertainty is reduced to ≈ 0.03 m Pa as compared to ≈ 0.4 mPa for Hg.

Oil, however, absorbs gases. And the low reflectivity makes interferometric measurements more difficult.

Oil manometers have almost disappeared in application [[60], p. 78].

Piston Gauges. Piston gauges have some advantages over manometers at low pressures. Because of the low coefficients of volume expansion no special provisions beyond standard laboratory controls are necessary for the temperature. There is also good short term repeatability in measurements even in the presence of vibrations.

The major disadvantage is in the limitation on pressures attainable because of the piston mass. The pressure range is of the order of 2–10 kPa.

Static Expansion. Static expansion is a technique first proposed by the Danish physicist Martin Knudsen (1871–1949). If an ideal gas, or a gas for which the thermal equation of state is adequately known, expands into a vacuum the final pressure can be computed from measurement of the initial and final volumes. The initial pressure can be chosen high enough for accurate measurement using a manometer or piston gauge.

This technique is not used for direct measurement of pressure. Rather a pressure transducer (i.e. PSG or CDG) is calibrated using a static expansion and the calibrated transducer is then used for the actual measurement. State-of-the-art measurements in the range $10^{-5} - 10^3$ Pa have been reported [[60], p. 87].

6.4 Density

In principle we can obtain a direct experimental determination of the thermal equation of state from fluid density measurements at specific temperatures and pressures.

The most obvious approach to the determination of density is through application of Archimedes' principle. For liquids we measure the volume displaced by a known mass. If the fluid is a gas we measure the buoyant force on a mass of known volume suspended in the fluid.

6.4.1 Magnetic Suspension

In this approach a magnetic suspension is used to alter the effective mass of a body floating in a liquid. The floating body is usually a hollow glass vessel containing a magnet or soft iron core. A magnetic field from a solenoid surrounding the float is used to provide a levitating force to the float, which can be determined from a measurement of the current. The position of the float can be observed optically.

6.4.2   Two-Sinker Densimeter

The two-sinker densimeter is based on a novel approach in which buoyant forces on two masses are compared. The masses are of different shapes, one being a ring and the other a sphere, of different densities and volumes, but both have the same mass. The masses can each be placed on a support or suspended. The suspension is connected to an analytical balance which measures the buoyant forces. The arrangement is shown schematically in Fig. 6.4. Sinker 1 is a solid gold disk. Sinker 2 is a gold covered quartz glass sphere. The analytical balance is connected or released by an electromagnet.

Fig. 6.4 Two-sinker densimeter. Disk is solid gold. Sphere is gold covered quartz glass. Reprinted from Experimental Thermodynamics, Vol 6/J.V. Nicholas and D.R. White, Density, Fig. 5.1, 2003, with permission from Elsevier

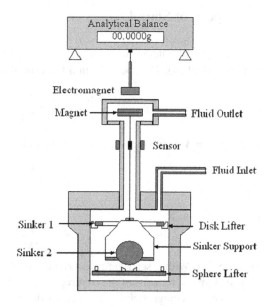

6.4.3   Single-Sinker Densimeter

For applications not requiring the accuracy of the two-sinker method, a simpler design is the single-sinker. The sinker is a cylinder of quartz glass. The single-sinker is a simpler approach, but requires a novel magnetic suspension coupling. This coupling allows taring of the balance system with the sinker decoupled from the magnetic suspension. A second measurement then provides the mass of the sinker in the fluid.

The single-sinker densimeter covers a range of 2–$2\,000\,\mathrm{kg\,m^{-3}}$ for temperatures from 233 to 523 K and pressures up to 30 MPa [[60], p. 144].

6.4.4 Vibrating Bodies

In a vibrating tube densimeter the density of a fluid is found from the resonant frequency of a U-tube containing the fluid. A vibrating wire densimeter is a standard densimeter in which the weight of the sinker is determined from the resonant frequency of the wire suspending the sinker.

The vibrating wire technique is applicable at high pressures (over 100 MPa) and temperatures (up to 473 K). The vibrating tube can be used at higher temperatures, but the pressure is limited to about 50 MPa [[60], p. 150].

6.5 Speed of Sound

We shall consider here only isotropic fluids. Sound waves are infinitesimal disturbances in the fluid. The speed of an infinitesimal disturbance u satisfies [[35], p. 5]

$$u^2 = \left(\frac{\partial P}{\partial \rho}\right)_S, \tag{6.2}$$

where $\rho = n/V$ is the molar density of the fluid. We may write (6.2) as

$$u^2 = \frac{1}{\rho\,\kappa_S} \tag{6.3}$$

using (5.30), or as

$$u^2 = \frac{\gamma}{\rho\,\kappa_T} \tag{6.4}$$

using (5.31). With a separate measurement of density, (6.3) provides the basis for almost all measurements of the isentropic compressibility κ_S [[60], p. 300]. Because the isothermal compressibility κ_T is easily determined by direct measurements, (6.4) provides a measurement of the ratio of specific heats γ.

If we express the thermal equation of state in terms of ρ we have (see Exercises 5.8 and 5.9)

$$u^2 = \left[\left(\frac{\partial \rho}{\partial P}\right)_T - \frac{1}{\rho^2}\frac{T}{C_P\,(T,P)}\left(\frac{\partial \rho}{\partial T}\right)_P^2\right]^{-1} \tag{6.5}$$

and

$$u^2 = \left(\frac{\partial P}{\partial \rho}\right)_T + \frac{n}{\rho^2}\frac{T}{nC_V\,(T,\rho)}\left(\frac{\partial P}{\partial T}\right)_\rho^2 \tag{6.6}$$

If the thermal equation of state is known the specific heats $C_P(T,P)$ and $C_V(T,\rho)$ can be obtained directly from (6.5) and (6.6) with measurements of the speed of sound as a function of (T,P) and (T,ρ). Alternatively we may obtain $C_P(T,P^0)$ and $C_V(T,\rho^0)$ from measurements of the speed of sound at reference pressure P^0 and molar density ρ^0 and calculate $C_P(T,P)$ and $C_V(T,\rho)$ from (5.49) and (5.50).

We may then obtain all the fundamental thermodynamic properties of fluids (gases or liquids) from the thermal equation of state and measurements of the speed of sound.

6.6 Calorimetry

Calorimetry remains the method of choice for measurements of mixing enthalpies and specific heats. Here we will only discuss certain specific calorimeters to provide a sense of the techniques available.

6.6.1 AC Calorimetry

In its simplest form the AC (alternating current) operates by heating a sample with a heat source composed of a steady plus a sinusoidal component. The sample is connected to a heat bath, which is a metal block held at constant temperature, via a mount with known thermal resistance R so that the loss of heat to the bath is $(T - T_{bath})/R$. The temperature sensor is a thermistor,[1] whose resistance is recorded using a Wheatstone bridge and a precision programmable resistor.

If the temperature of a sample of n_s mols rises by ΔT it gains a heat $n_s C_{P,s} \Delta T$. Conservation of energy produces the relationship

$$n_s C_{P,s} \frac{dT}{dt} = P_0 (1 + \sin(\omega t)) - \frac{T - T_{bath}}{R}. \tag{6.7}$$

After a transient, the steady state temperature in the sample varies sinusoidally at the frequency of the source and with a phase difference that is known in terms of R, $n_s C_{P,s}$, and ω. The specific heat $C_{P,s}$ is determined by the ratio of P_0 to the AC temperature amplitude in the sample.

Joule heating AC calorimeters can be used to determine specific heats to within $\pm 0.01\%$ [[60], p. 340]. They have the disadvantage of requiring relatively large samples.

With light irradiation AC calorimeters very small samples may be used. It is not necessary to attach the heater to the sample, which may then have dimensions of the order of $1 \times 1 \times 0.1$ mm. But the accuracy of measurements is limited by the difficulty in determining the heat absorbed by the sample.

---

[1] The electrical resistance of a thermistor varies with temperature. Thermistors are generally ceramic or polymer, in contrast to a resistance thermometer, which is metal.

Radiant heat sources such as halogen lamps with a mechanically chopped beam or diode lasers have been used. Thermocouples with very fine (μm diameter) wires are employed as temperature sensors. The bath is again a metal block (e.g. copper).

Applications to biology require accurate measurements of the specific heat of small amounts of liquid in, for example, studies of denaturation of proteins. H. Yao of Tokyo Institute of Technology and I. Hatta of Nagoya University reported a method in which the problems of a small AC heat source and connection to the constant temperature bath are solved by placing the liquid sample in a 5 cm metal tube of inner diameter 290 μm and 20 μm wall thickness [[60], p. 344]. The internal cell volume was then (about) 3.3 μl. Both ends of the tube extended out of the bath and were connected to an AC power source so that the tube wall became the Joule heating source. Temperature was sensed by a chromel-constantan thermocouple with 13 μm diameter wire. Three separate measurements were made. The first was with the sample cell empty, the second with the sample cell containing a liquid of known density and specific heat, and the third was with the sample of interest in the cell. From the three measurements of temperature differences and phase angles the measurements were repeatable to within $\pm 0.01\%$ with an uncertainty of $\pm 0.5\%$.

6.6.2 Differential Scanning Calorimetry

In differential scanning calorimetry (DSC) the sample and a reference are heated together in a furnace. The sample and reference are each small (1–30 mg) sealed in metal crucibles with volumes $\sim 10 \mu$l. The furnace temperature varies linearly and the temperatures of the sample and references are recorded as a function of furnace temperature. Any change in the state of the sample, for example a chemical or phase change, is observed as a relative temperature differential.

Power-Compensated DSC. In power-compensated DSC two separate furnaces are used. One furnace contains the sample under study and the other contains the reference, each in metallic crucibles. The furnaces are in the same constant temperature bath and are heated at the same rate. Any change in the physical condition of the sample is then noted as a differential change in the power ΔP required to maintain both sample and reference at the same heating rates. Temperature sensing is by a platinum resistance thermometer.

Heat Flux DSC. In heat flux DSC the reference and sample crucibles are identical and placed in single furnace. Temperature sensing is thermoelectric. Either the difference in heats exchanged between the furnace and the crucibles required to maintain the same heating rates or the temperature difference between crucibles is measured.

Modifications. Variations and modifications of DSC include single cell DSC and temperature modulated DSC. In single cell DSC the separate sample and reference cells are replaced by a single cell. Two separate measurements are then made. This places some severe requirements on temperature control, but results in mechanical simplicity. In temperature modulated DSC AC heating is used. The furnace temperature then varies sinusoidally.

Specialized DSCs have also been designed for applications such as high temperature gas measurements (800–1 900 K) and rapid reaction rates.

X-ray diffraction measurements can be performed during DSC measurements to study the relationship between thermal effects and structural changes.

6.6.3   Nanocalorimetry

Micro-system technology from the manufacture of integrated circuits has made possible the development of calorimeters with millimeter dimensions, heating elements with well-defined characteristics, mechanical mountings, and corresponding thermometry. The term 'nanocalorimeters' is currently accepted for these instruments, although it refers to the power detected rather than the dimensions. With the small sizes there are corresponding rapid response time and high sensitivity.

Micro Electromechanical Systems. This allows the construction of the nanocalorimeter on a single silicon substrate. In the example illustrated in Fig. 6.5 the substrate is a single silicon wafer about 100 mm in diameter and 0.5 mm thick. An individual calorimeter of this form has dimensions of the order of 1 mm. Etched into this wafer are the locations for the heater at the center and the thermopiles to measure the temperature difference between the heat source and the surrounding wafer serving as a heat sink. The heater and thermopiles are isolated by a dielectric. The silicon below the membrane has been removed by back etching. This cavity contains the sample.

Similar nanocalorimeters have been fabricated on thin plastic membranes such as mylar.

These nanocalorimeters find use particularly in biomedical and biochemical work, such as in the detection of glucose and urea in blood from measurements of the enthalpy of an enzyme reaction. They are also used in measurements made on the metabolism of living cells and their response to medication or toxins. The silicon membrane in the cavity can be coated with an enzyme to induce a reaction in a (biological) sample and the enthalpy can be measured from the temperature change.

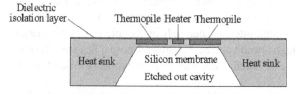

Fig. 6.5 Nanocalorimeter fabricated on a silicon wafer. The heater and thermopiles are in photoetched cavities and covered by a dielectric isolation layer. Reprinted from Experimental Thermodynamics, Vol 6/J.V. Nicholas and D.R. White, Calorimetry, Fig. 7.22, 2003, with permission from Elsevier

6.6.4 Cryogenics

At very low temperatures substances may undergo phase transitions. For example a sample may have a frozen-in disordered state as in a glass. Structural relaxation takes place at the glass transition temperature with release of enthalpy. Insight into the physics of the transition is gained from measurements of this enthalpy. The time domain covered in these studies is determined by the structural relaxation time of the sample. These longer time domain measurements are complementary to the short time measurements of NMR and dielectric studies.

To attain high precision the standard practice has been to use large samples (30–50 g). But the materials studied are often very expensive in a highly purified state. This has led to miniaturization of low-temperature calorimeters. Miniaturization also means that the heat released by resistance thermometers during measurements becomes an important consideration and thermometer location is critical. Differential methods have also been used in which the sample is compared to a reference, as discussed above.

Adiabatic demagnetization techniques (see Sect. 7.3) have been almost entirely replaced by ^{3}He cryostats or by ^{3}He-^{4}He dilution refrigerators [[60], p. 483].

6.6.5 High Temperatures

At the beginning of the twenty-first century there has been a great improvement in measurements at high temperatures. Solid state detectors such as Si- and InGaAs-photodiodes now serve as fast detectors in high speed pyrometry. New techniques for attaining high temperatures, such as laser pulse heating, are also available in addition to resistive pulse heating. Temperatures in the regime 1,000–10,000 K can be studied. The heating rates for pulse heating are between 10^{4}–10^{10} K s^{-1} so that the measurement techniques must be fast requiring energy storage in a bank of batteries (ms experiments) or in a bank of capacitors (μs experiments). The samples are wires, foils, or tubes and current is passed through them. Measurements cover the temperature range from 293 K to above the melting point of the sample.

Temperature of the sample is measured with a calibrated pyrometer. A pyrometer measures radiated intensity from the sample and the temperature is obtained from Planck's radiation law. The radiated intensity depends on sample size, which must be measured separately by a high speed camera. For this purpose a fast framing can be used. this permits pictures to be taken every 9 μs with an exposure time of about 300 ns.

Data reduction requires accurate values for the emissivity of the sample. For accurate results separate measurements must be performed and a comparison made to black body spectra.

Measurements in the millisecond time regime are presently accurate from 1 000 K to the sample melting point. Temperature measurements are uncertain to about 5 K and specific heat is uncertain to within 2%. Reproducibility is about 0.5% at 2 000 K.

Microsecond measurements cover the temperature range from 1,000–6,000 K.

6.7 Summary

In this chapter we have provided a very limited survey of the measurement techniques, limitations, and capabilities available at the beginning of the 21st century. In each case that we discussed the basic physical principles used were not themselves esoteric. For example Archimedes' principle formed the basis of many of the densimeters. Notable, rather, were the advances based on human ingenuity, such as the single- and two-sinker densimeters, and the advances based on micro electromechanical technologies.

None of the techniques we discussed is universal. Each instrument is intended for a specific range of temperature and pressure. In some instances, such as the temperature scale ITS-90, the instruments to be used are specified. In general, however, the engineer or scientist must simply become familiar with what is available and exercise judgement.

In certain applications it may not be necessary to use the most accurate instrument available, while for some applications it may be imperative to have as accurate a measurement as possible. The decision will be based on a thorough understanding of the implications of errors in measurements for the problem at hand.

The advances in laboratory measurement techniques arise from the necessities of research and engineering development. They are also a result of additional capabilities that result from advances in technology. We are often able to purchase commercial instruments intended for specific measurements. Advances in instrumentation are, however, the result of research.

Chapter 7
The Third Law

7.1 Introduction

Arnold Sommerfeld referred to the third law as the most far reaching generalization of classical thermodynamics in the 20th century [[150], p. 72]. This principle was established by Walther Nernst in his investigations of the thermodynamics of chemical reactions [37, 71].

A major problem facing chemists at the beginning of the 20th century was the quantitative understanding of equilibrium in chemical reactions. This problem was particularly acute in Germany because of the importance of the chemical industry to the German economy. As we shall find in a subsequent chapter (see Sect. 13.3) to predict the state of equilibrium in a chemical reaction requires knowledge of the Gibbs energy and, therefore, knowledge of the absolute value of the entropy. The issue we encountered in Sect. 5.3.4 is not, then, simply academic. This reference value for the entropy is determined as a consequence of the third law.

As with almost all of science, the path to the resolution of the original problem of predicting chemical equilibrium was not straightforward and certainly was not simple. And the third law, which is Nernst's hypothesis, is more foundational than the statement that absolute entropy vanishes at absolute zero.

Because of the consequences and implications of the third law for the behavior of matter near the absolute zero of thermodynamic temperature, we shall devote this entire chapter to the discussion of the third law. Our treatment will be brief, but complete in relevant details.

7.2 Nernst's Hypothesis

We consider a finite isothermal process in a closed system during which the state of the system changes from 1 to 2. This finite process may be the result, for example, of a chemical reaction occurring in the system. We do not require knowledge of the details of the process.

The internal energy and entropy change by the amounts $\Delta U = U_2 - U_1$ and $\Delta S = S_2 - S_1$ and the difference in the Helmholtz energy in this isothermal process is

$$\Delta F]_{T=\text{constant}} = \Delta U - T\Delta S. \tag{7.1}$$

Nernst designated this isothermal difference in the Helmholtz energy by the letter A. That is

$$A = \Delta U - T\Delta S. \tag{7.2}$$

From the first law the work done by the system in a finite reversible isothermal process is

$$W = T\Delta S - \Delta U. \tag{7.3}$$

If the finite isothermal process were reversible $\Delta F]_{T=\text{constant}}$ would be the negative of the maximum work that could be obtained from the system in the process. The process considered is, however, not necessarily reversible so that W is not the maximum work that could be obtained.

Since $S = -(\partial F/\partial T)_V$, we can always express the change in the entropy in a finite process in terms of the corresponding change in the Helmholtz energy as

$$\Delta S = -\left(\frac{\partial \Delta F}{\partial T}\right)_V, \tag{7.4}$$

with $\Delta F = F_2 - F_1$. For a finite isothermal process we have then

$$A - \Delta U = T\left(\frac{\partial A}{\partial T}\right)_V, \tag{7.5}$$

regardless of the nature of the isothermal process.

If the finite process is a chemical reaction, ΔU can be obtained from calorimetric measurements of the heat of reaction. If the properties of the constituents in the reaction are known, the isothermal change in Helmholtz energy $(= A)$ can be calculated. Nernst wrote, "It occurred to me that we have here a limiting law, since often the difference between A and ΔU is very small. It seems that at absolute zero A and ΔU are not only equal, but also asymptotically tangent to each other. Thus we should have

$$\lim_{T \to 0} \frac{dA}{dT} = \lim_{T \to 0} \frac{d(\Delta U)}{dT} = 0 \text{ for } T = 0." \tag{7.6}$$

[quoted in [150], p. 72]. This is Nernst's hypothesis.

From (7.2) we have,

$$\lim_{T \to 0} \frac{dA}{dT} = 0 = \lim_{T \to 0} \frac{d(\Delta U)}{dT} - \lim_{T \to 0} \Delta S - \lim_{T \to 0} T\frac{d(\Delta S)}{dT}. \tag{7.7}$$

The first term on the right of (7.7) is zero from (7.6) and the third term vanishes with T, regardless of the form of ΔS. As a consequence of Nernst's hypothesis, therefore,

$$\lim_{T \to 0} \Delta S = 0. \tag{7.8}$$

In his 1906 publication Nernst put forward a stronger position on the entropy change. There he stated that "For all changes occurring in pure condensed substances at the absolute zero, the change in entropy is zero." [120, 152] There is then a constant value for the entropy of any (pure condensed) substance at the absolute zero of thermodynamic temperature. Planck pointed out that this constant may be taken to be zero for all substances resulting in a universal definition of what is called absolute entropy [[132], p. 274].

A potential difficulty arises if we consider that elements in their naturally occurring states are often specific mixtures of isotopes. There is then a mixing entropy if we consider the formation of the naturally occurring state of the element from its isotopes. However, unless we are dealing with a situation in which the concentrations of the isotopes do in fact vary, this is of no consequence. We may then take the absolute entropy of naturally occurring mixture of isotopes to be zero [162].

The problem of the arbitrary constant in the entropy factor appearing in the Gibbs and Helmholtz energies is then resolved as a consequence of the third law. In practical terms this was what we sought in our consideration of the third law.

7.3 Unattainability of Absolute Zero

Temperatures in the mK range can be attained by adiabatic demagnetization. This was suggested by Peter J.W. Debye[1] in 1927. Although this can be exceeded at this time (see Sects. 6.2.3 and 10.3.4), adiabatic demagnetization provides a good example of the principle we wish to discuss here.

In Fig. 7.1 we have plotted the entropy of a paramagnetic salt, subjected to an external magnetic field, as a function of temperature in the neighborhood of $T = 0$. Each panel in Fig. 7.1 includes the variation of entropy with temperature when the sample is exposed to high and low values of the external magnetic field intensity. In panel (b) we have magnified the region indicated in the small box in panel (a). The data for the curves were calculated using the results of a statistical mechanical analysis of system of spins in an external magnetic field [[162], p. 308].

In the process of adiabatic demagnetization the sample is first immersed in a liquid helium bath and cooled to a temperature of about 1 K in the presence of a low magnetic field. This is the thermodynamic state 1 in Fig. 7.1(a). The sample is then subjected to an increasing magnetic field while the temperature is held constant. This is the vertical dashed line from state 1 to 2 in Fig. 7.1(a). The sample is then adiabatically isolated and the external magnetic field decreased. This is the dashed line from state 2 to 3 in Fig. 7.1(a). The sample temperature decreases as the applied field is decreased, since the curves approach one another. But there is a limit to the temperature attainable that is determined by the shape of the $S(T)$ curve for the lower value of the applied field. If this curve had the form indicated by the dashed line in Fig. 7.1(a) it would be possible to attain the temperature $T = 0$. This would be the case if the entropy at $T = 0$ were dependent on the value of the external field. But the third law denies this possibility.

---

[1] Peter Joseph William Debye (1884–1966) was a Dutch physicist and physical chemist.

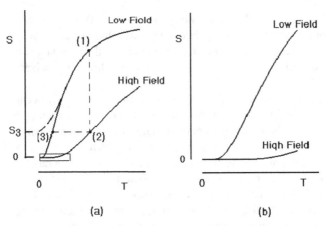

Fig. 7.1 Entropy $S(T)$ for adiabatic demagnetization of a paramagnetic salt. $S(T)$ is plotted for high and low values of magnetic field (**a**). Enlargement of boxed area (**b**)

We can repeat the process of adiabatic demagnetization systematically as shown in Fig. 7.2, which is an enlargement of Fig. 7.1(b) with sets of isothermal and adiabatic processes drawn. Because the two curves for $S(T)$ at the two values of magnetic field intensity asymptotically approach one another, we cannot attain the temperature $T = 0$ in a finite number of steps.

The technique of laser cooling has been used recently in studies of Bose-Einstein condensation (see Sect. 10.3.4) [3, 32]. Using laser cooling and magnetic trapping temperatures in the n K range have been attained. The technique of laser cooling is not an equilibrium technique. And at this time its use is in the investigation of fundamental physics not related directly to the third law. Nevertheless, no violation of the third law limitation on the attainability of $T = 0$ has been found.

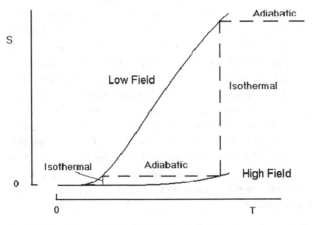

Fig. 7.2 Repetition of the isothermal and adiabatic steps shown in Fig. 7.1(b)

7.3.1 Limits on H and G

From the definitions of the Helmholtz and Gibbs energies we have

$$U - F = H - G.$$

Therefore, in a finite process,

$$\Delta F - \Delta U = \Delta G - \Delta H.$$

Nernst's hypothesis that

$$\lim_{T \to 0} \Delta F = \lim_{T \to 0} \Delta U = 0 \qquad (7.9)$$

and that ΔF and ΔU are asymptotically tangent to each other is then equivalent to

$$\lim_{T \to 0} \Delta G = \lim_{T \to 0} \Delta H = 0. \qquad (7.10)$$

Either of these relationships results in a limit identical to (7.7) with the consequence that $\lim_{T \to 0} \Delta S = 0$. Then we may deal with either ΔF and ΔU, or with ΔG and ΔH. Either pair of potentials is equivalent for discussions of the consequences of the third law.

7.4 Consequences of the Third Law

7.4.1 Coefficient of Expansion

The coefficient of expansion is, using the MR (4.55),

$$\beta = \frac{1}{V} \left(\frac{\partial V}{\partial T} \right)_P = -\frac{1}{V} \left(\frac{\partial S}{\partial P} \right)_T. \qquad (7.11)$$

In the limit of zero temperature the entropy is independent of pressure. Therefore $\lim_{T \to 0} (\partial S / \partial P)_T = 0$, and

$$\beta \, (T = 0) = -\frac{1}{V} \lim_{T \to 0} \left(\frac{\partial S}{\partial P} \right)_T = 0. \qquad (7.12)$$

The coefficient of expansion also then vanishes at the absolute zero of temperature.

7.4.2 Specific Heats

From the Gibbs equation we can express the specific heats as

$$nC_V = T \left(\frac{\partial S}{\partial T} \right)_V \quad \text{and} \quad nC_P = T \left(\frac{\partial S}{\partial T} \right)_P. \qquad (7.13)$$

Integrating these equations with $S(T = 0) = 0$,

$$S(T,V) = \int_0^T \frac{nC_V}{T'} dT' \tag{7.14}$$

and

$$S(T,P) = \int_0^T \frac{nC_P}{T'} dT'. \tag{7.15}$$

The arbitrary functions of V and of P that are required in these integrals vanish because the entropy vanishes at $T = 0$ independently of pressure or volume.

Since S vanishes at $T = 0$, the Eqs. (7.14) and (7.15) require that C_P and C_V must vanish at least as linear functions of T in the limit $T = 0$. This form for C_V has been verified experimentally for solids with the specific heat vanishing as T for metals and as T^3 for insulators and is well understood in the theoretical (statistical mechanical) picture near absolute zero. The electron contribution to the specific heat vanishes as T (see Sect. 10.4.2, (10.87)) and the contribution from the lattice vibrations vanishes as T^3 when the temperature approaches absolute zero [[79], p. 72]. The coefficient of expansion, β, has the same T^3 dependence as C_V in the limit of absolute zero [[79], p. 76].

Since β vanishes at absolute zero, the general equation

$$C_P - C_V = T \frac{\beta^2}{\kappa_T} v \tag{7.16}$$

(see (5.32)) guarantees that C_P remains greater than C_V provided $T > 0$ and that C_P is equal to C_V in the limit of zero temperature.

7.5 Summary

The third law was originally a hypothesis about the asymptotic values of changes in the Helmholtz and internal energies resulting from an isothermal process as the temperature at which the process takes place approaches absolute zero. Nernst termed this hypothesis his heat theorem and made it clear that this was fundamentally a law and not simply another mathematical recipe. The discovery of this law earned him the 1920 Nobel Prize in chemistry [[37], pp. 129–131]. The consequences of the third law have been shown to be correct in the most modern studies of the behavior of matter near absolute zero.

As we have seen in this chapter, the asymptotic character of the third law is what results in the unattainability of absolute zero of temperature, even though we have been able to attain temperatures in the nanokelvin regime (see Sect. 10.3.4). The asymptotic character of the law is also the source of the decision to set absolute entropy equal to zero at $T = 0$.

In 1922 Max Planck noted, in his *Treatise on Thermodynamics*, how remarkable was the conclusion that C_P must approach zero at $T = 0$ as a consequence of the third law. The experimental and theoretical verification of this, since Planck's writing, are indications of the validity of Sommerfeld's claim that this is the most far reaching generalization of classical thermodynamics in the twentieth century.

Exercises

7.1. Show that the ideal gas, defined by $PV = nRT$ and C_V =constant violates the third law of thermodynamics. Show this for the actual third law, not just for the consequences of the third law.

7.2. In a solid the internal energy consists of the energy of the electrons and the energy of the lattice formed by the sites of the nuclei. the nuclei vibrate in a manner similar to harmonic oscillators. Correspondingly there are two contributions to the specific heat of a solid. in the low temperature regime these are $C_V^{\text{electron}} = \frac{\pi^2}{2} n k_B \frac{T}{T_F}$ and $C_V^{\text{lattice}} = 3 n k_B \frac{4\pi^4}{5} \left(\frac{T}{T_D}\right)^3$. Assume that the specific heat of a solid is a linear sum of these two contributions as $T \to 0$. The differential of the entropy of a certain solid has been reported as

$$dS = nC_V \frac{dT}{T} + \Gamma C_V \frac{dV}{V}$$

where Γ is a constant for all temperatures. Will this solid satisfy the third law?

7.3. The entropy of an ideal paramagnetic salt with a total spin, S, per atom, and a number of atoms, N, located in a region where the magnetic field intensity is H is given by [[162], p. 308]

$$S = Nk_B \ln \frac{\sinh\left[(2S+1)\mu_0 H/(k_B T)\right]}{\sinh\left[\mu_0 H/(k_B T)\right]}$$
$$- (2S+1)\frac{N\mu_0 H}{T} \coth \frac{(2S+1)\mu_0 H}{k_B T} + \frac{N\mu_0 H}{T} \coth \frac{\mu_0 H}{k_B T}.$$

Does this form of the entropy satisfy the third law?

Chapter 8
Models of Matter

8.1 Introduction

The goal of this chapter is to provide some details of our kinetic molecular picture
of matter before we undertake a development of the modern approach to this subject provided by statistical mechanics. Our intention is to consider the consequences
of molecular motion and interaction while still requiring a basic understanding in
terms of identifiable interactions and the concept of average behavior of molecules.
We will discuss models of fluids, which include gases and vapors, fluids near condensation, and, to some extent, liquids. These models will provide thermal equations
of state from which we will be able to obtain insight into molecular behavior based
on the kinetic and interactive potential energies of the molecules. But we will not be
able to obtain a thermodynamics. For that we require a more fundamental approach,
which will engage the consequences of microscopic structure beyond probabilities.
The more fundamental approach is statistical mechanics.

We begin with the simplest ideas of molecular motion that lead to the concept
of the ideal gas. To this we will add the complexities of molecular interactions.
The introduction of repulsive and attractive forces between molecules results in the
semitheoretical thermal equation of state of Johannes Diderik van der Waals [156],
which is as far as we can go while maintaining a basically understandable connection with molecular dynamics. Our final considerations of the Redlich-Kwong thermal equation of state will be indicative of the complexities that must be considered
in any attempt to approximate the thermal behavior of matter.

We shall use a basically historical approach. We do this in part to reveal the roots
and the age of some of the molecular ideas. Understanding how some of the greatest
scientists of the 18th and 19th centuries engaged the critical question of the molecular structure of matter also reveals the complexity of the questions.

8.2 The Ideal Gas

We introduced the ideal gas in an earlier chapter (see Sect. 2.3.1) as a fictional substance which can be used to model the behavior of real gases at low densities and

C.S. Helrich, *Modern Thermodynamics with Statistical Mechanics*,
DOI 10.1007/978-3-540-85418-0_8, © Springer-Verlag Berlin Heidelberg 2009

moderate temperatures. Because of its simplicity, particularly in light of what is presently known about the molecular structure of matter, we shall begin our considerations of models of matter with the kinetic theory of the ideal gas. In kinetic theory we consider a gas to be a specific collection of particles (atoms or molecules) moving rapidly in otherwise empty space and assume that the measured properties, such as pressure and temperature, can be identified from an average of the mechanical behavior of these particles.

The form of the constant appearing in (A.10), obtained from ideal gas thermometry, may be found experimentally. But it is insightful to establish the value based on the arguments Daniel Bernoulli (1700–1782) presented in 1738 [16, 75, 83, 128].

8.2.1 Ideal Gas Kinetic Theory

Bernoulli supposed that a gas consists of an enormous number of particles moving rapidly and separated by great distances compared to the size of the particles. Based on the fact that gases are compressible, he assumed that most of the gas was empty space. The pressure was the result of collisions of the gas particles with the vessel walls. This remains our basic picture of the ideal gas.

An atomic theory of matter was not new in 1738. At that time Robert Boyle's (1627–1691) picture of a gas, in which the particles were considered to behave as springs exerting forces directly on one another, was widely accepted [75]. And Boyle's result that the pressure of an ideal gas is inversely proportional to the volume at constant temperature,

$$P = \frac{F(T)}{V}, \tag{8.1}$$

where $F(T)$ is a function only of the temperature and the amount of gas present, was known [16].

Isaac Newton (1642–1727) discussed forces between gas particles in his *Philosophiae Naturalis Principia Mathematica* (1687) apparently to place Boyle's law on a mathematical footing. And many scientists in the seventeenth century were able to speak of heat in terms of the motion of particles, but not as free translation. Bernoulli's hypothesis that the gas particles were in rapid translational motion and that the pressure on the walls of the containment vessel was not a constant force but the result of a large number of impacts of extremely small particles was a completely new idea.

Bernoulli's idea led him easily to Boyle's law (8.1). We can see this using a simple verbal argument. First we assume that the average speed of the particles in a container remains constant. Then the number of collisions of the rapidly moving particles with the vessel walls is proportional to the density of particles present in the volume. And in a system with a constant number of particles, the particle density is inversely proportional to the volume. Therefore, if the speed of the particles doesn't change, the pressure is inversely proportional to the volume

$$P \propto \frac{1}{V}.$$ (8.2)

This is almost Boyle's law. We need only the function of temperature in (8.1) to complete the argument.

The force exerted on the wall by the collision of a particle increases with the speed of the particle. That is

$$P = \frac{\text{function of particle speed}}{V}$$ (8.3)

If we assume, with Bernoulli, that the particle speed increases with an increase in temperature,[1] then we can replace the function of particle speed appearing in (8.3) with a function of temperature and we have the form of Boyle's law in (8.1). That is, Boyle's law follows simply from Bernoulli's idea without any complicated discussion of the forces between the particles (atoms or molecules) making up the gas. Indeed the forces between the particles do not enter the argument at all. And we have pressure on the container walls even if there is no interaction between the particles.

Bernoulli also pointed out that if we keep the volume and the temperature constant, we can increase the number of collisions with the walls by introducing more particles. That is the number of collisions is simply proportional to the number of particles in the volume. Therefore, the function of temperature $F(T)$ in (8.1) must include a factor proportional to the number of particles in the volume. In modern terminology we would say $F(T)$ is proportional to the number of mols n of gas present. That is $F(T) = nf(T)$, where $f(T)$ is a function only of temperature. Then Boyle's law becomes

$$PV = nf(T).$$ (8.4)

We recognize (8.4) as the thermal equation of state of an ideal gas if we can show that the function of temperature $f(T)$ is proportional to the ideal gas temperature.

Bernoulli came close to doing this. He cited the 1702 experiments of Guillaume Amonton (1663–1705) which indicated that the pressure of a gas is proportional to temperature if the volume is held constant. And Bernoulli claimed that his experiments of a similar sort were more accurate than Amonton's [16].

With a simple argument Bernoulli was able to relate the gas pressure to the square of the speed of the gas particles. He pointed out that an increase in the speed of the particles had two consequences. Both the force imparted to the container walls in each collision and the number of collisions each increased in direct proportion to the increase in speed. The pressure is proportional to the product of the number of collisions per unit time and the force on the walls due to each collision. That is, the pressure is proportional to the square of the speed of the particles, which is a measure of the kinetic energy of the particles. The increase in pressure is then proportional to an increase in the kinetic energy of the particles.

---

[1] Bernoulli actually wrote that if the heat is intensified then the internal motion of the particles increases.

Bernoulli did not frame the conclusion in such modern terms and the reader of the 18th century would not have reached the same conclusions that may seem obvious to us [[16], footnote on p. 62]. It is, nevertheless, notable how close Bernoulli came to the thermal equation of state of the ideal gas and a modern understanding of the relation between temperature and the kinetic energy of the gas particles.

We need only take a few steps to add flesh to the rather solid bones Bernoulli left us. Let us consider that the gas is contained in a cubic vessel with smooth walls. We shall assume that the collisions with the walls are elastic. An elastic collision of a gas particle with a smooth wall is illustrated in Fig. 8.1. We take the direction x to be horizontal and the direction y to be vertical. The particle velocity perpendicular to the wall (y-direction) reverses in the collision and the velocity parallel to the wall (x-direction) is unchanged. The particle then moves at a constant velocity in the x-direction independently of any and all collisions with walls parallel to the x-direction. If the particle has a velocity component v_x in the x-direction and the length of each side of the container is ℓ then the time between collisions with walls perpendicular to the x-direction is ℓ/v_x and the number of such collisions per unit time is v_x/ℓ. On each collision with a wall perpendicular to the x-direction the momentum transferred to the wall is $2mv_x$ where m is the mass of the particle. The rate of transfer of momentum resulting from collisions with the walls perpendicular to the x-direction is then

$$2mv_x \times \frac{v_x}{\ell}.$$

Similar expressions result for collisions with walls perpendicular to the y- and z-directions with corresponding velocity components.

The total area of the six sides of the cube is $6\ell^2$. The pressure is the average rate of momentum transferred to all walls per unit area, or

$$P = \frac{2mN}{6\ell^3}\overline{v^2}, \tag{8.5}$$

where N is the number of particles in the container and $\overline{v^2} = \overline{\left(v_x^2 + v_y^2 + v_z^2\right)}$ is the average of the square of the velocity of a particle. The average kinetic energy of a gas particle is then $(m/2)\,\overline{v^2}$. If there are no interactions among the gas particles the energy of the particles is only kinetic. We may then write

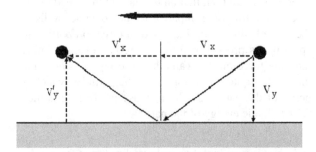

Fig. 8.1 Elastic collision with a smooth wall. Particle velocity is unchanged in the horizontal direction. Velocity is reversed in direction but unchanged in magnitude in the vertical direction

$$U = \frac{1}{2}Nm\overline{v^2} + U_0, \tag{8.6}$$

where U_0 is the reference energy. Then (8.5) becomes

$$P = \frac{2}{3}\frac{U}{V}, \tag{8.7}$$

$$\frac{1}{V} = \frac{3}{2}\frac{P}{\frac{3}{2}nf(T)} = \frac{P}{nf(T)}$$

where $V = \ell^3$ is the volume of the container.

If we introduce Boyle's law, as Bernoulli had it in (8.4), then (8.7) becomes

$$U = \frac{3}{2}nf(T). \tag{8.8}$$

Adding a few steps to Bernoulli's results produces, then, an expression for the internal energy of a gas made up of point-like particles.

For Bernoulli the formulation of thermodynamics was more than a century in the future. But we can use thermodynamics to find $f(T)$.

The general dependence of internal energy on volume, for any substance, is (see Sect. 5.2, (5.17))

$$\left(\frac{\partial U}{\partial V}\right)_T = T\left(\frac{\partial P}{\partial T}\right)_V - P \tag{8.9}$$

Using (8.7) and (8.8) we have

$$\left(\frac{\partial U}{\partial V}\right)_T = 0$$

$$= \frac{T}{f(T)}\left(\frac{df(T)}{dT}\right)P - P.$$

or

$$\frac{T}{f(T)}\left(\frac{df(T)}{dT}\right) = 1. \tag{8.10}$$

Integrating (8.10) we have

$$\ln f(T) = \ln(\text{constant} \times T), \tag{8.11}$$

or

$$f(T) = \text{constant} \times T \tag{8.12}$$

in which the constant is mathematically arbitrary. Its value must be determined experimentally.

The internal energy in (8.8) is then

$$U = \frac{3}{2}n \times \text{constant} \times T. \tag{8.13}$$

Statistical mechanics results in the identification of the constant in (8.13) as the universal gas constant R.

The dependence of the internal energy of an ideal gas on thermodynamic temperature T has been determined from experimental measurements made on real gases at low densities. Experiments performed on monatomic gases show that the internal energy is, to a high degree of accuracy, a linear function of the temperature [162]. For diatomic gases a linear dependence of the internal energy on temperature holds over limited ranges. We may then claim that the internal energy of an ideal gas is a linear function of the (thermodynamic) temperature. For n mols of a particular gas, then

$$U = nC_V T + nu_0. \tag{8.14}$$

Here u_0 is the reference value for the specific (molar) internal energy. We have been able to understand the microscopic basis of this result with some amplification of the Bernoulli kinetic theory of gases in which particles move freely with no interaction.

Our amplification has included a reach across the centuries into Gibbs' thermodynamics and developments of the twentieth century in our use of (8.9). But the physical picture contained in Bernoulli's idea remains in tact. It provides us with a simple physical picture of the fictitious ideal gas that we first encountered as a limit in measurements using a gas thermometer. And the result (8.14) provides insight into the meaning of temperature in physical terms. For the ideal gas thermodynamic temperature is a measure of the kinetic energy of the particles (atoms or molecules). For condensed matter we may expect results that are not quite as simple as this. But Bernoulli had provided a deep insight into the meaning of temperature.

The ideal gas is then a substance in which the particles (atoms or molecules) are widely separated, infinitesimal in size, and moving very rapidly. Because they are infinitesimal in size and widely spaced, most of the gas particle's existence is spent moving in a straight line and we may neglect interactions between particles. Interactions with the walls cause thermodynamic pressure on the walls. The internal energy of the ideal gas is directly proportional to the thermodynamic temperature T. And if the gas particles can be considered to be structureless point particles, with no rotational or vibrational energies, then the thermodynamic internal energy results only from the translational energy of the gas particles and is equal to $(3/2) nRT$.

We also recognize the points at which this picture fails for a nonideal gas. We should anticipate that interactions will become important in real gases and that these will eventually lead to condensation. The fictitious ideal gas never condenses and so is a very unrealistic model for low temperatures. We should also anticipate that molecular structure will play a role in the behavior of real gases. And we recognize that in Bernoulli's work the meaning of average is not clear. We conducted an average in our amplification of Bernoulli's work (see (8.5) and (8.6)). But we made no effort to explain how the average was to be performed. Each of these points will be considered in the next sections and in subsequent chapters.

8.2.2 Kinetic Theory with Collisions

There was a remarkable silence regarding gas kinetic theory between the work of Bernoulli and the reconsideration of this in the middle of the nineteenth century. Then in 1856 August Karl Krönig (1822–1879) revived the investigations in kinetic theory with a brief paper in *Annalen der Physik*. The paper contained no advance over the ideas of Bernoulli and the influence it had is apparently attributable to the fact that Krönig was, at that time, a well-known scientist in Germany [16]. But Krönig's paper induced a publication by Clausius on the kinetic theory of gases in 1857. There Clausius acknowledged the priority of Krönig, and of Joule, but claimed that he had held these basic ideas on the kinetic theory of gases at the time of his 1850 publication on the second law. His desire had been, however, to keep the experimental basis of the second law separate from the more speculative kinetic theory.

The principal contribution Clausius made in 1857 was to point out that an equilibration must occur among the various forms of kinetic energy (at that time called *vis viva* or "living force"), which included rotation and vibration as well as translation. He also discussed evaporation and specific heats.

In 1858 Clausius published a discussion of diffusion and heat conduction in gases in response to a critique by C. H. D. Buijs-Ballot (Buys-Ballot) that appeared at the beginning of the year. Buijs-Ballot had questioned the ability of a kinetic hypothesis, with the molecules moving rapidly, to account for the slow diffusion of smoke into air or of gases into one another. Clausius' discussion made clear the effect of molecular collisions and provided estimates of the mean free path, which is the average distance a molecule travels between collisions. He also claimed that molecular interactions include a long range, weak attractive force and a short range stronger repulsive force. Although Clausius was careful to avoid basing thermodynamics on kinetic hypotheses, he was the first to show that a connection could be made between actual molecular models and macroscopic thermodynamics [16].

Maxwell is acknowledged as the first to introduce statistical ideas into kinetic theory in a publication of 1860. Prior contributions, including those of Bernoulli,[2] considered that the molecules all moved at the same velocity and that equilibration among the energies would result in uniform velocities for each component. Maxwell realized that equilibration (by collisions) meant that molecular velocities would be statistically distributed. In 1867 he introduced an intermolecular repulsive force that varied inversely as the fifth power of the distance between molecular centers to simplify the treatment of intermolecular collisions.

Maxwell considered that his work in the kinetic theory of gases was an exercise in mechanics. He confessed to becoming rather fond of this mathematical effort and required, as he put it, to be "snubbed a little by experiments [16]." One of the results he obtained, that he expected to be incorrect, was that the viscosity of a gas was independent of the density and depended only on temperature. Maxwell's own experiments, conducted with his wife, Katherine, later showed that at constant

---

[2] Bernouli considered mean values and not averages.

temperature the viscosity was actually independent of pressure (i.e. independent of density) [106].

Much of our present kinetic theory originated with Ludwig Boltzmann (1844–1906). Boltzmann considered that the number of gas molecules in a small volume having velocities within limits could be expressed in terms of a statistical distribution function that depended upon the position of the volume, the velocity of the molecules, and the time. The equation he obtained for this distribution function, called the Boltzmann equation, considers the dependence of the distribution function on intermolecular collisions as well as differential changes in the coordinates, velocities and time [9–12].

The Boltzmann equation is still used in much of kinetic theory [31, 116]. So we could claim that with Boltzmann there was a triumphant end to the quest for an atomic and kinetic description of gases [10]. And Einstein had finally established the validity of the atomic theory based on experimental observations of Brownian motion [48]. But the intellectual triumph is marked by tragedy. In 1906 Boltzmann hanged himself while on vacation with his family in the village of Duino near Trieste. Many have speculated that Boltzmann's unpleasant arguments with the energeticists, who denied the existence of atoms, led to his suicide. But Boltzmann also suffered from manic depression [23]. And this may have contributed to his state of mind. There is no way of establishing the reason for the suicide based on what is known. We can only say that this is one of the greatest tragedies in the history of science.

8.2.3 Collisional Models

As we noted above Maxwell studied the affect of short range repulsive forces, specifically forces that varied as r^{-5} where r is the distance between molecular centers. These are called Maxwell Molecules and were chosen by Maxwell to eliminate the greatest difficulty in the Boltzmann equation, which is the appearance of the relative velocity between colliding molecules as a factor in the integrand in the collision integral. David Mintzer points out that as interesting as it is mathematically, the r^{-5} force law represents only the interaction between permanent quadrupoles when angles of orientation are neglected [116]. In Fig. 8.2 we represent a collision between two Maxwell molecules. In Fig. 8.2 one molecule is considered to be fixed and is represented by a star. The trajectories of the colliding molecules are represented by dotted paths. These were obtained by numerical integration of the equations of motion for scatter in a central force field. If there were no repulsive force the molecules would pass one another at the distance b. This is called the impact parameter. A similar illustration was published by Maxwell in 1866 [109] and is reproduced in the monograph by Chapman and Cowling [25] and by Jeans [83].

In his *Lectures on Gas Theory* Ludwig Boltzmann[3] [10] pointed out that a very similar picture would result from the scattering of elastic spheres with radii equal

---

[3] Ludwig Eduard Boltzmann (1844–1906) was an Austrian theoretical physicist and philosopher. Boltzmann held professorships at the unversities of Graz, Vienna and Leipzig.

Fig. 8.2 Scattered trajectories in bimolecular collisions. The stationary molecule is indicated by a *star*, the scattered trajectories by *dots*, b is the impact parameter

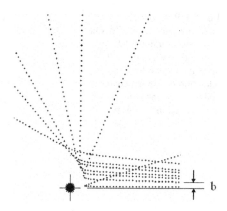

to twice the distance of closest approach when $b = 0$, i.e. the head-on collision. To illustrate this we draw the corresponding hard spheres for three of the collisions of Fig. 8.2 with the smallest impact parameters and one grazing collision in Fig. 8.3. The collisions with very small impact parameters are well represented by collisions between hard spheres. The representation is less appropriate as the impact parameter increases. The hard elastic sphere then models a situation in which the most important collisions are those for which the approach of the molecules to one another is close to the minimum. For these collisions the momentum and energy transfer among the collision partners is a maximum.

Clausius discussed this (almost) impenetrable sphere surrounding the molecules in his 1858 paper on the mean free path [16]. He termed this the *sphere of action* of the molecule. Because the molecules are not impenetrable spheres, we shall use Clausius' designation of the sphere of action in descriptions of repulsive molecular collisions .

The distance of closest approach is a function of the collisional energy. We can see this graphically by considering the interaction potential between molecules with attractive and repulsive forces as proposed by Clausius. We consider again a fixed

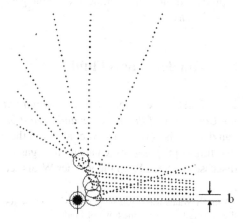

Fig. 8.3 The hard sphere approximation to the collision of Maxwell molecules

Fig. 8.4 Potential energy of
molecular interaction. This is
a plot of the Lennard-Jones
potential (see (13.69)). It
accounts for the long range
attraction of the molecules
and the short range repulsion
as indicated in the insert

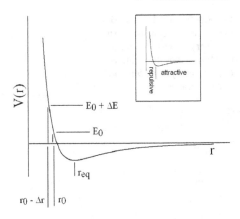

scattering center as in Figs. 8.2 and 8.3. In Fig. 8.4 is a plot of the potential energy
$V(r)$ of the interaction between two molecules separated by a distance r. The inset
in the upper right corner shows the regions of repulsive and attractive force. The
force is $-dV(r)/dr$. The minimum of the potential energy is at the intermolecular
distance $r = r_{eq}$ where $dV(r)/dr = 0$. There is no force between the molecules with
centers separated by this distance and it is possible for two molecules to remain in
equilibrium at this distance from one another. We consider an incoming (scattered)
molecule with the energy E_0. The distance of closest approach for this molecule is
r_0. At the point r_0 the total energy of the incoming molecule is equal to the potential
energy. If the collision energy is increased by ΔE the distance of closest approach
is decreased by an amount Δr as shown.

We should then assume that each molecule occupies a spherical volume with radius equal to half the distance of closest approach at the average of the collisional
energy. The centers of two such hard elastic sphere molecules cannot be closer than
twice this radius, which we shall designate as σ. The volume occupied by this (almost) impenetrable sphere is $(4/3)\pi\sigma^3$ [10]. Each molecule may then be considered
as occupying a spherical volume of radius equal to the average distance of closest
approach. If there are N molecules in the gas, the volume occupied by the molecules
is $(4/3)N\pi\sigma^3$.

8.3   Van der Waals Fluid

In his Ph.D. dissertation, *Over de continuiteit van den gas- en vloeistofoestand* (*On
the Continuity of the Gaseous and Liquid State*) (Leiden, 1873), Johannes Diderik
van der Waals[4] (1837–1923) considered the consequences of the long range attractive forces proposed by Clausius in a gas made up of hard elastic spheres. As we
discussed in Sect. 8.2.3 the van der Waals' choice of hard elastic spheres with weak

---

[4] Van der Waals was a self-taught individual who took advantage of the opportunities afforded by
the university in his home town of Leiden.

long range attractive forces for the molecules of the fluid is a simplified model of a more complete description. The van der Waals (thermal) equation of state, which he obtained for this gas, provides a qualitative description of the behavior of gases in regions for which the ideal gas assumption is no longer valid. However, the actual quantitative successes of the van der Waals equation of state have been modest. And in practical applications the van der Waals equation of state has been replaced by empirical equations representing the behavior of specific gases over limited regions. Nevertheless the qualitative success of the van der Waals equation of state in describing the behavior of a substance in the very difficult region in which the phase transition between gas and liquid states occurs is remarkable. We may then accept the van der Waals description as one of a fluid and not simply a gas.

As we shall see the van der Waals description of the state of a gas just at the onset of condensation or of a liquid just at the onset of boiling is one of a metastable state that exists only as long as there is no (very slight) disturbance in the system that will precipitate the phase change. The possibility of such small disturbances, that are amplified in the system, are indicative of the molecular basis of matter. That is the van der Waals equation of state, which is finally a description of a continuous fluid state of matter, retains the molecular character of its origin in predicting the metastable conditions that are observed experimentally. For our purpose, which is to gain an insight into the molecular picture of matter, a study of the van der Waals equation of state is, therefore, important. We begin by considering the molecular interactions treated by van der Waals.

8.3.1   Van der Waals Model

Clausius assumed that the effect of long range forces between molecules was small. Van der Waals claimed that he did not neglect these forces in his dissertation. Boltzmann, however, pointed out that he was unable to derive the van der Waals thermal equation of state without this assumption [10]. The attractive force itself is not negligible. But the average effect on a molecule surrounded by other molecules vanishes.

To see that this is the case we note first that the attractive force does not change rapidly with distance (see Fig. 8.4). Therefore the total attractive force on a particular molecule in a specific direction is proportional to the number density of neighboring molecules in that direction. The situation, in this regard, is different for molecules near the container wall and those in the bulk of the gas. In the bulk of the gas the number density is uniform in all directions. Therefore, in the bulk gas, the average of the attractive force is zero and the molecules move, as Clausius pointed out, in straight lines punctuated by billiard ball-like collisions. The molecules near the container wall, however, experience a net attractive force back into the bulk gas that is proportional to the number density of the molecules in the bulk gas. The net attractive force on the molecules in the region next to the wall is also proportional to the number density of molecules in this region (next to the wall). The net attractive force back into the bulk gas is then proportional the square of the molecular density of the gas, if we neglect any difference in the density of the bulk gas and the gas at the wall. The net force on the molecules near the wall is then proportional to

$$\left(\frac{N}{V}\right)^2 f_m. \tag{8.15}$$

where f_m is an average of the attractive force per molecule. This causes a reduction of the pressure on the container wall, which we shall designate as P_i. This reduction in pressure has the general form

$$P_i = \left(\frac{N}{V}\right)^2 \alpha, \tag{8.16}$$

in which α proportional to f_m.

At a molecular level the situation in the bulk gas is then very similar to the kinetic picture of ideal gas (Bernoulli) except that the volume available to the molecules is modified by the size of the sphere of action surrounding each molecule. We shall then assume that the pressure of the bulk gas is related to the kinetic energy in the same way as in Bernoulli's ideal gas (see Sect. 8.2.1, (8.5)). If the sphere of action is introduced to limit the volume available to molecules colliding with a surface, the calculation of the pressure in the bulk gas, P_g, results in

$$P_g = Nm\frac{\overline{v^2}}{3(V-B)}, \tag{8.17}$$

where

$$B = N\frac{2\pi\sigma^3}{3} \tag{8.18}$$

is half the volume occupied by the spheres of action of the molecules[5][10].

The pressure of the gas on the vessel wall is the measured quantity. This is equal to the bulk gas pressure, P_g, minus the reduction in pressure due to the molecular attraction, P_i. That is

$$P = P_g - P_i$$
$$= Nm\frac{\overline{v^2}}{3(V-B)} - N^2\frac{\alpha}{V^2}, \tag{8.19}$$

using (8.16) and (8.17).

For the ideal gas we also have $(Nm/3)\overline{v^2} = nRT$ (see Sect. 8.2.1, (8.6) and (8.13)). Then (8.19) becomes

$$\left(P + n^2\frac{a}{V^2}\right)(V - nb) = nRT. \tag{8.20}$$

---

[5] The calculation, although straightforward, is involved.

This is the van der Waals thermal equation of state. Here

$$b = \frac{2\pi\sigma^3}{3}N_A \tag{8.21}$$

is half the total volume of a mol of the spheres of action (N_A is Avogadro's number $6.0221367 \times 10^{23}\,\text{mol}^{-1}$) and

$$a = N_A^2\alpha$$

is proportional to the intermolecular attractive force.

The van der Waals equation of state is, therefore, a mathematical expression of three ideas, all of which are directly identifiable as aspects of the molecular constitution of matter. (1) The molecules move rapidly in almost straight lines between collisions, which are very well modeled as hard sphere encounters. (2) The molecules take up space. (3) There is an attractive force between the molecules. These three aspects of the molecular picture of matter are rather clearly identified in (8.20). The constant b is proportional to the volume of the molecular spheres of action. And the average effect of the attractive force is present in the constant a. The similarity of the van der Waals equation of state to that of the ideal gas is based in the kinetic picture of molecular motion. These are aspects of the molecular picture of matter that could be tested in the laboratory because van der Waals had proposed an equation of state that embodied them separately. There are other semiempirical equations of state, such as that of Dieterici [41, 42], which in some cases fit the data better than the van der Waals model [[83], p 68]. But here we shall consider only the van der Waals model because of its clarity.

8.3.2 Condensation

How well does the van der Waals equation of state represent real fluid behavior? To answer this question we shall compare the isotherms of a van der Waals fluid with those obtained for oxygen. The oxygen isotherms are plotted in Fig. 8.5. In the region below the dashed line the vapor and liquid phases coexist in equilibrium. In this region the isotherm is a horizontal line as the phase change from vapor (gas) at (a) to liquid at (b) takes place. The small black diamond is the critical point ($T_c = 154.58\,\text{K}$, $P_c = 5.043\,\text{MPa}$). The very steep isotherms to the left of the critical point in Fig. 8.5 are those of liquid oxygen. Oxygen does not exist in the liquid state at temperatures above that of the critical point.

8.3.3 Van der Waals and Maxwell

In Fig. 8.6 we have plotted a number of isotherms for a general van der Waals fluid contained in a closed system. These may be compared to the isotherms of the oxygen shown in Fig. 8.6. We see that above the critical point the general shape of the

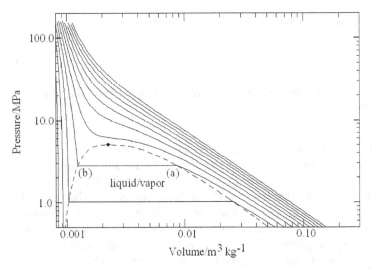

Fig. 8.5 Isotherms of oxygen [102]. In the region under the *dashed line* liquid and vapor coexist in equilibrium. At (**a**) is pure vapor at (**b**) is pure liquid. The critical point is the small *black diamond* at the *top* of the *dashed line*

isotherms is similar to that of the oxygen isotherms. The van der Waals isotherms below the critical point do not, however, have the correct form. The horizontal two phase region, shown in Fig. 8.6 for oxygen, is lacking in the van der Waals isotherms of Fig. 8.6. We may correct this defect with a simple thermodynamic argument. But we are not interested in simply adjusting the defects in the van der Waals equation of state. We are interested in understanding models of the molecular basis of matter. So we shall consider the remedy in detail.

The remedy was first proposed by Maxwell [107]. But his proposed remedy was presented almost as an after thought. Maxwell was primarily interested in how well the van der Waals model of molecules and their interactions reproduced measurements that had been made by Thomas Andrews in 1869 [2]. Maxwell was writing for readers not necessarily convinced of the molecular kinetic theory being developed. We are not those same readers. But we shall turn to Maxwell's insight and clarity to provide the understanding we seek of molecular models.

Fig. 8.6 Isotherms of a van der Waals fluid. The fluid is homogeneous. There is no two-phase region

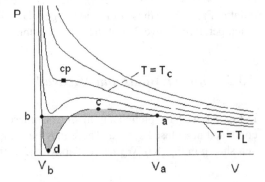

As the thermodynamic temperature increases above the critical point the kinetic energy of the molecules increases. The closest approach of the molecules during collisions and the collision times decrease. The weak, long range attractive forces have less effect and the infrequent, hard sphere collisions dominate. Therefore gas behavior becomes more ideal as the thermodynamic temperature increases. For high temperatures the van der Waals isotherms closely resemble those of the ideal gas, except for volumes small enough for the molecular volume to be noticed (high pressures). Van der Waals fluids become incompressible at small finite volumes, as do real fluids. The behavior of van der Waals fluids at high temperatures is not, however, as interesting as that at lower temperatures. It was the behavior of the van der Waals fluid near and below the critical point that attracted the attention of Maxwell.

In Fig. 8.6 we have indicated the critical isotherm at the temperature $T = T_c$, the critical temperature. The critical isotherm passes through the critical point cp that we have designated by a small black square. At the critical point the critical isotherm has a horizontal tangent and no curvature. That is both $(\partial P/\partial V)_T$ and $(\partial^2 P/\partial V^2)_T$ vanish at cp. Along all isotherms above the critical isotherm the slope $(\partial P/\partial V)_T$ is negative. This is a mechanical stability criterion required of any real physical substance. For each isotherm below the critical isotherm there is a portion, such as $c \rightarrow d$ on the isotherm at T_L, along which the slope of the isotherm $(\partial P/\partial V)_T$ is positive. Where $(\partial P/\partial V)_T > 0$ the van der Waals equation of state does not describe the behavior of any real substance.

Below the critical point real fluids (vapors) begin to condense as the pressure is increased isothermally. When this occurs the system will have vapor and liquid present and will no longer satisfy the assumption of homogeneity inherent in the van der Waals equation of state.

Maxwell, however, claimed that van der Waals' bold approach to the very difficult problem of the continuity of the liquid and gaseous forms of matter would not fail to have a notable effect on the development of a molecular theory of matter. This was the basis of Maxwell's interest in van der Waals' results, which he discussed in a publication [107] and in the revised edition of his *Theory of Heat* [108], that was published after his death. Maxwell was more interested in the continuity of the gas state at a into the region $a \rightarrow c$ and the continuity of the liquid state at b into the region $b \rightarrow d$ than he was in the nonphysical region $c \rightarrow d$.

Specifically Maxwell considered whether one might (isothermally) compress the vapor into the region $a \rightarrow c$ or decompress the liquid into the region $b \rightarrow d$. He answered affirmatively in both instances, pointing particularly to the phenomenon called *bumping*, in which a liquid may be heated in a smooth container beyond the normal boiling point with bubbles forming spontaneously in the bulk liquid. This failure of a liquid to boil without nucleation sites is a common occurrence, which is avoided by introducing boiling beads into distillation columns. Subcooling a gas below the condensation point is also a known phenomenon [cf. [123], p. 314]. The existence of these metastable states, and particularly their spontaneous destruction by the formation vapor (bubbles) or of condensed (droplets) in the bulk fluid indicates that matter is particulate, i.e. molecular.

The thermodynamic equilibrium state is not an absolute state imposed upon continuous matter. It is an average of kinetic molecular states. There will always be deviations (fluctuations) from the average if matter is particulate. For example if U_0 is the average energy of a system, then the actual energy of the collection of molecules making up the physical system will be

$$U = U_0 + \delta U, \tag{8.22}$$

in which δU is the fluctuation in the energy. There are similar fluctuations in any physical quantity if the system is actually particulate in nature. These fluctuations are not measurable on a macroscopic scale. They have averages equal to zero, as we see from the average of both sides of (8.22). But fluctuations are generally position and time dependent and may produce macroscopic effects depending upon conditions in the system. For example a correlation between fluctuations in energy and particle density δN may result in spontaneous condensation or boiling at some point in the bulk fluid. The growth of these fluctuations is related to correlations among them and averages of correlations are macroscopic quantities, such as $\overline{(\delta U \delta N)}$. These phenomena are properties only of particulate matter and a result of the fact that a particulate description of matter cannot be based simply on averages. The molecular theory of matter brings with it a rich basis for exploration.

We then have a more detailed graphical picture what is happening in the regions of the van der Waals fluid where condensation, or the phase transition between gas and liquid occur. This is illustrated in Fig. 8.7. The metastable states in Fig. 8.7 encompass portions of isotherms such as $a \rightarrow c$ and $d \rightarrow b$ in Fig. 8.6. The unstable states are those for which $(\partial P / \partial V)_T > 0$. The line separating these regions is called a spinodal [123].

Maxwell cited what he called the masterful experiments of Thomas Andrews [2]. In these experiments Andrews considered the transition of carbonic acid from a gas to a liquid state along an isotherm at a temperature above T_c. He showed that under these conditions the transition occurs with no separation into two phase regions within the system. The system remained homogeneous throughout, although in the initial state the system contained only a gas and in the final state it contained only a liquid. This was in agreement with the predictions of the van der Waals equation

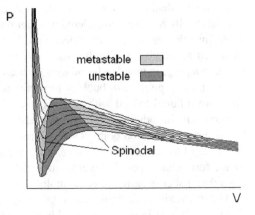

Fig. 8.7 Metastable (*light shading*) and unstable (*dark shading*) regions of the van der Waals model fluid. The line separating metastable and unstable regions is the spinodal

of state. The conclusion was that there is essentially no physical difference between the gas and the liquid state. We can understand this in terms of a molecular picture of matter. The gas and liquid differ only in the distance between the molecules. The exact role of the critical temperature in this dissolution of the phase separation remained obscure, however.

We then have at least qualitative agreement with the van der Waals ideas about the behavior of molecules in a fluid as long as the system remains homogeneous. This condition is met for compression above the critical temperature and into the metastable regions in Fig. 8.7 below the critical temperature. Metastable states are, however, not the subject of classical thermodynamics. The van der Waals equation of state may be adjusted to make it applicable in classical thermodynamics. To do this we must exclude metastable states and replace the transition between the gas in state a and the liquid of state b in Fig. 8.6 by a straight horizontal (isobaric) line, which agrees with experiment. How to construct the horizontal line for the heterogeneous two phase transition is then of some practical interest, but is not of fundamental interest in our understanding of the behavior of the molecules. Maxwell proposed a simple graphical construction to accomplish this that is now known as the Maxwell construction.

The first section of Gibbs' paper on the equilibrium of heterogeneous substances appeared in the October, 1875, edition of the Transactions of the Connecticut Academy. In this work Gibbs established the basis of thermodynamic stability of exactly the type that interested Maxwell here. But this work by Gibbs was not available when Maxwell published his journal treatment of the van der Waals equation. So our present treatment of this problem as simply one of requiring that the Gibbs energy is continuous between the states a and b is based on a thermodynamic stability requirement not available to Maxwell. The Maxwell construction is based on the second law.

Maxwell asked us to assume that the system passed from state a to state b through the hypothetical sets of states on the van der Waals isotherm and then returns to the state a along the straight line, two phase process $a \rightarrow b$. The cycle from a back to a is then isothermal, and the second law denies the possibility of doing work in this cycle. The work done is the algebraic sum of the areas shaded in Fig. 8.6. This sum is zero if the areas are identical. The Maxwell construction then requires that these areas be equal. Using the thermodynamic stability requirements of Gibbs (see Sect. 12.4) we realize that this requires that the Gibbs energy at state a is identical to that at b. The equality of the areas then follows from (4.39) with $dT = 0$. Our mathematical treatment is then extremely simple. We should not, however, allow that to obscure the point that Maxwell is trying to preserve the van der Waals construct in the unshaded regions of Fig. 8.7. To treat the van der Waals equation as a noble but inadequate attempt at an equation of state is to miss the subtleties of the molecular picture it reveals, as we have shown in our discussion above. To go beyond this, however, requires the methods of statistical mechanics.

As we shall see in Sect. 8.4, we can obtain much better agreement with experiment with more modern equations of state. But we lose a physical understanding of the parameters involved.

8.3.4 Critical Point and Constants

The isotherm passing through the critical point in both the real fluid (oxygen) of Fig. 8.5 and the van der Waals fluid of Fig. 8.6 has a horizontal tangent and no curvature at the critical point. That is

$$\left(\frac{\partial P}{\partial V}\right)_T\bigg|_{T=T_c,\ P=P_c} = \left(\frac{\partial^2 P}{\partial V^2}\right)_T\bigg|_{T=T_c,\ P=P_c} = 0 \qquad (8.23)$$

for both the real fluid and the van der Waals fluid. These are two algebraic equations from which we can obtain the two constants in the van der Waals equation of state. The result (see Exercises) is the set of relationships

$$V_c = 3nb, \qquad (8.24)$$

$$P_c = \frac{1}{27}\frac{a}{b^2}, \qquad (8.25)$$

and

$$T_c = \frac{1}{R}\left(\frac{8}{27}\frac{a}{b}\right). \qquad (8.26)$$

Because there are only the two van der Waals constants, and there are three eqs. (8.24), (8.25) and (8.26), we can combine these to yield

$$a = \frac{27}{64}\frac{(RT_c)^2}{P_c}, \qquad (8.27)$$

$$b = \frac{R\,T_c}{8\,P_c}, \qquad (8.28)$$

and

$$T_c = \frac{8}{3nR}P_c V_c. \qquad (8.29)$$

The eqs. (8.27) and (8.28) can be used to calculate a and b from data for the critical point (T_c, P_c) [102]. We provide the results for some common gases in Table 8.1.

If we normalize T, P, and V to the critical point values, defining the reduced pressure and temperature as

$$\overline{P} = \frac{27b^2 P}{a},$$

$$\overline{T} = \frac{27RbT}{8a},$$

and the reduced molar volume as

$$\overline{v} = \frac{V}{3nb},$$

we find that the van der Waals equation of state takes the form

Table 8.1 Van der Waals constants

Gas	$a/\mathrm{Pa\,m^6\,mol^{-2}}$	$b/\left(10^{-5}\,\mathrm{m^3\,mol^{-1}}\right)$
He	0.00346	2.37
Ne	0.0216	1.73
H_2	0.0247	2.66
A	0.136	3.22
N_2	0.137	3.86
O_2	0.138	3.19
CO	0.147	3.95
CO_2	0.366	4.28
N_2O	0.386	4.44
NH_3	0.423	3.72
H_2O	0.553	3.05
SO_2	0.686	5.68

Source: D.R. Lide (ed.): CRC Handbook of Chemistry and Physics, 84th ed. (CRC Press, Boca Raton).

$$\left(\overline{P}+\frac{3}{\overline{v}^2}\right)(3\overline{v}-1)=8\overline{T}. \tag{8.30}$$

That is, all fluids that can be represented accurately by the van der Waals model have the same equation of state. This is known as the law of corresponding states.

8.4 Beyond van der Waals

Modern applications have required better equations of state than that of van der Waals [cf. [123]]. And basic scientific investigations into molecular interactions have led us to more general formulations of equations of state based on statistical mechanics [cf. [52, 139]]. These statistical mechanical formulations lead directly to the virial equation of state, which is an expansion in terms of the density. Since density is small for real gases and approaches zero for the ideal gas, we may consider an expansion in terms of the density to be intuitively logical without any reference to statistical mechanics. In this section we shall pursue a density expansion in order to place the van der Waals equation of state in a more general context. This will provide a physical basis for possible extensions of the van der Waals equation of state.

8.4.1 Compressibility

We may extend the ideal gas equation of state to nonideal situations by writing a general gas equation of state as

$$PV = nzRT,$$

where z is the compressibility. The compressibility

$$z = \frac{Pv}{RT}.$$ (8.31)

is then a dimensionless quantity that conveniently expresses the deviations of the gas from ideality. For the ideal gas $z = 1$.

The compressibility depends on two thermodynamic variables, which we may choose in any convenient fashion. In Fig. 8.8 we present the compressibility $z = z(T,P)$ as a set of isotherms in the (z,P) plane. The dashed line is the saturation line enclosing the two-phase liquid-vapor region. The lowest isotherm in Fig. 8.8 is at the critical temperature ($T = 126.19\,\text{K}$) and the highest is at 600 K. These two isotherms are labeled in Fig. 8.8. The isotherm at the critical temperature is tangent to the saturation line at the critical pressure (3.3958 MPa). The temperatures of the other isotherms are those between 130 and 250 K in steps of 20 K. Each isotherm approaches $z = 1$ as the pressure decreases to zero and the gas becomes ideal.

We note that the value of z for the isotherm at 600 K is always ≥ 1 for all pressures, while for all other isotherms shown in Fig. 8.8 the value of z is less than 1 over some of the range of pressure. Also as the pressure increases the value of z for all isotherms becomes greater than 1 and continues to increase. The positive curvature of the isotherms is only less as the temperature of the isotherm increases. In general terms the initial drop in the value of z for low temperature isotherms is a result of the attractive component of the intermolecular force. The increase of the value of z for any isotherm, as pressure increases, indicates the dominance of the repulsive

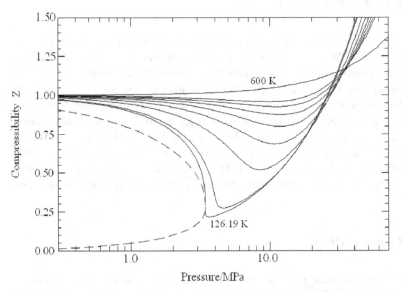

Fig. 8.8 Compressibility z of nitrogen [102]. The *dashed (saturation) line* separates the liquid/vapor two phase region. The critical isotherm (126.19 K) is tangent to the *saturation line*

component of the intermolecular force as the pressure (and, therefore density) increases, while the temperature remains constant. We shall see this below for the van der Waals model.

Since the specific volume of the ideal gas is RT/P, we may write the compressibility as

$$z = \frac{v}{v^{ig}},$$

where v^{ig} is the ideal gas specific volume. We have seen that the van der Waals fluid behaves like an ideal gas as the temperature increases and the molecular interactions are dominated by the repulsive (hard sphere) forces. But in the van der Waals fluid the molecules still occupy a volume. Therefore, in high temperature regions the van der Waals fluid behaves like an ideal gas for which the specific volume is $(v - b)$. The compressibility of the van der Waals fluid (gas) is then

$$z_{hs} = \frac{v}{(v-b)}, \tag{8.32}$$

where the subscript hs designates the hard sphere approximation.

From (8.20) we obtain

$$z = z_{hs} - \frac{a}{RTv}. \tag{8.33}$$

The contribution to the compressibility neglected by approximating the intermolecular interaction by hard spheres of finite volume is then $-a/RTv$.

8.4.2   Virial Expansion

Using (8.32) in (8.33) we have

$$z = 1 + \frac{b\rho}{(1 - b\rho)} - \left(\frac{a}{RT}\right)\rho, \tag{8.34}$$

where $\rho = 1/v$ is the density of the fluid (gas). Expanding (8.34) in powers of the density we obtain

$$z = 1 + \left(b - \frac{a}{RT}\right)\rho + b^2\rho^2 \cdots \tag{8.35}$$

for the van der Waals fluid. Equation (8.35) for the compressibility in terms of powers of the density is the virial equation of state, or simply virial expansion, referred to above. The virial equation of state is written generally as

$$z = 1 + B(T)\rho + C(T)\rho^2 + \cdots. \tag{8.36}$$

The coefficient $B(T)$ of the first power in ρ, is called the second virial coefficient and the coefficient $C(T)$ of the second power in ρ is called the third virial coefficient. For the van der Waals fluid the second virial coefficient

$$B_{vdW}(T) = b - \frac{a}{RT} \qquad (8.37)$$

depends on the temperature and all other virial coefficients are positive constants. The $n+1$st virial coefficient is equal to b^n.

In the van der Waals model the second virial coefficient $B_{vdW}(T)$ is negative for low temperatures when the attractive force contribution a/RT dominates, and becomes positive as the thermodynamic temperature increases. With an increase in thermodynamic temperature the contribution of the hard sphere (repulsive) component of the intermolecular force, which is characterized by the constant b, dominates the second virial coefficient (see Sect. 8.3.1, (8.21)). For nitrogen, using $a = 0.137\,\mathrm{Pa\,m^6\,mol^{-2}}$ and $b = 3.86 \times 10^{-5}\,\mathrm{m^3\,mol^{-1}}$ (see Table 8.1), we find that the value of $B_{vdW}(T)$ changes sign at $T = 425.89\,\mathrm{K}$. The actual temperature at which the second virial coefficient of nitrogen changes sign from negative to positive is 326.28 K [102]. We present a comparison of the actual second virial coefficient and the van der Waals prediction of (8.37) in Fig. 8.9.[6]

The third virial coefficient of nitrogen is almost constant for high temperatures and very nearly equal to the value

$$C_{vdW}(T) = b^2 = 1.9 \times 10^{-6}\,\left(\mathrm{m^3\,kg^{-1}}\right)^2$$

predicted by the van der Waals model for nitrogen. The actual value of $C(T)$ for nitrogen is between $1.9 \times 10^{-6}\,\left(\mathrm{m^3\,kg^{-1}}\right)^2$ at 273 K and $1.5 \times 10^{-6}\,\left(\mathrm{m^3\,kg^{-1}}\right)^2$ at 800 K [102].

In general we may obtain a virial expansion for the compressibility of any non-ideal gas as a Taylor series expansion of $z(T,\rho)$ in ρ

$$z = 1 + \rho \left(\frac{\partial z}{\partial \rho}\right)_T\bigg|_{\rho=0} + \frac{1}{2!}\rho^2 \left(\frac{\partial^2 z}{\partial^2 \rho}\right)_T\bigg|_{\rho=0} + \cdots \qquad (8.38)$$

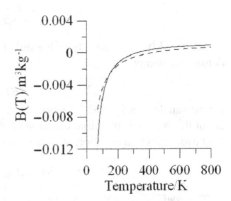

Fig. 8.9 Second virial coefficient for nitrogen. *Solid line* data from reference [102], *dashed line* calculated from (8.37)

---

[6] Actual data are at constant density.

since $z = 1$ in the ideal gas limit ($\rho \to 0$). In general the second and third virial coefficients are then

$$B(T) = \left(\frac{\partial z}{\partial \rho}\right)_T\bigg|_{\rho=0},\tag{8.39}$$

and

$$C(T) = \frac{1}{2!}\left(\frac{\partial^2 z}{\partial^2 \rho}\right)_T\bigg|_{\rho=0}.\tag{8.40}$$

Another useful extreme is the high temperature limit $1/RT \to 0$. In this limit the attractive forces between molecules may be neglected and the molecules modeled as hard spheres. The general high temperature approximation to z is then a Taylor series in $(1/RT)$ around the point $(1/RT) = 0$, which is

$$z = z_{hs} + \frac{1}{RT}\left(\frac{\partial z}{\partial(1/RT)}\right)_\rho\bigg|_{\frac{1}{RT}=0}$$
$$+ \frac{1}{2!}\left(\frac{1}{RT}\right)^2\left(\frac{\partial^2 z}{\partial(1/RT)^2}\right)_\rho\bigg|_{\frac{1}{RT}=0} + \cdots,\tag{8.41}$$

since in the high temperature limit $z \to z_{hs}$. Using the virial expansion (8.38) for z we have, to first order in the density,

$$\left(\frac{\partial z}{\partial(1/RT)}\right)_\rho = \rho\frac{dB}{d(1/RT)}.$$

Therefore (8.41) can be written, to first order in the density, as

$$z = z_{hs} + \frac{\rho}{RT}\frac{dB}{d(1/RT)}\bigg|_{\text{high }T},\tag{8.42}$$

where we have written the limit $1/RT = 0$ as "high T." This simple relationship provides a bridge on which we can build better approximate equations of state while still claiming to have some basis in the physics of molecular interactions.

From (8.42) we see that

$$\frac{1}{RT}\frac{dB}{d(1/RT)}\bigg|_{\text{high }T} = \lim_{\rho\to 0}\text{ slope of }z(T,\rho)\text{ for high }T.$$

The value of this slope may be determined from isothermal measurements of $z(T,\rho)$ extrapolated to zero density. These provide empirical approximations for $dB/d(1/RT)$, which may then be integrated to obtain the second virial coefficient $B(T)$ and an empirical equation of state from (8.38).

For the van der Waals equation of state we have an analytical expression for the second virial coefficient from (8.33), which is

$$B_{\mathrm{vdW}}\left(T\right) = -a\left(\frac{1}{RT}\right).$$

Then

$$\frac{1}{RT}\frac{\mathrm{d}B_{\mathrm{vdW}}}{\mathrm{d}\left(1/RT\right)} = -\frac{a}{RT} \tag{8.43}$$

for the van der Waals fluid at high temperatures.

8.4.3 Redlich-Kwong Equation of State

Experimentally $B\left(T\right)$ is not a linear function of $1/RT$ for real gases, which indicates deviations from the assumptions in the van der Waals treatment. One of the most successful equations of state, which takes this deviation into account is the Redlich-Kwong equation of state [137]. Redlich and Kwong did not present a derivation of their equation in their original publication. Rather they argued for its validity based on the results obtained. We can, however, obtain their result as an extension of (8.43) [123].

Experimentally the dependence of $(1/RT)\mathrm{d}B(T)/\mathrm{d}(1/RT)$ for real gases is more accurately represented as proportional to $(1/RT)^{3/2}$ [137] than $(1/RT)$ as found for the van der Waals fluid. Then

$$\frac{1}{RT}\frac{\mathrm{d}B}{\mathrm{d}\left(1/RT\right)} = -\frac{a'}{(RT)^{\frac{3}{2}}},$$

or

$$B(T) = 2a'\left(\frac{1}{RT}\right)^{\frac{1}{2}}.$$

The integration constant is zero, since $\lim_{T\to\infty} B\left(T\right) = 0$. The virial equation of state in the form appearing in (8.42) is then, to first order,

$$z = z_{\mathrm{hs}} - \frac{a'\rho}{(RT)^{\frac{3}{2}}}. \tag{8.44}$$

If we now accept the van der Waals form of z_{hs}, we have

$$z = \frac{1}{(1-b\rho)} - \frac{a'\rho}{(RT)^{\frac{3}{2}}}. \tag{8.45}$$

The final equation of state presented by Redlich and Kwong has an additional factor of $(1+b\rho)$ in the denominator of the second term in (8.45). That is

$$z = \frac{1}{(1-b\rho)} - \frac{a\rho}{RT^{\frac{3}{2}}(1+b\rho)} \tag{8.46}$$

in the form it was originally presented by Redlich and Kwong. Here a may still be interpreted in terms of an attractive force. However, the interpretation cannot be equivalent to that of van der Waals. There is no theoretical justification for the inclusion of the factor $(1 + b\rho)$ [138], which affects the physical interpretation of a in (8.46). Modern treatments also include a dependence of a on temperature [123]. Nevertheless, the Redlich-Kwong equation of state, and modifications of it, provide some of the best representations of the actual behavior of fluid substances that we have. On the one hand the fact that we can still base these equations of state in some fashion on the basic aspects of intermolecular interactions provides something of the microscopic insight we are seeking in this chapter. On the other hand, however, the fact that we are not able to derive an accurate equation of state based on simple aspects of molecular behavior indicates the complexity inherent in the behavior of matter.

Our treatment here has also been only of a pure homogeneous substance. We have chosen to ignore the difficulties introduced when interactions among molecules of different components must be considered [52]. These situations are, of course, of great interest in almost any application. And considerable effort has been devoted to developing methods for treating mixtures of fluid substances. Our goal in this chapter, however, has been to provide insight into the microscopic, molecular treatment of matter before undertaking a systematic study of microscopic thermodynamics. Our treatment of the Redlich-Kwong equation of state has taken us a step beyond this goal. And, in the development of the Redlich-Kwong equation of state we have seen something of the sort of approach that is required to treat real systems. Statistical mechanics will provide a systematic basis for studies of the molecular basis of matter. But there will still be no simple path revealing the complexities underlying the behavior of matter [cf. [97]].

8.5 Summary

In this chapter we have seen that the basic kinetic description of matter in a gas or vapor state retains the foundational ideas of Bernoulli. We showed that interactions among the molecules explained some of the observations, such as diffusion in gases. And we considered the general form of the molecular interactions as a combination of attraction and repulsion. The very important concept of a statistical average was introduced by Maxwell and Boltzmann. And we saw the consequences of this in what the van der Waals equation of state reveals, provided we know what to look for. But there we also saw the failure of some aspects of this approach. To go beyond the van der Waals equation of state, such as in the discussion of the Redlich-Kwong equation of state, it was necessary to abandon the clarity of the van der Waals model and accept temperature and density dependencies that could not be readily explained. We need a more systematic approach to the problem of basing our science on the molecular structure of matter. This will be the topic of Chap. 9.

Exercises

8.1. Show that the internal energy of a van der Waals gas is $U = nC_V T - n^2 a/V +$ constant, where C_V is a constant specific heat. (Use (5.17))

8.2. With the internal energy of the van der Waals fluid in Exercise 8.1 show that the fundamental surface for the van der Waals fluid is $S = nR \ln \left[(V - nb)(U + n^2 a/V)^{C_V/R} \right]$ (Hint: first obtain the entropy as a function of (T,V) and then eliminate the temperature algebraically).

8.3. From the form of the fundamental surface for the van der Waals fluid in Exercise 8.2 obtain the Helmholtz energy for the van der Waals fluid and obtain the equation of state in the normal fashion from the Helmholtz energy.

8.4. Using the condition (8.23), find the van der Waals constants a and b in terms of the volume V_C and pressure P_C at the critical point. With these obtain the nondimensional form of the van der Waals equation of state (8.30).

8.5. Unfortunately our assumption that the specific heat C_V for the van der Waals fluid is a constant is not universally valid. Using (5.32) and the van der Waals equation of state to obtain a universal relationship between C_P and C_V for the van der Waals fluid.

8.6. We will want to know good our original approximation of constant specific heats really is if we want to use the van der Waals model. Obtain an approximation to $C_P - C_V$ in (8.5) by expanding in terms of the interaction parameter is a. Obtain a final form in terms of the compressibility factor $z(P,T) = PV/RT$ and the grouping an^2/PV^2, which gives a direct indication of the dependence on the van der Waals force.

8.7. Using (6.6) obtain the speed of sound in a van der Waals fluid. Does this reduce to the speed of sound in an ideal gas (see Exercise 5.10)?

8.8. In Exercise 2.19 we encountered an initial experimental attempt to ascertain the dependence of the internal energy of a gas on volume in 1807 by Gay-Lussac. Neither Gay-Lussac nor Joule could obtain reasonable results from this experiment. Why this would have been the case can be deduced if we consider the results we should expect from such an experiment based on the van der Waals model of a gas.

Determine the change in the temperature of the water bath in the course of the experiment if the gas being investigated is a van der Waals fluid. Use the internal energy of the van der Waals fluid as given in Exercise 8.1. You may assume that the specific heat of the water far exceeds that of the gas, and you may neglect (at least initially) the specific heat of the glass vessel. The experimental apparatus is shown in Fig. 8.10.

Fig. 8.10 Gay-Lussac
apparatus

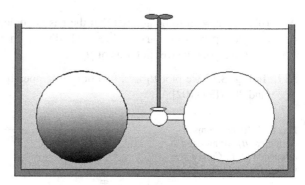

8.9. We encountered the Joule-Thomson in Exercises 2.20 and 3.7. We realize
that the experiment determined $(\partial H/\partial P)_T$ from a measurement of the Joule
coefficient $\mu_J = (\partial T/\partial P)_H$. We now have some reasonable models of a gas.

Find the Joule-Thomson coefficient for the van der Waals fluid. [Hint: use
(5.18)]

8.10. In the Joule-Thomson experiment the initial state of the gas and the final pressure are chosen. The final temperature is measured. Experimental data for
nitrogen are shown in Fig. 8.11 [102]. The slope of a smooth curve drawn
through these data is the Joule coefficient $\mu_J = (\partial T/\partial P)_H$. In the experiment
the initial pressure always must exceed the final pressure. So we may begin
at any one of the states (A), (B), or (C) in Fig. 8.11 and the final state in the
experiment may be any of the states to the left of the initial state.

   (a) Describe what happens if you begin a Joule-Thomson experiment at any
one of the states (A), (B), or (C) and expand to any of the allowed states
(B)–(D).

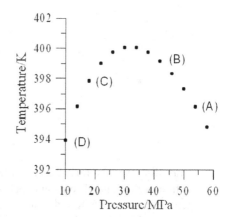

Fig. 8.11 Joule-Thomson
experiment. *Single points*
represent possible initial and
final states

(b) If you want to be assured that the gas is cooled on expansion where (on
what part of the curve in Fig. 8.11) do you choose your initial condition?
Give your answer in terms of μ_J.

8.11. In Fig. 8.12 we plot μ_J as a function of temperature for pressures of 10, 15,
and 20 MPa [102].

Fig. 8.12 Joule-Thomson
coefficient μ_J for nitrogen as
a function of T. *Curves* are
isobars for 10, 15, and 20 MPa

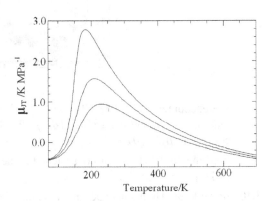

The point at which $\mu_J = 0$, is called the inversion point. We see in Fig. 8.12
that the inversion point is a function of (T, P). A plot of the inversion points
is called an inversion curve.

In Fig. 8.13 we present the inversion curve for nitrogen [102]. We have indicated the region for which $\mu_J > 0$ in Fig. 8.13.

Fig. 8.13 Inversion curve for
nitrogen

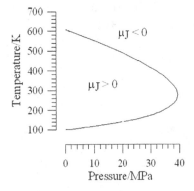

Throttling (see Exercise 2.21) is a term used to describe an industrial application of the Joule-Thomson experiment. We can liquefy air if we begin at a
temperature of 175 K and a pressure of 20 MPa and throttle to atmospheric
pressure. Find the final state you need to attain to accomplish this on Fig. 8.13.

8.12. This is the basis of Carl von Linde's (1842–1934) liquid air machine. At the
core of the machine is a throttling process. High pressure, low temperature

gas is throttled from a tube into a vessel. The liquid nitrogen accumulates at the bottom of the vessel, where it can be bled off. Because we want a continuous process, we recirculate the very cold gas remaining in the vessel after throttling. Recalling that we have to cool our compressed gas before throttling, we realize that this very cold, recirculated gas can be used before feeding back into the process.

Based on this plan, design your own liquid nitrogen machine.

8.13. What is the Joule-Thomson coefficient of the ideal gas?

8.14. Find the inversion temperature T_i (the temperature for which $\mu_J = 0$) for the van der Waals fluid.

8.15. Are there both heating and cooling regions for the van der Waals gas in throttling? Use your answer to Exercise 8.9.

8.16. Find the Helmholtz energy as a function of characteristic variables for the van der Waals fluid. Begin by writing down the steps as they were outlined in Chap. 5. This is always good practice. Then proceed with the steps you have outlined.

8.17. (a) Obtain an explicit form of the general thermodynamic relationship for $C_P - C_V$ in (5.32) for the van der Waals fluid.

(b) Show that if we neglect molecular size this reduces to $C_P - C_V \approx R(1 + 2aP/R^2T^2)$.

(c) How large is $2aP/R^2T^2$ for some familiar gases (see Table 8.1)? Can it be neglected for temperatures of the order of $300\,\mathrm{K}$?

8.18. Find the internal energy as a function of characteristic variables for the van der Waals fluid. Begin by writing down the steps as they were outlined in Chap. 5. This is always good practice. Then proceed with the steps you have outlined. [Hint: look carefully at your results in Exercises 8.16 and 8.17.]

8.19. For the virial equation written in the form $Pv/RT = 1 + B'(T)P + C'(T)P^2 + \cdots$ where B', C', $\cdots$ are functions of T (see (13.66)), find the Joule-Thomson coefficient.

8.20. (a) Find β for a van der Waals fluid.

(b) Show that it reduces to the value $1/T$ for the ideal gas.

Chapter 9
Statistical Mechanics

*The laws of thermodynamics ... express
the laws of mechanics ... as they appear
to beings ... who cannot repeat their
experiments often enough to obtain any
but the most probable results.*
Josiah Willard Gibbs

9.1 Introduction

Statistical mechanics is based on the particulate (atomic or molecular) structure of matter. These particles obey the laws of rational mechanics, which may be the classical mechanics of Newton, Lagrange, and Hamilton, or quantum mechanics. Our formulation must be insensitive to the choice we may make for the mechanics of the particles.

A rational thermodynamics, based on this particulate structure of matter, will result if we can find an expression for one of the thermodynamic potentials that contains only a mechanical description of the particles and requires no additional hypotheses regarding probabilities. The first objective of this chapter will be to identify a function, based solely on mechanical principles, with a thermodynamic potential. This identification will form our bridge between the microscopic and thermodynamic pictures of matter.

We shall base our development on the statistical mechanics of Gibbs. This is the foundation for all modern work based on a microscopic particulate description of matter. We shall demonstrate this through applications to fluids under wide ranges of temperature. In the subsequent chapter on quantum statistical mechanics our demonstration will include extreme conditions encountered in present research.

9.2 Gibbs' Statistical Mechanics

Between autumn of 1900 and summer of 1901 Gibbs wrote a summary of the final fourteen years of his scientific work, which was published by Yale University Press in 1902 as *Elementary Principles in Statistical Mechanics* [56]. With the title of this

C.S. Helrich, *Modern Thermodynamics with Statistical Mechanics*,
DOI 10.1007/978-3-540-85418-0_9, © Springer-Verlag Berlin Heidelberg 2009

book he gave the new science, based on microscopic physics, the name *statistical mechanics* [141].

Gibbs claimed that the lack of progress in developing a microscopic basis for thermodynamics resulted from focusing our efforts on the problem of explaining the laws of thermodynamics based on mechanical principles. Much work had been devoted to studies of the statistical behavior of individual molecules. Gibbs proposed instead to consider the statistical behavior of systems of molecules. This was not a theory of gases, or even of fluids; it was a theory of rational mechanics.

Central to Gibbs' statistical mechanics is the fact that macroscopic thermodynamics is an experimental science, which is based on the measurement of only a few parameters (e.g. temperature, pressure and volume), while the mechanical picture of the molecular system necessarily requires knowledge of positions, orientations, velocities, angular velocities, and vibrations of an extremely large number of particles.

Previous attempts to go from a detailed mechanical description of the behavior of the individual constituents of matter to a thermodynamic description required assumptions or postulates regarding probabilities. Examples are Boltzmann's assumption about the rate of collisions in a gas (see Sect. 14.3) and the postulates present in the concept of a single particle distribution function.

In the statistical mechanics Gibbs accepted that measurements we make on the system before us on the laboratory bench will be the same as those made on a vast number of such systems having the same few macroscopic parameters. This is true independent of the exact mechanical description of the particles. The macroscopic thermodynamic description is then common to a vast number of possible mechanical systems.

Although we now understand the power of statistical mechanics, Gibbs' claims were modest. In 1900 we were unable to account for the fact that our measurements of specific heats did not reflect the expected seven quadratic energy terms for diatomic molecules (see Sects. 8.2.1 and 9.10). This deterred Gibbs from, in his words, "attempting to explain the mysteries of nature" and forced him to be content with a more modest aim. In retrospect, however, it is precisely this modesty that produced one of the greatest strengths in the resulting theory: independence of the form of the mechanics.

In his classic work, *The Principles of Quantum Mechanics*, Paul A. M. Dirac shows that the Gibbs theory emerges naturally from the elements of the quantum theory [[44], pp.130–135]. The Pauli Principle will require the introduction of modifications in the counting procedures for quantum particles. But Gibbs' statistical mechanics remains unscathed by the quantum theory.

9.3   Ensembles of Systems

A rigid complex molecule can translate and rotate. The description of this motion requires three coordinates to locate the center of mass and three angles to specify the orientation of the molecule about the center of mass. Each of these coordinates has associated with it a translational or angular momentum. That is 12 variables, each of which depends on the time, are required to specify the exact mechanical state of each rigid complex molecule. If there are N molecules in the system then $12N$ variables are required to specify the exact mechanical state of the system.

To obtain the coordinates and momenta for all of the N particles of the system as functions of the time we must solve the $12N$ coupled mechanical equations of motion for those N particles. Even if we have the capability of numerically integrating this set of equations we still require values for the $12N$ initial positions and momenta at some instant, to use as the required initial conditions. Knowledge of these initial conditions for the system requires measurements beyond our physical capabilities. The exact state of the mechanical system of molecules is forever vastly underdetermined.

Systems of molecules may differ from one another by something as small as the orientation of a single molecule. Generally systems of molecules, which cannot be distinguished from one another by any measurement we can make, may have large numbers of molecules with energies, locations, and orientations that differ among systems. These mechanical systems are, however, identical in nature because they are indistinguishable from one another by any measurement we are able to perform. Such systems are members of what is called an ensemble of systems.

We shall refer to the physical properties of a single system in the ensemble as *microscopic properties*. These are determined by the *microscopic state* or *microstate* of the system, which is known if the all positions and all momenta of all of the particles are known. From these positions and momenta values of properties such as the total energy, momentum, and density may be calculated. These are the respective microscopic energy, momentum, and density of the single system.

Each ensemble is specified by the choice of a certain set of properties with the same values for all of the systems in the ensemble. For example we may choose to consider an ensemble of all systems having a certain thermodynamic temperature and a certain number of molecules. Each system in this ensemble may not have, however, exactly the same value of the energy. The energy of the σth system in the ensemble is the value of the *Hamiltonian*[1] for that system, which we shall designate as H_σ. This is the sum of the individual energies of the molecules. A change in the states of a subset of those molecules will not be sufficient to affect the measured temperature. But that change will result in a different system in the ensemble with a different numerical value for the Hamiltonian.

Gibbs said that we should identify thermodynamic quantities with averages of the corresponding microscopic quantities over all systems in the ensemble. So the thermodynamic internal energy of the systems in our constant temperature ensemble is the average of the system Hamiltonian over all systems in the ensemble. This is termed the *ensemble average* of the Hamiltonian and is designated as $\langle H \rangle$. The deviation of the value of a microscopic property for a single system from the ensemble average of that property is termed a *fluctuation*.

---

[1] The Hamiltonian is named in memory of William Rowan Hamilton (1805–1865) the Irish physicist and mathematician who developed the form of Newtonian mechanics commonly used in modern physics.

We, therefore, focus our attention on the behavior of the ensemble, rather than on the most probable behavior of the molecules in a particular system. The relationships among these ensemble averaged quantities constitutes what Gibbs called *rational thermodynamics*, or *statistical thermodynamics*.

The relationships among ensemble averaged quantities are a consequence of the fact that matter is particulate and that those particles behave according to certain laws reflected in the form of the Hamiltonian. The object of statistical mechanics is to discover which ensemble averaged quantities correspond to thermodynamic functions. When these are discovered we gain an understanding of how the behavior of large numbers of particles result in thermodynamic behavior of those systems. And we can discover the consequences of changes in the Hamiltonian on that thermodynamic behavior. In this we do not attempt to explain the laws of thermodynamics. They are valid independently of our microscopic picture of matter. And the relationships among the ensemble averaged quantities are a result of rational mechanics and independent of the laws of thermodynamics. Discovering a correspondence provides between certain ensemble averaged quantities and thermodynamic potentials provides a profound insight which continues to form the basis for fruitful research. But this does not explain the laws of thermodynamics on mechanical principles.

9.4 Phase Space

9.4.1 Concept

The mechanical state of a system is most easily represented in the phase space of classical mechanics. This representation may also be used for quantum systems as long as we are concerned only with variables having classical analogs.

Example 9.4.1. As an example we consider the simple harmonic oscillator, which is a mass on a spring. This is a conservative mechanical system for which the Hamiltonian H is the sum of kinetic and potential energies

$$H = \frac{p^2}{2m} + \frac{1}{2}k_S x^2, \tag{9.1}$$

where $p = m\,dx/dt$ is the momentum of the oscillator and k_S is the spring constant. The plot of p versus x is the phase space trajectory of the oscillator. Since the energy of the oscillator is constant ($H = $ constant) the phase space trajectory is an ellipse. This is shown in Fig. 9.1. The mechanical state of the oscillator is represented by a point on the phase space trajectory, as shown in Fig. 9.1. We call this point the representative point of the oscillator. If we know the location of the representative point we know all the mechanical properties of the oscillator. This representative point moves along the trajectory as the mechanical state of the oscillator changes with time.

Fig. 9.1 Phase space
trajectory of harmonic
oscillator

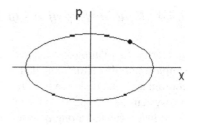

A particle moving in two dimensions requires a phase space with four dimensions: two position and two momentum coordinates. We can speak of a trajectory in this four dimensional phase space. But we can not draw the phase space trajectory for this simple system, or even construct a mental picture of it. We can construct any number of orthogonal axes mathematically, but we are not able to translate this into even a mental image of four mutually perpendicular axes.

9.4.2 μ - Space

The problem is no more difficult mathematically for more realistic situations. If the molecules are rigid we require 6 coordinates and 6 momenta. The trajectory of the representative point is then a curve in a phase space of 12 dimensions. And the mechanical state of our molecule is represented by a point in this phase space of the single molecule. We cannot draw on paper nor picture in our minds a space of 12 dimensions. We can, however, refer to and understand this representation.

This single molecule phase space is called μ - space, where the Greek letter μ designates molecule. The state of our system of N molecules is then represented by a collection of N points in this μ - space; one for each molecule. The representative point of each molecule follows a trajectory in μ - space and the collection of moving representative points resembles a cloud or a swarm of bees.

9.4.3 Γ -Space

We want to extend this concept to an ensemble of systems. To do this we require a phase space in which the state of each system is represented by a point. Because the microstate of our system of rigid molecules is specified if we have values for each of the 12 coordinates and momenta for each of the N molecules in the system, the state of the system is represented by a point in a phase space of $12N$ dimensions. This space is called Γ - space, where the Greek letter Γ designates gas. Modern applications of statistical mechanics include more than gases, but the name has remained.

9.4.4 Relationship of μ - to Γ -Space

To clarify the relationship between μ - and Γ - spaces we shall consider one of the simplest possible situations: that for which the particle state is specified by a single coordinate. This is shown in Fig. 9.2. We designate this coordinate as q and confine our system to the space $0 \le q \le L$. We may suppose, for example, that there is a force field oriented along q so that the particle energy depends on the value of the coordinate q. The μ - space for the particle then has one dimension. We partition this μ - space into two one dimensional *cells* of length $L/2$. These are labeled as cells (1) and (2) in the μ - space of Fig. 9.2

We assume that our system consists of two particles, which we designate as a and b with the coordinates q_a and q_b. We consider the situation in which one particle is located in cell (1) and the other in cell (2) and represent the particles by filled circles in the μ - space of Fig. 9.2. We have purposely placed the particles at arbitrary positions within the cells to emphasize the fact that specification of a cell in μ - space implies that the particle state is known only to within the limits of the cell dimension. We shall designate this state, which is determined by the number of particles in each of the μ - space cells, as the state Z^* [[46], p. 27].

The Γ - space for this system has two coordinates: one position for each particle. This is shown in Fig. 9.2. The partitioning of Γ - space reflects the original partitioning of μ - space. This we have emphasized by placing the drawing of Γ - space immediately below that of μ - space with partitions aligned.

If the particles are distinguishable from one another we can specify the cell in which each particle is located. If particle a is in cell (1) and b is in cell (2) the system occupies the lightly shaded region in Γ - space. If particle a is in cell (2) and b is in cell (1) the system occupies the darkly shaded region. For distinguishable particles the region occupied by the system in Γ - space is either the lightly or the darkly shaded region, but not both. If particles a and b are indistinguishable the

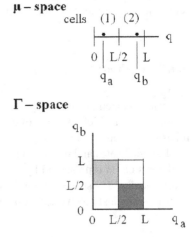

Fig. 9.2 Representation in μ - and Γ - space of particles described by a single coordinate q. The particles are located in separate one dimensional μ - space cells at q_a and q_b. In Γ - space there is one particle in each of two areas

system occupies both the shaded regions of Γ - space. The area occupied by the system in Γ - space then depends on whether the particles are distinguishable or indistinguishable.

We shall designate the area (or volume) of regions in Γ - space corresponding to a Z^* state as $W(Z^*)$. That is, for the situation we are considering, there are two possibilities. These are

$$W(Z^*) = \begin{cases} (L/2)^2 & \text{for distinguishable particles} \\ 2(L/2)^2 & \text{for indistinguishable particles.} \end{cases}$$

We may generalize this by partitioning our one-dimensional μ - space into a large number m of one dimensional cells of of length ℓ and considering N particles with representative points distributed in these cells. The state Z^* is now that in which there are a number N_i^* particles in the ith cell in μ - space. The Z^* state is then specified by the set of occupation numbers $\{N_i^*\}_{i=1}^m$ of the m cells in μ - space. Our Γ - space now has N dimensions and can no longer be drawn.

A more general situation is one in which μ - space has at least 6 dimensions: three position and three momentum coordinates for point particles. We partition this more general μ - space into equal cells each with a 6 dimensional volume ω. This volume ω in $\mu-$ space corresponds to the length $L/2$ of the one-dimensional cells in the μ - space of Fig. 9.2

The partitioning, or coarse-graining, of μ - space is a mathematical exercise and is independent of the physics, except that it reflects the accuracy we can claim in determining the particle state. The final limits of the cell size are provided by the Heisenberg indeterminacy (uncertainty) principle, which forbids a mathematical limit of zero cell volume. We may then consider that all particles with representative points inside of the cell have the same state.

We consider a particular state of the system in which m of the μ - space cells contain particle representative points and the rest are empty. If the particles are distinguishable the volume $W(Z^*)$ in Γ - space is ω^N. This corresponds to the area $(L/2)^2$ in the Γ - space of Fig. 9.2. If the particles are indistinguishable the volume in Γ - space occupied by the state Z^* increases by a factor equal to the number of ways we can distribute N particles among the m cells with N_i particles in the ith. This is $N!/\prod_{i=1}^m N_i!$. The volume occupied by the state Z^* for a system of N indistinguishable particles in Γ - space is then

$$W(Z^*) = \frac{N!}{\prod\limits_{i=1}^m N_i!} \omega^N. \tag{9.2}$$

This corresponds to the area $2(L/2)^2$ in the Γ - space of Fig. 9.2.

Two separate Z^* states Z_k^* and Z_ℓ^* correspond to two separate sets of occupation numbers and result in two separate disjoint volumes in Γ - space $W(Z_k^*)$ and $W(Z_\ell^*)$. To see this we consider our example of two indistinguishable particles in

Fig. 9.3 Representation in μ - and Γ - space of particles described by a single coordinate q. The particles are located in the same one dimensional μ - space cell. In Γ - space there are two particles in each of two areas

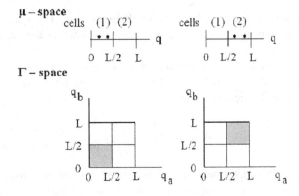

one (μ - space) dimension. Placing one particle in each cell results in $W(Z^*)$ equal to the sum of the shaded regions in Fig. 9.2. In Fig. 9.3 we have represented the two possible Z^* states in which there are two particles in single cells. The two separate Γ - space volumes are shown. Then we have three possible Z^* states for two indistinguishable particles: $Z_1^* \Longrightarrow \{N_1^* = 1, N_2^* = 1\}$, $Z_2^* \Longrightarrow \{N_1^* = 2, N_2^* = 0\}$, and $Z_3^* \Longrightarrow \{N_1^* = 0, N_2^* = 2\}$. These result in three disjoint volumes in Γ - space:

$W(Z_1^*)$ = sum of both shaded volumes in Fig. 9.2
$W(Z_2^*)$ = lower left shaded volume in Fig. 9.3
$W(Z_3^*)$ = upper right shaded volume in Fig. 9.3

An ensemble average may then be separated into a sum of integrals over all possible (separate) Z^* states with disjoint volumes. This is particularly helpful if we deal with the microcanonical ensemble.

9.4.5 Volumes in Phase Space

We shall assume that the coordinates and momenta take on a continuous set of values. This is true for quantum systems with closely spaced energy levels as well as for all classical systems. We then have continuous sets of coordinates $\{q_\alpha\}_{\alpha=1}^{12N}$ and momenta $\{p_\alpha\}_{\alpha=1}^{12N}$ which specify the state of the system. We shall use the designation Ω for a point in Γ - space. Specifically

$$\Omega = \left(\{q_\alpha\}_{\alpha=1}^{12N}, \{p_\alpha\}_{\alpha=1}^{12N}\right). \tag{9.3}$$

For simplicity we number the coordinates and momenta sequentially so that molecule 1 has the sets $(q_1 \dots q_{12})$ and $(p_1 \dots p_{12})$, molecule 2 has $(q_{13} \dots q_{24})$ and $(p_{13} \dots p_{24})$, and so on. We may then define an infinitesimal volume in this Γ - space as

$$d\Omega = C_N^{\text{dim}} dq_1 \dots dq_{12N} dp_1 \dots dp_{12N}. \tag{9.4}$$

We have introduced the constant C_N^{dim} in order to make the Γ - space differential $d\Omega$ non-dimensional. The form of C_N^{dim} comes from Heisenberg's indeterminacy principle, which requires that the product of two differentials $dq_\mu dp_\mu$ can be no smaller than $\hbar/2 = h/4\pi$, where h is Planck's constant equal to $6.6260755 \times 10^{-34}$ J s or $4.1356692 \times 10^{-15}$ eV s.

We then divide Γ - space into cells of volume h^{fN} where f is the number of pairs of coordinates (g_μ, p_μ) required to define the state of a molecule. Then

$$C_N^{\text{dim}} = \frac{1}{h^{fN}}. \tag{9.5}$$

If the N particles are indistinguishable we introduce a factor of $1/N!$. For indistinguishable particles we have then

$$C_N^{\text{dim, indis}} = \frac{1}{N! h^{fN}}. \tag{9.6}$$

Inclusion of the factor $1/N!$ gives us what is called "correct Boltzmann counting." This is also a mathematical resolution of the so-called Gibbs paradox. Referring to this as a paradox is something of a misnomer, as we pointed out in Sect. 4.7.2 resulting from historical confusion. But this serves as a standard identifier.

9.5 Ensemble Averages

The representation of an ensemble in Γ - space is the collection of all of the representative points for the systems in the ensemble. The representative point of a particular system moves along a trajectory in Γ - space as the values of the coordinates and momenta of each of the individual molecules change. The ensemble is then a dynamic cloud of moving points in Γ - space. This cloud will be more or less dense in various regions of Γ - space depending on the system trajectories.

We define the density of the cloud of representative system points in Γ - space as $D(\Omega, t)$ where t is the time. This density is defined by the requirement that $D(\Omega, t) d\Omega$ is the number of systems with representative points in the volume $d\Omega$ at the time t. The fraction of the systems in the ensemble contained in $d\Omega$ at the time t is $D(\Omega, t) d\Omega / \mathfrak{N}$, where $\mathfrak{N}$ is the total number of systems in the ensemble. The difficulty is that the number of systems in the ensemble is actually infinite, as is the number of systems in the infinitesimal volume $d\Omega$. However, the ensemble average of the microscopic, single system property $\Phi(\Omega)$, given at the time t by

$$\langle \Phi(t) \rangle = \frac{\int_\Gamma \Phi(\Omega) D(\Omega, t) d\Omega}{\mathfrak{N}}, \tag{9.7}$$

in which the integral is over all of Γ - space, is finite. That is the function

$$P_\Gamma(\Omega, t) = \frac{D(\Omega, t)}{\mathfrak{N}} \tag{9.8}$$

remains finite and is integrable. Indeed

$$\int_\Gamma P_\Gamma\left(\Omega,t\right)\mathrm{d}\Omega = 1. \tag{9.9}$$

The function $P_\Gamma\left(\Omega\right)$ is the *coefficient of probability*[2] of the phase space volume $\mathrm{d}\Omega$. $P_\Gamma\left(\Omega,t\right)\mathrm{d}\Omega$ is a statistical measure of the number of system representative points in $\mathrm{d}\Omega$. With the definition in (9.8) the ensemble average in (9.7) becomes

$$\langle \Phi\left(t\right)\rangle = \int_\Gamma \Phi\left(\Omega\right)P_\Gamma\left(\Omega,t\right)\mathrm{d}\Omega. \tag{9.10}$$

To proceed we must have a functional form for $P_\Gamma\left(\Omega,t\right)$.

9.6   Coefficient of Probability

9.6.1   Nonequilibrium

The theory we have at this point is valid for equilibrium or for nonequilibrium systems. We shall limit ourselves, in this text, to equilibrium systems. Gibbs also limited himself to equilibrium considerations and "especially to those cases and properties which promise to throw light on the phenomena of thermodynamics." Modern investigations have not, however, stopped at this level.

In the general case the coefficient of probability obeys what is now known as the *Liouville equation* (see appendix) [154]. The Liouville equation is based on the canonical equations of Hamilton and contains all the difficulties of the entire mechanical problem. We may, however, introduce reduced distribution functions for 1, 2, ... particles and obtain, from the Liouville equation, a hierarchy of equations for these reduced distribution functions. This is known as the *BBGKY* (Born, Bogoliubov, Green, Kirkwood, and Yvon[3]) Hierarchy [[95], pp. 176–185], [[117], pp. 41–49], [[139], pp. 310–313]. The BBGKY hierarchy is a mathematical reduction of the original problem and alone is no simplification. To simplify the problem we write the reduced distribution functions in terms of single particle distributions and 2, 3, ... particle correlation functions [111], [[95], p. 178], [[117], p. 45]. We may then argue for the dropping of correlation functions above a certain level to obtain a truncated set of equations. The truncated set yields the Boltzmann equation if correlations are dropped [31].

---

[2] This quantity is defined in terms independent of additional postulates. The name is an identifier.
[3] The original references that resulted in identifying these names with this hierarchy are given in [Montgomery 1964] on p. 49.

9.6.2   Equilibrium

If our thermodynamic system is in equilibrium the Liouville equation is satisfied if $P(\Omega)$ depends on Ω only through the Hamiltonian. That is $P_\Gamma(\Omega) = P_\Gamma(H)$. This is true for either classical or quantum systems. Therefore any integrable function of the Hamiltonian is a perfectly acceptable candidate for the coefficient of probability $P_\Gamma(H)$. There is no other explicit requirement imposed on $P_\Gamma(H)$.

Our model of the molecules of the system determines the actual form of the Hamiltonian. For example for point atoms with an interaction potential $\phi_{\mu\nu}$ between particles μ and ν (see Fig. 8.4) and an interaction potential Φ_{boundary} with the system boundary, the Hamiltonian is

$$H(\Omega,V) = \sum_{\mu=1}^{3N} \frac{p_\mu^2}{2m} + \frac{1}{2} \sum_{\mu \neq \nu} \phi_{\mu\nu} + \Phi_{\text{boundary}}. \tag{9.11}$$

The Hamiltonian then depends on the Γ - space point Ω (see (9.3)) and, because of Φ_{boundary}, also on the system volume V. We then designate the Hamiltonian as $H(\Omega,V)$.

To guarantee that (9.9) will be satisfied $P_\Gamma(H)$ must remain finite for all possible values of the Hamiltonian. This excludes the simple choice $P_\Gamma(H) = \text{constant}$, except if we limit $P_\Gamma(H)$ in such a way that it vanishes everywhere except on a thin shell. That is

$$P_\Gamma(H) = \begin{cases} \text{constant for} \leq H \leq E + \Delta E \\ 0 \text{ otherwise.} \end{cases} \tag{9.12}$$

Where the energy limit ΔE is arbitrarily small. The coefficient of probability defined by (9.12) is that for the microcanonical ensemble. All systems in the microcanonical ensemble have essentially the same energy, to within the band of width E.

The exponential function

$$P_\Gamma(H) = \exp\left(\frac{\Psi - H}{\Theta}\right), \tag{9.13}$$

in which Ψ and Θ are functions of thermodynamic parameters and independent of the coordinates of Γ - space, is also a logical candidate for $P_\Gamma(H)$. Both Ψ and Θ have dimensions of energy and Θ is positive. Gibbs claimed that the function in (9.13) seems to be the simplest functional form conceivable for $P_\Gamma(H)$. The coefficient of probability defined by (9.13) is that for the canonical ensemble.

There is another ensemble beyond the canonical: the grand canonical ensemble. In this ensemble the requirement of a known number of particles is relaxed. Perhaps surprisingly this is the simplest ensemble to use for quantum systems. We will introduce the grand canonical ensemble in the next chapter.

9.7   Thermodynamics of Ensembles

9.7.1   Canonical Ensemble

We begin with the canonical ensemble because the direct mathematical treatment of the canonical ensemble is considerably easier than that of the microcanonical ensemble. Since Θ and Ψ are the only macroscopic parameters in (9.13), a connection with thermodynamics can only be formed from them. We shall consider each of these separately.

The Modulus Θ. We can determine the identity of the modulus Θ in (9.13) if we consider our system to be composed of two subsystems A and B separated by a diathermal wall. The combined system is in thermal equilibrium and the two subsystems are in thermal equilibrium with one another. The coefficients of probability for the systems taken separately are

$$P_\Gamma\left(H_A\right) = \exp\left(\frac{\Psi_A - H_A}{\Theta_A}\right) \tag{9.14}$$

and

$$P_\Gamma\left(H_B\right) = \exp\left(\frac{\Psi_B - H_B}{\Theta_B}\right) \tag{9.15}$$

For the total system the thermodynamic function Ψ is $\Psi_A + \Psi_B$ and the Hamiltonian H is $H_A + H_B$. The coefficient of probability for the combined system is then

$$P_\Gamma\left(H_A + H_B\right) = \exp\left[\left(\frac{\Psi_A + \Psi_B - H_A - H_B}{\Theta_{AB}}\right)\right]. \tag{9.16}$$

The only way we can obtain the coefficient of probability defined in (9.16) from a combination of the separate coefficients of probability in (9.14) and (9.15) is by requiring that $P_\Gamma\left(H_A + H_B\right) = P_\Gamma\left(H_A\right)P_\Gamma\left(H_B\right)$ and $\Theta_A = \Theta_B = \Theta_{AB}$. That is the moduli of the subsystems and the combined system must all be identical. If the two systems are in equilibrium across a diathermal wall we know, from the zeroth law, that the temperatures of the subsystems are the same. Therefore the modulus Θ of the canonical distribution is a function only of the thermodynamic temperature.

The Function Ψ. Using (9.9) we have an expression for Ψ as a function of Θ and V.

$$\Psi\left(\Theta, V\right) = -\Theta \ln\left(\int_\Gamma \exp\left[-\frac{H\left(\Omega, V\right)}{\Theta}\right] d\Omega\right). \tag{9.17}$$

If we now define the canonical partition function as

$$Q(\Theta,V) = \int_\Gamma \exp\left[-\frac{H(\Omega,V)}{\Theta}\right] d\Omega, \qquad (9.18)$$

(9.17) takes on the simple form

$$\Psi(\Theta,V) = -\Theta \ln Q(\Theta,V), \qquad (9.19)$$

and the canonical ensemble average in (9.10) becomes

$$\langle \Phi \rangle = \frac{1}{Q(\Theta,V)} \int_\Gamma \Phi(\Omega) \exp\left[-\frac{H(\Omega,V)}{\Theta}\right] d\Omega. \qquad (9.20)$$

Specifically the ensemble averaged energy can be obtained from the canonical partition function as

$$\langle H \rangle = \frac{1}{\Theta^2}\left(\frac{\partial}{\partial \Theta} \ln Q\right)_V. \qquad (9.21)$$

Because our fundamental definitions of the thermodynamic functions appeared first in differential form, we are led to consider the differential of $\Psi(\Theta,V)$. Using (9.20) and (9.19) we have

$$d\Psi(\Theta,V) = -\left[\left(\frac{1}{\Theta}\right)\langle H \rangle - \frac{1}{\Theta}\Psi(\Theta,V)\right]d\Theta + \left\langle \frac{\partial H}{\partial V}\right\rangle dV. \qquad (9.22)$$

From the Gibbs equation (2.25) we know that

$$P = -\left(\frac{\partial U}{\partial V}\right).$$

Since $H(\Omega,V)$ is the energy of the single system, the pressure for a single system in the ensemble is $-\partial H(\Omega,V)/\partial V$. The thermodynamic pressure is then the ensemble average of $-\partial H(\Omega,V)/\partial V$. That is

$$P = -\left\langle \frac{\partial H}{\partial V}\right\rangle. \qquad (9.23)$$

With thermodynamic pressure defined by (9.23), (9.22) becomes

$$d\Psi(\Theta,V) = -\frac{1}{\Theta}\left[\langle H \rangle - \Psi(\Theta,V)\right]d\Theta - PdV. \qquad (9.24)$$

We have not yet explicitly specified the form of the dependence of the modulus Θ on the thermodynamic temperature. If Θ is directly proportional to T, i.e. if $\Theta = k_B T$, where k_B is a constant, then (9.24) becomes

$$d\Psi(T,V) = -\left(\frac{1}{T}\right)\left[\langle H \rangle - \Psi(T,V)\right]dT - PdV. \qquad (9.25)$$

From (4.16) we recognize (9.25) as the Pfaffian of the Helmholtz energy provided the entropy can be identified as

$$S = \left(\frac{1}{T}\right)\left[\langle H\rangle - \Psi(T,V)\right].\qquad(9.26)$$

And, if we identify $\Psi(T,V)$ as the Helmholtz energy then (9.26) is

$$F = \langle H\rangle - TS,$$

which is the correct thermodynamic expression for the Helmholtz energy. Therefore $\Psi(T,V)$ is the correct statistical mechanical formulation of the Helmholtz energy and is the bridge we sought between the microscopic, rational mechanical description of matter, and the thermodynamic description.

From (9.19) we have

$$F = -k_B T \ln Q(T,V),\qquad(9.27)$$

which, is a concise and general formulation of the Helmholtz energy in terms of the canonical partition function. The constant k_B is Boltzmann's constant defined such the product of k_B and Avogadro's number is the gas constant,

$$R = N_A k_B.\qquad(9.28)$$

Entropy. Using $\Theta = k_B T$ and taking the ensemble average of the logarithm of the coefficient of probability defined in (9.13), we have

$$\left(\frac{1}{T}\right)[\langle H\rangle - \Psi] = -k_B \langle \ln P_\Gamma\rangle.$$

Our general expression for the entropy in (9.26) is then

$$S = -k_B \langle \ln P_\Gamma\rangle.\qquad(9.29)$$

This is the expression for the thermodynamic entropy obtained by Gibbs for the canonical ensemble. We shall consider (9.29) to be the statistical mechanical formulation of the entropy.

We recall that P_Γ is a measure of the density of systems in Γ - space (see (9.8)). And that P_Γ involves all the interactions among the molecules of the individual systems, as we see in (9.13). Therefore, although (9.29) is very simple in form, a simple interpretation of the entropy based on (9.29) is not possible, as it was for the internal energy (see (9.21)) or the thermodynamic pressure (see (9.23)).

9.7.2 *Microcanonical Ensemble*

The microcanonical ensemble is the ensemble of systems all of which have the same energy and the same number of particles. To avoid possible confusion we shall designate the coefficient of probability for the microcanonical ensemble as $P_\Gamma^m (H)$. As in the case of the canonical ensemble, the Hamiltonian for the microcanonical ensemble is also a function of Ω and V. The ensemble average of a single system, microscopic property Φ is then

$$\langle \Phi \rangle = \int_{E < H < E + \Delta E} \Phi P_\Gamma^m (H) \, d\Omega, \tag{9.30}$$

where the integral is over the shell in Γ - space for which the Hamiltonian has values between an energy E and $E + \Delta E$.

Our treatment of the microcanonical ensemble is immensely simplified if we assume that we can also use the Gibbs statistical form of the entropy in (9.29) for the microcanonical ensemble. That is

$$S = -k_B \int_{E < H < E + \Delta E} P_\Gamma^m (H) \ln \left[P_\Gamma^m (H) \right] d\Omega. \tag{9.31}$$

For a thermodynamic system the entropy is a maximum in systems for which the energy and the volume are constants. This will be developed rigorously in a later chapter (see Sect. 12.3). Therefore the entropy S in (9.31) must be a maximum at equilibrium for the correct form of $P_\Gamma^m (H)$. If (9.31) is the correct statistical form for the entropy then this maximization should yield the result in (9.12).

The function $P_\Gamma^m (H)$ must also satisfy the normalization condition (9.9), which is then a constraint on the problem of maximizing the entropy. For extremum problems with constraints we use the method of Lagrange underdetermined. multipliers (see Sect. A.4). The subsidiary function (A.22) is

$$h = \int_{E < H < E + \Delta E} P_\Gamma^m (H) \left\{ \alpha - k_B \ln \left[P_\Gamma^m (H) \right] \right\} d\Omega. \tag{9.32}$$

The first variation of h in (9.32) is

$$\delta h = 0 = \int_{E < H < E + \Delta E} \delta P_\Gamma^m (H) \left\{ \alpha - k_B \ln \left[P_\Gamma^m (H) \right] - k_B \right\} d\Omega. \tag{9.33}$$

Then, for arbitrary variations $\delta P_\Gamma^{(m)} (H)$ we must have

$$P_\Gamma^m (H) = \exp \left(\alpha / k_B - 1 \right), \tag{9.34}$$

which is a constant. So the systems in the microcanonical ensemble are uniformly distributed over the energy shell.

To find the value of this constant we use the constraint in (9.9). This results in

$$P_\Gamma^m (H) = \begin{cases} \frac{1}{\Omega_{\Delta E}} & \text{for } E - \Delta E \leq H \leq E + \Delta E \\ 0 & \text{otherwise.} \end{cases} , \qquad (9.35)$$

where $\Omega_{\Delta E}$ is the dimensionless volume of the energy shell containing the system representative points for the microcanonical ensemble. With (9.35) the entropy in (9.31) is

$$S = k_B \ln (\Omega_{\Delta E}) . \qquad (9.36)$$

In the so-called thermodynamic limit for which $N \to \infty$ and $V \to \infty$ with $N/V =$ constant the entropy in (9.36) can be simplified [[139], p. 347]. In the appendix we show that

$$\lim_{N,V \to \infty, \frac{N}{V} = \text{const}} \ln [\Omega_{\Delta E} (E, V, N)] = \ln [\Omega (E, V, N)] \qquad (9.37)$$

where

$$\Omega (E, V, N) = \int_{H < E} d\Omega . \qquad (9.38)$$

The integral appearing in (9.38) is a much simpler integral to perform than one over a thin energy shell.

The statistical mechanical entropy for the microcanonical ensemble in (9.36) appears simpler than that for the canonical ensemble in (9.29). But the simplification is only apparent and results from the constancy of the distribution of system states on the energy shell. The general Hamiltonian still retains all of the interactions.

Quantum Mechanical Result. To obtain a correct quantum mechanical description of the system we must begin by considering that the particles in the system posses quantum mechanical spin and that the mechanical description is based on the Schrödinger equation. This we do in the next chapter. For our purposes here, however, we assume that the introduction of quantum theory changes only the energy description. This is valid provided we avoid low temperatures, extreme densities, as in neutron stars, or detailed considerations of the solid state.

A quantum system in a microcanonical ensemble may exist in a number of possible states, each of which has the same energy E. We shall designate this set of quantum states by the numbers 1, 2, ..., $N_{qs} (E)$, where $N_{qs} (E)$ is the number of quantum system states with the energy E. The probability that the system is in the nth of these quantum states is P_n. For the quantum system the entropy formulation corresponding to (9.31) is

$$S = -k_B \sum_{n=1}^{N_{qs}(E)} P_n \ln (P_n) . \qquad (9.39)$$

The summation in (9.39) is over all states n in the set of possible system quantum states for which the energy is constant. The sum of the probabilities of the system quantum states must be unity.

$$\sum_{n=1}^{N_{qs}(E)} P_n = 1. \tag{9.40}$$

We then maximize the entropy in (9.39) subject to the constraint (9.40) using the method of Lagrange underdetermined. multipliers as in the preceding section. The result is

$$P_n = \frac{1}{N_{qs}(E)}. \tag{9.41}$$

If we use (9.41) in the Gibbs formulation for the entropy (9.39) we have

$$S = k_B \ln\left(N_{qs}(E)\right). \tag{9.42}$$

That is, the statistical mechanical entropy of quantum systems distributed micro-canonically is simply proportional to the logarithm of the number of system quantum states.

Equation (9.42) is a strikingly simple expression for the entropy of the quantum microcanonical ensemble. In principle it is not different from the classical result (9.35). The volume of the energy shell in Γ - space is simply replaced by the number of available system quantum states. Both (9.35) and (9.42) result from the assumption that the Gibbs entropy, formulated for the canonical ensemble in (9.29), can be used for the more restrictive microcanonical ensemble. Our derivation of (9.42) also excludes considerations of the Pauli principle.

Maxwell-Boltzmann Distribution. To perform the actual integration over an energy shell to calculate the ensemble average of a microscopic quantity Φ for systems in a microcanonical ensemble is a formidable problem. If, however, we can show that almost all the systems in the microcanonical ensemble fit a certain category, then we can avoid the difficulties of the exact calculation by calculating averages for single systems that fit this category. We shall now show that almost all the systems in the microcanonical ensemble have particles distributed in energies according to a Maxwell-Boltzmann distribution. The Maxwell-Boltzmann distribution is the single particle equivalent of the microcanonical distribution for systems and the integrations are of the same form for each.

There is a large number of possible sets of occupation numbers $\{N_i^*\}$ that will result in a value for the system Hamiltonian within the limits required for the microcanonical ensemble, i.e. $E < H < E + \Delta E$. Using the notation of Sect. 9.4.4 each of these is a particular Z^* state for the system. Because each state Z^* produces a separate volume in Γ - space, we will take in all of Γ - space if we sum over all such Z^* states. Using (9.35) in (9.30) the ensemble average becomes

$$\langle \Phi \rangle = \left(\frac{1}{\Omega_{\Delta E}} \right) \sum_{\text{all } Z^*} \int_{W(Z^*)} \Phi_{Z^*} d\Omega, \tag{9.43}$$

where Φ_{Z^*} is the value of the system property Φ when the state of the system is Z^*.

For sufficiently small cells in μ - space all particles with representative points in the cell (i) have a *single particle property* φ_i corresponding to the system property Φ. The quantity Φ_{Z^*} is then the sum

$$\Phi_{Z^*} = \sum_i N_i^* \varphi_i \tag{9.44}$$

over all cells that are occupied when the system is in the state Z^*. Then for the state Z^*, we have

$$\int_{W(Z^*)} \Phi_{Z^*} d\Omega = \left(\sum_i N_i^* \varphi_i \right) W(Z^*), \tag{9.45}$$

where the Γ - space volume $W(Z^*)$ is given by (9.2). With (9.45) the microcanonical average in (9.43) becomes

$$\langle \Phi \rangle = N \sum_{\text{all } Z^*} \left(\frac{W(Z^*)}{\Omega_{\Delta E}} \right) \sum_i \frac{N_i^*}{N} \varphi_i. \tag{9.46}$$

This is exactly true regardless of the distribution of the particles in each of the systems in the microcanonical ensemble.

We now ask for the state Z^* which occupies the greatest volume in Γ - space. That is we maximize $W(Z^*)$ subject to the constraints that the total number of particles N and the system energy E are constants using the method of Lagrange undetermined multipliers. Because of the factorials in (9.2) it is much easier mathematically to deal with the logarithm of $W(Z^*)$ than with $W(Z^*)$ itself. The logarithm of $W(Z^*)$ is a monotonic function of $W(Z^*)$ so maximization of $\ln W(Z^*)$ is an equivalent problem to that of maximizing $W(Z^*)$.

The simplification is brought about by the use of Stirling's approximation[4]

$$N! \approx \sqrt{2\pi N} \left(\frac{N}{e} \right)^N, \tag{9.47}$$

which for $N = 10$ is accurate to within 0.83% and for $N = 100$ is accurate to within 0.083%. With (9.47) we have

$$\ln N! \approx N \ln N - N + \ln \sqrt{2\pi N}. \tag{9.48}$$

Since $N \gg \ln \sqrt{2\pi N}$ for large values of N, e.g. for $N = 1000$ the ratio $\ln \sqrt{2\pi N}/N = 0.004$, Stirling's approximation takes the form

---

[4] Stirling's approximation was derived by James Stirling (1692–1770), a Scottish mathematician. At the time of this writing, a proof appears in *Wikipedia*. Daniel Schroeder [144] has an elementary discussion.

$$\ln N! \approx N \ln N - N. \tag{9.49}$$

We will use (9.49) in $\ln W(Z^*)$ to simplify our problem of finding the extremum.

Using (9.49) the subsidiary function for the problem of maximizing $\ln W(Z^*)$ subject to the constraints

$$\sum_{i=1}^{m} N_i = N \text{ and } \sum_{i=1}^{m} N_i \varepsilon_i = E$$

is

$$h_N = N \ln \omega - \sum_{i=1}^{m} N_i \ln \frac{N_i}{N} + \alpha_N \left(\sum_{i=1}^{m} N_i - N \right) - \beta_N \left(\sum_{i=1}^{m} N_i \varepsilon_i - E \right), \tag{9.50}$$

The first variation of h_N in (9.50) is

$$\delta h_N = - \sum_{i=1}^{m} \left(\delta N_i \ln \frac{N_i}{N} + N_i \frac{N}{N_i} \frac{\delta N_i}{N} - \alpha_N \delta N_i + \beta_N \delta N_i \varepsilon_i \right). \tag{9.51}$$

The requirement that $\delta h_N = 0$ results in the distribution of occupation numbers

$$\frac{N_i}{N} = \frac{1}{Z} \exp(-\beta_N \varepsilon_i), \tag{9.52}$$

where

$$Z = \sum_{i=1}^{m} \exp(-\beta_N \varepsilon_i) \tag{9.53}$$

is the "sum over states" (*Zustandsumme* in German) and is the μ - space equivalent of the partition function $Q(\Theta, V)$ for the canonical ensemble (9.18). Z is often also called the molecular partition function function and designated as q (see Sect. 9.12). The average of a single particle property φ_i over all the particles in the system is then

$$\langle \varphi \rangle = \frac{1}{Z} \sum_{i}^{m} \varphi_i \exp(-\beta_N \varepsilon_i). \tag{9.54}$$

With $\beta_N = 1/k_B T$ (see Exercises) we can obtain the ratio of the relative numbers of particles in distinct energy states (μ - space cells) from (9.52) as

$$\frac{N_i}{N_j} = \exp\left[-\frac{(\varepsilon_i - \varepsilon_j)}{k_B T} \right]. \tag{9.55}$$

This is the *Maxwell-Boltzmann distribution* of the particles among the energy states. In practical terms this result tells us that the numbers of particles in high energy states increases as the temperature increases because the factor $1/k_B T$ decreases with temperature.

How much of the energy shell volume is covered by systems with occupation numbers given by (9.52)? We answer this question by considering the form of the extremum of h_N. In the appendix we show that the extremum is very peaked and that system states with (μ - space) occupation numbers given by the Maxwell-Boltzmann distribution of (9.52) cover almost the entirety of the constant energy sheet in Γ - space. Therefore for almost all systems in a microcanonical ensemble the particles are distributed among energies according to a Maxwell-Boltzmann distribution. We can then replace the sum in (9.46) with a single term and write the microcanonical ensemble average of the system property as

$$\langle \Phi \rangle = N \left(\frac{1}{Z} \right) \sum_i^m \varphi_i \exp \left(-\frac{\varepsilon_i}{k_B T} \right), \tag{9.56}$$

using (9.52).

If we define the single particle (μ - space) average over equilibrium (Maxwell-Boltzmann) particle states as

$$\langle \Phi \rangle_\mu = \frac{1}{Z} \sum_i^m \varphi_i \exp \left(-\frac{\varepsilon_i}{k_B T} \right), \tag{9.57}$$

Then (9.56) is simply

$$\langle \Phi \rangle = N \langle \Phi \rangle_\mu \tag{9.58}$$

The result in (9.58) is an important and very useful result. It allows us to treat the microcanonical ensemble with the same ease that we treat the canonical ensemble. We require only that the property for which we are seeking an average value is a property that can be defined for each particle. Energy and momentum are examples. The entropy is not. The entropy is obtained from the distribution of system states in Γ - space according to (9.36).

Ergodic Theorem. What we have presented h.ere is related to the ergodic theorem. The ergodic theorem is historically interesting and has produced a mathematical literature, much of which seems irrelevant to physics [[67], p. 9], [[23], p. 146], [[24], p. 58], [[91], pp. 44–62]. The problem is, however, real and very logical.

Any laboratory measurement determines the time average of a system property. In statistical mechanics we replace this with a Γ - space average over the ensemble. The ergodic theorem justifies this replacement.

The measurement in the laboratory is over a span of the system's lifetime, if we can use such terminology. This translates into the trajectory of the system representative point in Γ - space. The ergodic theorem is then valid if this trajectory passes through every point in Γ - space and if the part of the trajectory covered during the laboratory measurement is identical for all parts of all trajectories. What we have shown here is that the second part of this claim is valid for the microcanonical ensemble. Almost all system representative points on the energy shell have the same occupation numbers $\{N_i^*\}$ given by the Maxwell-Boltzmann distribution (9.55). For almost all points in Γ - space the contribution to the ensemble average is then identical. We have not, however, established the Ergodic theorem.

Gibbs was aware of the difficulty [23]. But ensemble theory is not based on the ergodic theorem and has no need of it. Ensemble theory also provides an accurate representation of the behavior of real systems (see Sects. 9.11 and 9.13.5). So in practical terms we may consider the ergodic theorem to be primarily only of historical or mathematical interest.

Thermodynamics. If we use the results of Sect. 9.7.2 we may obtain a thermodynamic potential from the sum over states Z in (9.53). Analogously to (9.19) we define the function $\psi(\beta, V)$ as

$$\psi(\beta, V) = -\frac{1}{\beta} \ln Z(\beta, V). \tag{9.59}$$

where $\beta = 1/k_B T$. Using (9.53) and (9.57) we have from (9.59)

$$\left[\psi - T \left(\frac{\partial \psi}{\partial T} \right)_V \right] d\beta + \beta \left(\frac{\partial \psi}{\partial V} \right)_T dV = \langle \varepsilon \rangle_\mu d\beta + \beta \langle d\varepsilon \rangle_\mu, \tag{9.60}$$

where

$$\langle d\varepsilon \rangle_\mu = \frac{1}{Z} \left[\sum_{i=1}^m \left(\frac{\partial \varepsilon_i}{\partial V} \right) \exp(-\beta \varepsilon_i) \right] dV$$

$$= -\frac{1}{N} P dV \tag{9.61}$$

is the microcanonical equivalent of (9.23) in Sect. 9.7.1. With (9.61) the differential expression in (9.60) becomes

$$\left[N\psi - T \left(\frac{\partial N\psi}{\partial T} \right)_V \right] d\beta + \beta \left(\frac{\partial N\psi}{\partial V} \right)_T dV = \langle H \rangle d\beta - \beta P dV \tag{9.62}$$

where $\langle H \rangle$ is the ensemble averaged energy for the system. Equation (9.62) is satisfied if $N\psi = F$, the Helmholtz energy, since then $(\partial N\psi / \partial T)_V = -S$ and $(\partial N\psi / \partial V)_T = -P$. That is

$$F(T, V) = -N k_B T \ln Z(T, V) \tag{9.63}$$

for the microcanonical ensemble. Equation (9.63) is the equivalent of (9.27).

Boltzmann's Entropy. Gibbs' work is based, at least in part, on Boltzmann's work on gas theory (see Sect. 8.2.2). A result similar to (9.29) had been obtained

by Boltzmann for more restrictive conditions [9, 11, 12, 46]. And Gibbs cites
Boltzmann in his discussion of the entropy [56].

Boltzmann introduced a single particle distribution function $f(\mathbf{v},t)$ defined by
the requirement that the number of gas atoms or molecules with a velocity in the
range $\mathbf{v} \rightarrow \mathbf{v}+d\mathbf{v}$ at the time t is $f(\mathbf{v},t)d\mathbf{v}$. This is a generalization of Maxwell's
probability density function (see Sect. 8.2.2). Based on $f(\mathbf{v},t)$ Boltzmann was able
to show that a quantity he designated as $H(t)$[5,6], defined by [25, 46, 116]

$$H(t) = \int f(\mathbf{v},t) \ln (f(\mathbf{v},t)) \, d\mathbf{v} \qquad (9.64)$$

underwent rapid variations with time, but had a greater probability of decreasing
than of increasing. That is $H(t)$ decreases statistically in time. This is illustrated in
Fig. 9.4, which is a plot of $H(t)$ from a Monte Carlo simulation of the relaxation
of a gas from a microstate in which all particles are in a single energy state to a
Maxwell-Boltzmann distribution. Steps in the Monte Carlo (Metropolis) algorithm
are considered as representative of time steps in Fig. 9.4 [70].

Boltzmann then identified the entropy as

$$S(t) = -Nk_B H(t). \qquad (9.65)$$

Boltzmann's analysis was based on his formulation of the rate of collisions between particles (atoms) in the gas (see Sect. 8.2.2). The result provides a formulation
for the entropy as a function of time, but only near equilibrium. At equilibrium $S(t)$
in (9.65) becomes a maximum. From his kinetic description Boltzmann was able to
show the distribution function $f(\mathbf{v},t)$ was Maxwellian at equilibrium

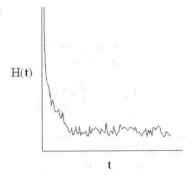

H(t)

Fig. 9.4 The Boltzmann $H(t)$
function. A Monte Carlo
simulation of a gas relaxing
from a microstate in which all
particles have the same energy
to a Maxwell-Boltzmann
distribution

t

---

[5] Boltzmann originally used the letter E for this function [12] and changed the notation to H in
1895. The use of H was introduced by S.H. Burbury [18], who makes no suggestion of the origin
of this term. There is no specific indication that H is capital Eta, as some modern pysicists are
claiming [16].

[6] We will designate Boltzmann's $H(t)$ with a time dependence so that there is no confusion with
the Hamiltonian.

$$f(\mathbf{v},t) \propto \exp\left(-\frac{1}{2}mv^2/k_\mathrm{B}T\right),$$

and that $S(t)$ in (9.65) exhibited statistical variations around the equilibrium value. Expressed in the terms of Sect. 9.7.2, (9.36), Boltzmann's entropy is

$$S = k_\mathrm{B} \ln W, \tag{9.66}$$

where W is the volume in Γ - space covered by most probable, i.e. the Maxwell-Boltzmann, distribution. In this sense W is a measure of the probability of the distribution. The origin of the notation W is the German *Wahrscheinlichkeit* (probability). Equation (9.66) appears on Boltzmann's tombstone [158].

Equation (9.66) is essentially the same as (9.36) and (9.42). We must only assume that the probability that a system representative point will be at a particular location on the microcanonical energy shell is inversely proportional to the shell volume, and that the probability that a quantum system will be in a particular state is inversely proportional to the number of states available. This is the assumption of equal á priori probabilities [[54], p. 202].

Equation (9.66) is the source of the interpretation of entropy as a measure of disorder, based on the assumption that disorder is more probable than order. As we have seen, however, the most probable distribution at equilibrium is a Maxwellian in the energy. There is no implication of disorder. Structures may be the result of complex interactions, which are common to biology [cf. [26, 70, 78]].

9.8 Information Theory

The basic mathematical concepts of information theory applied to problems in communication were first presented by Claude Shannon[7] [146], [147], [148]. A concise development of the mathematical foundations of the theory is available in the book by Aleksandr I. Khinchin[8] [90].

Central to information theory is the mathematical representation of a measure of the uncertainty we have regarding the outcome of an event. This measure of our uncertainty must be such that (1) if all events are equally likely it is a maximum, (2) The uncertainty of a combined event is equal to the sum of the uncertainties of the separate events, and (3) Adding an impossible event to the scheme does not affect the uncertainty. Such a function is

$$H = -K\sum_i p_i \ln p_i, \tag{9.67}$$

---

[7] Claude Elwood Shannon (1916–2001) was an American electrical engineer and mathematician. In 1948 Shannon published the seminal work "A Mathematical Theory of Communication" in the *Bell System Technical Journal*.

[8] Aleksandr Yakovlevich Khinchin (1894–1959) was a Russian mathematician.

where K is a positive constant[9] [[90], p. 9]. From (9.39) we recognize this uncertainty as the statistical mechanical entropy provided we identify p_i as a system state and K as k_B, or it is the Boltzmann entropy (9.65) if we identify p_i as the distribution function $f(v,t)$ and K as Nk_B.

The application of information theory to statistical mechanics was presented in two fundamental papers by Edwin T. Jaynes[10] [81], [82]. There Jaynes pointed out that, although the mathematical facts concerning the maximization of the statistical entropy were known to Gibbs, Shannon's contribution was to show that the expression for the statistical entropy has a deeper meaning independent of that supplied by thermodynamics. We can then formulate the problem of obtaining the distribution of particles in μ - space as one in which we seek to maximize our uncertainty subject to our limited knowledge about the system (number of particles, energy, average velocity, ...) as constraints. This maximization of uncertainty (statistical entropy), Jaynes said, is not an application of a law of physics. It is merely a method of reasoning that ensures that no unconscious arbitrary assumptions have been introduced.

Information theory remains of interest in particularly artificial intelligence and studies of the human mind [cf. [49, 62, 65, 151]] as well as studies of nonlinear systems [6]. And discussions of information theory appear in some modern thermodynamics texts [4, 22].

9.9 Potential Energies

Generally the potential energy of the molecules depends only on the coordinates and is independent of the momenta. We may then write the Hamiltonian for a system of N particles (molecules) as a sum of separate kinetic and potential energy terms

$$H(\Omega,V) = \sum_{i}^{N} \mathscr{E}_{i,kin} + \Phi. \qquad (9.68)$$

Here $\mathscr{E}_{i,kin}$ is the kinetic energy of the ith molecule, and Φ is the potential energy of the entire system.

Interactions with the container walls are intermolecular collisions because the walls are composed of atoms or molecules. Inclusion of such details becomes very important in modern plasma physics. However, here we shall only consider that the wall confines the particles in our system. This is accomplished by introducing a potential energy which is zero inside the wall and infinity outside. This potential mathematically forbids the passage of particles to the region outside the container.

---

[9] A proof is given by Khninchin.

[10] Edwin Thompson Jaynes (1922–1998) was Wayman Crow Distinguished Professor of Physics at Washington University in St. Louis.

The greatest complication results from the potential energy of interaction between the moving molecules. If the forces of interaction between the particles are short range then we can assume that collisions are well modeled by hard-sphere interactions (see Sect. 8.2.3). The Coulomb (r^{-2}) force, of interest in studies ionized gases, is, however, not short range. A shielded Coulomb potential is introduced to produce a more realistic model. This shielded potential is the same as the Thomas-Fermi shielding introduced to deal with free electrons in solids. Detailed treatments of chemically interacting molecules must include interactions as well.

The difficulty in treating these situations comes from the integration over particle coordinates required for the ensemble average. If we include the long range portion of the intermolecular potential there will also be multibody collisions among the particles, such as we treated in a simplified form in the van der Waals model (see Sect. 8.3.1). The only realistic, systematic approach to the treatment of the effect of the details of molecular interactions is the BBGKY hierarchy mentioned in Sect. 9.6.

An alternative approach to interparticle interactions, which avoids the difficulties of integration, is Monte Carlo simulation. Monte Carlo simulation was developed to treat neutron diffusion in fast neutron fission. It has since been adapted to the treatment of essentially any system of interacting particles [115]. This includes application to biological systems and particularly to cancer therapy, in which the studies are of motion of charged particles in biological tissue. Personal computers are presently capable of Monte Carlo simulation of complex systems [cf. [70]].

In this text we will, however, treat neither the BBGKY hierarchy nor Monte Carlo simulations. Although neither approach is discouragingly difficult, their consideration would simply take us too far from our goal of obtaining a basic understanding of microscopic thermodynamics. We shall then explicitly include only the potential energies internal to the molecules of the system and the potential energy barrier of the wall in the term Φ of (9.68). That is we shall write

$$\Phi = \Phi_{\text{boundary}} + \sum_{i}^{N} \phi_{i,\text{internal}}, \tag{9.69}$$

where $\phi_{i,\text{internal}}$ is the internal energy of the ith molecule and may be considered to be related to vibrational motion. The boundary interaction potential Φ_{boundary} vanishes within the system and is infinite outside.

9.10 Equipartition Principle

The equipartition principle is a general classical result for ideal gas systems in thermodynamic equilibrium. It provides a very simple relationship between the thermodynamic internal energy of a substance and the number of quadratic energy terms in the Hamiltonian for the molecules of the ideal gas [83].

We base our proof of the equipartition principle on (9.57) with φ_i equal to the single molecule Hamiltonian. We shall assume that the single molecule Hamiltonian may be written as sum of N_{KE} quadratic kinetic energy terms and N_{PE} quadratic potential energy terms, which are independent of one another. We write the single particle Hamiltonian as

$$H_i = \sum_j^{N_{KE}} \alpha_j p_j^2 + \sum_k^{N_{PE}} \lambda_k q_k^2$$

for the molecular coordinates q_k and momenta p_j.

The μ - space average of the single particle Hamiltonian is then (see (9.57))

$$\langle H \rangle_\mu = \frac{1}{Z} \sum_{i=1}^m H_i \exp\left(-\frac{H_i}{k_B T}\right)$$

$$= k_B T^2 \frac{\partial}{\partial T} \ln Z.$$

For continuous variation in the coordinates q_k and momenta p_j the sum over states Z may be replaced by integrals. Then

$$Z = \frac{1}{h^f} \left\{ \int_\infty^\infty dp_1 \cdots dp_{N_{KE}} \prod_j^{N_{KE}} \exp\left(-\frac{\alpha_j p_j^2}{k_B T}\right) \right\} \cdots$$

$$\cdots \left\{ \int_\infty^\infty dq_1 \cdots dq_{N_{PE}} \prod_k^{N_{PE}} \exp\left(-\frac{\lambda_k q_k^2}{k_B T}\right) \right\}, \tag{9.70}$$

where we have used (9.5).

All integrals in (9.70) are of the form

$$\int_{-\infty}^\infty dx \exp\left(-\frac{ax^2}{k_B T}\right) = \sqrt{\frac{\pi k_B T}{a}}. \tag{9.71}$$

With (9.71) Z in (9.70) is

$$Z = \prod_j^{N_{KE}} \left(\frac{\pi k_B T}{h^2 \alpha_j}\right)^{\frac{1}{2}} \prod_\ell^{N_{PE}} \left(\frac{\pi k_B T}{\lambda_\ell}\right)^{\frac{1}{2}}. \tag{9.72}$$

Then

$$\langle H \rangle_\mu = k_B T^2 \frac{\partial}{\partial T} \ln Z$$

$$= \frac{1}{2}(N_{KE} + N_{PE}) k_B T, \tag{9.73}$$

and the ensemble averaged molar internal energy is

$$U = \frac{1}{2}(N_{KE} + N_{PE}) RT. \tag{9.74}$$

From the internal energy in (9.74) we can obtain the molar specific heat at constant volume as

$$C_V = \frac{1}{2}(N_{KE} + N_{PE})R.$$

Equations (9.73) and (9.74) are equivalent statements of the equipartition principle.

The equipartition principle is a completely general mathematical result, which was first proposed by Maxwell in 1867, on the basis of a separation of linear and rotational motion of gas molecules. It was extended by Boltzmann to include all separable energies.

Atoms in molecules are linked to one another by chemical bonds. The equilibrium bond length is the distance for which the binding energy is a minimum. We may then model the chemical bond energy as an attractive potential with a minimum. This is illustrated in Fig. 9.5. The nuclei of atoms in a molecule may vibrate relatively to one another. For small vibrations the potential energy between the nuclei may be approximated as a quadratic function of the displacement from equilibrium. The quadratic approximation to the potential is shown as a dashed line in Fig. 9.5. Motion in a quadratic potential is simple harmonic motion (see Sect. 9.4.1). Vibrational motion then contributes two quadratic terms to the Hamiltonian (see (9.1)).

For small vibrations the moments of inertia of the molecule may be assumed constant and the kinetic energies of rotation are quadratic functions of the rotational angular momenta. For a diatomic molecule there are two principal axes of rotation and for a general polyatomic molecule there are three.

A diatomic molecule then possesses two quadratic energies of vibration and two of rotation. To these are added three translational kinetic energies, which are quadratic in the translational momenta. We should then, according to the equipartition principle, expect that dipole gases will have a molar internal energy of $U = (7/2)RT$ and a molar specific heat of $C_V = (7/2)R$.

At the end of the nineteenth century measurements only revealed a specific heat of $(5/2)R$. This discrepancy caused Gibbs to recognize that the microscopic model of matter was incomplete when he wrote *Elementary Principles in Statistical Mechanics*. In his Baltimore Lectures of 1884, William Thomson (Lord Kelvin) referred this problem of the equipartition principle as one of the nineteenth century clouds over the dynamical theory of heat.

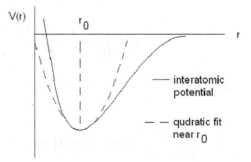

Fig. 9.5 Interatomic potential due to chemical bond as a function of interatomic spacing r (*solid line*). Quadratic fit near $r = r_0$ (*dashed line*)

There were even deeper problems. Measurements on the specific heat of hydrogen showed that C_V dropped from $(5/2)R$ at room temperatures to $(3/2)R$ below $80\,\text{K}$ [150]. The rotational energies seemed to have died out. Based on this, in 1911 Arnold Sommerfeld[11] proposed that quadratic energy terms (Sommerfeld referred to these as degrees of freedom) should be weighted, not counted.

This unsatisfactory situation is now understood in terms of the quantum theory. For that reason we shall turn to the quantum theory for our final description of molecules with structure.

9.11   Applications

We are now in a position to consider applications of statistical mechanics to structureless gases, which are those made up of particles that are accurately represented as point particles. These are the noble gases: helium (He), neon (Ne), argon (Ar), krypton (Kr), xenon (Xe), and radon (Rn). For these gases we can obtain a quantitative test of our statistical mechanical analysis by studying the entropy.

9.11.1   Sackur-Tetrode Equation

We could use (9.37) in (9.36) to obtain the statistical mechanical entropy. The calculation is not difficult, but is somewhat nuanced [[139], pp. 299–300]. We therefore turn to the calculation of the Helmholtz energy from either (9.27) or (9.63). We carry out the calculation of the canonical partition function in the example here.

Example 9.11.1. The ideal gas consists of point particles that do not interact with one another. For N identical, non-interacting point particles each of mass m the Hamiltonian is

$$H = \sum_{i=1}^{N} \frac{p_{xi}^2 + p_{yi}^2 + p_{zi}^2}{2m} \tag{9.75}$$

where $p_{xi, yi, zi}$ are the x, y, z−components of the momentum of the ith particle. The partition function is

$$Q = \int_{\Gamma} \exp\left[-\sum_{i=1}^{N} \frac{\left(p_{xi}^2 + p_{yi}^2 + p_{zi}^2 \right)}{2mk_B T} \right] d\Omega. \tag{9.76}$$

where, from (9.6)

---

[11] Arnold Johannes Wilhelm Sommerfeld (1868–1951) was professor of theoretical physics at the University of Munich.

$$d\Omega = \frac{1}{N! h^{3N}} \prod_{i=1}^{N} dp_{xi} dp_{yi} dp_{zi} dx_i dy_i dz_i. \tag{9.77}$$

The partition function is then

$$Q = \frac{1}{N! h^{3N}} \prod_{i=1}^{N} \left[\int_{p_{xi}=-\infty}^{\infty} \int_{p_{yi}=-\infty}^{\infty} \int_{p_{zi}=-\infty}^{\infty} \cdots \right.$$

$$\cdots \exp \left[-\frac{\left(p_{xi}^2 + p_{yi}^2 + p_{zi}^2 \right)}{2mk_B T} \right] dp_{xi} dp_{yi} dp_{zi} \Bigg] \cdots$$

$$\cdots \prod_{i=1}^{N} \left[\int \int \int_{\text{Volume}} dx_i dy_i dz_i \right]. \tag{9.78}$$

The integral over the coordinates extends over the volume of the system for each particle. That is

$$\prod_{i=1}^{N} \int \int \int_{\text{Volume}} dx_i dy_i dz_i = V^N.$$

Noting that

$$\int_{p=-\infty}^{\infty} \exp\left(-\frac{p^2}{2mk_B T} \right) dp = \sqrt{2\pi m k_B T}, \tag{9.79}$$

(9.78) becomes

$$Q = \frac{1}{N!} \left(\frac{2\pi m k_B T}{h^2} \right)^{\frac{3N}{2}} V^N, \tag{9.80}$$

Using (9.80) in (9.27) we have the Helmholtz energy for a gas of indistinguishable point particles. Using Stirling's approximation, this is

$$F = -Nk_B T \left\{ \ln \left[\left(\frac{2\pi m k_B T}{h^2} \right)^{\frac{3}{2}} \frac{V}{N} \right] + 1 \right\}$$

$$= -Nk_B T \left[\frac{3}{2} \ln T + \frac{3}{2} \ln \left(\frac{2\pi m k_B}{h^2} \right) + \ln \frac{V}{N} + 1 \right]. \tag{9.81}$$

For a gas of indistinguishable point particles the entropy is then

$$S = -\left(\frac{\partial F}{\partial T} \right)_V$$

$$= nR \left[\frac{3}{2} \ln T + \ln \frac{V}{N} + \frac{3}{2} \ln \left(\frac{2\pi m k_B}{h^2} \right) + \frac{5}{2} \right] \tag{9.82}$$

$$= nR \left\{ \frac{5}{2} \ln T - \ln P + \ln \left[\left(\frac{2\pi m}{h^2} \right)^{\frac{3}{2}} k_B^{\frac{5}{2}} \right] + \frac{5}{2} \right\}. \tag{9.83}$$

These are two forms of the Sackur-Tetrode equation for the absolute entropy of an ideal gas [142, 143, 153].

9.11.2 Mixing

As a result of the introduction of corrected Boltzmann counting, which provides the factor $1/N!$ in (9.80), the term

$$N\left(\ln\left(V/N\right)+\frac{5}{2}\right)=N\ln V+\frac{5}{2}N-N\ln N$$

appears in (9.82) rather than simply $N\ln V$. An application of the Sackur-Tetrode equation in the form (9.82) then results in the correct value of the mixing entropy whether the molecules are distinguishable or indistinguishable (see Sect. 4.7.2).

Clausius' original definition of entropy (see Sect. 1.3, (1.3)) provides dS for a closed system, i.e. for a system with the number of particles held constant. The correct integration of (1.3) for the ideal gas must then include an arbitrary function of the number of particles. That is

$$S(T,V,N) = Nk_B\left[\frac{3}{2}\ln T + \ln V\right] + k_B\phi(N), \tag{9.84}$$

where $k_B\phi(N)$ is the arbitrary function of N.

Wolfgang Pauli[12] recognized this problem and imposed the additional requirement that the entropy must be extensive regarding the numbers of particles [129]. Specifically Pauli required that

$$S(T,qV,qN) = qS(T,V,N), \tag{9.85}$$

where q is a scaling factor $0 < q < \infty$. Following Pauli we obtain

$$\phi(N) = N\phi(1) - N\ln N,$$

and (9.84) becomes

$$S(T,V,N) = Nk_B\left[\frac{3}{2}\ln T + \ln\frac{V}{N} + \phi(1)\right]. \tag{9.86}$$

For the ideal gas this results in (9.82) [80].

The form of the Sackur-Tetrode equation (9.82) results then from a recognition that the Clausius definition of entropy leaves open the question of extensivity of the

---

[12] Wolfgang Ernst Pauli (1900–1958) was an Austrian theoretical physicist. Pauli developed the spin algebra for electrons, the Pauli principle for symmetry of the quantum wave function, and postulated the existence of the neutrino.

entropy. Demanding extensivity produces (9.82) and resolves the presumed paradox. This has required no statistical argument and is not based in the quantum theory.

9.11.3 Experiment

The Sackur-Tetrode equation (9.82) or (9.83) contains no arbitrary constants and should provide an absolute entropy for an ideal gas. Does the Sackur-Tetrode equation, give an accurate formulation for the absolute entropy of a real gas in regions for which the ideal gas assumption is valid? To answer this question we make a direct comparison of the entropy obtained from the Sackur-Tetrode equation with the entropies available from NIST for real gases [102]. The NIST results are based on the most accurate pure fluid and mixture models available.

Because the Sackur-Tetrode equation is only valid for structureless molecules we choose one of the noble gases[13] (He, Ne, Ar, Kr, Xe and Rn) for our comparison. As we recall from the preceding chapter, the behavior of a real gas is well approximated by the ideal gas equation of state provided the compressibility $z \approx 1$. In Fig. 9.6 we have plotted z for argon as a function of specific volume v for the temperatures $T = 200$ K and $T = 800$ K. We have included approximate pressure ranges in Fig. 9.6 corresponding to $T = 200\,\mathrm{K}$ (low pressure) and $T = 800\,\mathrm{K}$ (high pressure).

Because we learn more about the limits of the Sackur-Tetrode results by considering conditions for which z deviates from unity, we shall compare entropies obtained from (9.82) with NIST entropies for specific volumes of $v = 0.03\,\mathrm{m^3\,kg^{-1}}$ and $v = 0.004\,\mathrm{m^3\,kg^{-1}}$.

In Fig. 9.7 we have plotted the entropy of argon calculated from (9.82) (dashed curves) and that obtained from NIST (solid curves) as functions of temperature for $v = 0.03\,\mathrm{m^3\,kg^{-1}}$ (a1) and $v = 0.004\,\mathrm{m^3\,kg^{-1}}$ (b). For low density $\left(v = 1\,\mathrm{m^3\,kg^{-1}}\right)$

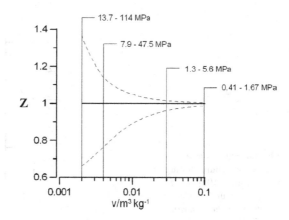

Fig. 9.6 Compressibility z of argon as a function of specific volume v. $T = 200\,\mathrm{K}$ (*lower dashed curve*), $T = 800\,\mathrm{K}$ (*upper dashed curve*), $z = 1$ (*solid line*) [102]

---

[13] The term noble gas is a direct translation of the German *Edelgas*, which was first used by H. Erdmann in a text on inorganic chemistry (*Lehrbuch der Anorganischen Chemie*) in 1898.

Fig. 9.7 Entropy as a
function of temperature for
argon. NIST results [102]
(*solid curve*), Sackur-Tetrode
equation (*dashed curve*).
Specific volume
$v = 0.03\,\mathrm{m^3\,kg^{-1}}$ **(a)**,
$v = 0.004\,\mathrm{m^3\,kg^{-1}}$ **(b)**

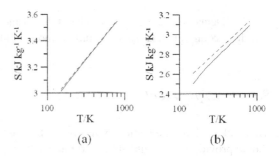

the Sackur-Tetrode and NIST entropies for argon agree to the fourth decimal place,
e.g. $3.7609\,\mathrm{kJ\,kg^{-1}\,K^{-1}}$ from the Sackur-Tetrode equation compared to
$3.7606\,\mathrm{kJ\,kg^{-1}\,K^{-1}}$ from NIST at $T = 150\,\mathrm{K}$. In Fig. 9.7 we see that the agreement between Sackur-Tetrode and NIST entropies for argon is quite good for
$v = 0.03\,\mathrm{m^3\,kg^{-1}}$ (a), which corresponds to a pressure range of 1.3–$5.6\,\mathrm{MPa}$ for
the temperature range $200 - 800\,\mathrm{K}$. Here the compressibility of argon varies between $z = 0.96$ and $z = 1.02$. There is, however, noticeable disagreement for
$v = 0.004\,\mathrm{m^3\,kg^{-1}}$ (b), which corresponds to a pressure range of 7.9–$47.5\,\mathrm{MPa}$ for
the temperature range $200 - 800\,\mathrm{K}$. Here the compressibility of argon varies between
$z = 0.77$ and $z = 1.15$.

In Fig. 9.8 we have plotted the entropy of argon as a function of volume for a
single isotherm at $T = 500\,\mathrm{K}$. The Sackur-Tetrode equation predicts a linear dependence on the logarithm of the volume (dashed curve in Fig. 9.8). The NIST results
(solid curve in Fig. 9.8) begin to deviate from the predictions of the Sackur-Tetrode
equation for volumes below about $0.01\,\mathrm{m^3\,kg^{-1}}$ (for $T = 500\,\mathrm{K}$). For argon the critical point is at a temperature of $150.69\,\mathrm{K}$ and a specific volume of $0.00187\,\mathrm{m^3\,kg^{-1}}$.
Particularly for low specific volumes the slope of S vs T is greater for the NIST
result than for the Sackur-Tetrode result.

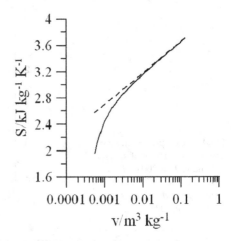

Fig. 9.8 Entropy as a
function of volume for argon.
NIST results [102] (*solid
curve*), Sackur-Tetrode
equation (*dashed curve*).
$T = 500\,\mathrm{K}$

From (4.19) and (4.18) we have

$$\left(\frac{\partial S}{\partial V}\right)_T = \left(\frac{\partial P}{\partial T}\right)_V.$$

The difference in the slopes of the curves in Fig. 9.8 then reflects a difference in the thermal equation of state of argon and that of the ideal gas. In the ideal gas the particles are geometrical points with no interaction. We know from our studies of the van der Waals equation of state (see Sect. 8.3.1, (8.20)) that the pressure dependence on temperature is primarily due to the finite volume of the particles. Collisional effects will be reflected by the way in which the van der Waals coefficient a depends on the temperature (see Sect. 8.4.3). So the deviation exhibited in Fig. 9.8 is primarily a result of the finite size of the atoms.

At constant volume the rate of change of entropy with thermodynamic temperature is proportional to the ratio $C_V(T,V)/T$ (see Sect. 5.3.4, (5.46)). We can then understand the deviation of the slope of S vs T in Fig. 9.7(b) by considering the form of $C_V(T,V)$ for argon. In Fig. 9.9 we have plotted $C_V(T)$ for argon with the specific volume as a parameter. When $v \geq 1\,\mathrm{m^3\,kg^{-1}}$ the specific heat is a constant equal to $(3/2)R = 0.3122\,\mathrm{kJ\,kg^{-1}\,K^{-1}}$ over the temperature range of 200–800 K. For specific volumes $v < 1\,\mathrm{m^3\,kg^{-1}}$ C_V depends upon the temperature over the range 200–800 K. In Fig. 9.9 the top curve is for $v = 0.004\,\mathrm{m^3\,kg^{-1}}$ and the bottom is for $v = 0.03\,\mathrm{m^3\,kg^{-1}}$.

In statistical mechanical terms we obtain

$$\frac{1}{n}\left(\frac{\partial S}{\partial T}\right)_V = \frac{1}{T}C_V(T,V) = \frac{k_B}{n}\left(\frac{\partial^2}{\partial T^2}\ln Q\right)_V$$

from Sect. 5.3.4, (5.46), and using (9.27). A decrease in specific volume (increase in density) requires that interaction energies between atoms (molecules) must be considered in the Hamiltonian. The partition function Q will then contain contributions from these interactions (collisions). The source of the disagreement between the Sackur-Tetrode and real gas entropies in the low specific volume regime (Fig. 9.7(b)) is then attributable to collisional effects. Treating these interparticle interactions is not beyond the scope of statistical mechanics. And we will consider the results of

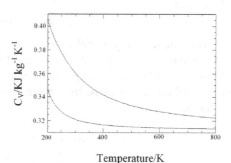

Fig. 9.9 C_V as a function of temperature for argon. Top curve is for $v = 0.004\,\mathrm{m^3\,kg^{-1}}$. Bottom curve is for $v = 0.03\,\mathrm{m^3\,kg^{-1}}$ [102]

collisions in our studies of chemical reactions in a later chapter. The detailed treatment of collisions is, however, beyond the scope of an introductory text.

Although there is substantial disagreement between the entropies obtained from the Sackur-Tetrode equation and those from NIST, the agreement is very satisfactory in ranges of thermodynamic variables for which the thermal equation of state of argon is well approximated by the ideal gas. The fact that the Sackur-Tetrode equation gives such good results in the ideal gas region is an indication of the validity of the statistical mechanics of Gibbs and of the ideas of Pauli, which resulted in the corrected Boltzmann counting. We have also been able to understand the sources of the deviations of the Sackur-Tetrode equation from the NIST results based on the assumptions in the atomic model used in the Sackur-Tetrode equation.

9.12 Molecular Partition Function

We may continue to avoid intermolecular interactions and still obtain a more realistic model of the gas if we add structure to the molecules. Diatomic molecules have rotational kinetic energies about two axes and internuclear vibration resulting from stretching and compression of the interatomic bond. Polyatomic molecules generally have rotational kinetic energies about three axes and may vibrate in a number of modes.

The classical treatment of each of these internal molecular energies is an analogue of the quantum treatment. We shall then first outline the classical treatment and carry out the details for the quantum treatment for each case. Our quantum treatment here will neglect the requirements of the Pauli principle resulting from the quantum mechanical spin. We reserve a correct quantum statistical treatment for a subsequent chapter.

Separation of Energies. The description of a body in motion, can be written in terms of coordinates of the center of mass (CM) of the body and coordinates describing the motion of the body around that center of mass. The proof of this is carried out in any text on mechanics [cf. [59]]. This results in a separation of $\mathscr{E}_{i,\text{kin}}$ in (9.68) into a kinetic energy of the CM and a kinetic energy of motion about the CM. That is

$$\mathscr{E}_{i,\text{kin}} = \mathscr{E}_{i,\text{kin-CM}} + \mathscr{E}_{i,\text{kin about CM}}. \tag{9.87}$$

The kinetic energy of molecular motion about the CM consists of kinetic energies of rotation and vibration. We shall consider the energy of the electrons separately. If we neglect the effect of vibrations on the moments of inertia the vibrational and rotational energies may be treated separately and the kinetic energy about the CM is a sum of these two

$$\mathscr{E}_{i,\text{kin about CM}} = \mathscr{E}_{i,\text{kin rot}} + \mathscr{E}_{i,\text{kin vib}}. \tag{9.88}$$

The internal potential energy is associated only with vibration. Then, using (9.69), (9.87), and (9.88), the Hamiltonian for a gas consisting of non-interacting structured molecules (9.68) becomes

$$H(\Omega,V) = \sum_{i}^{N} [\mathscr{E}_{i,\text{kin-CM}} + \mathscr{E}_{i,\text{kin rot}} + (\mathscr{E}_{i,\text{kin vib}} + \phi_{i,\text{internal}}) + \mathscr{E}_{i,\text{el}}] + \Phi_{\text{boundary}}.$$

(9.89)

We shall treat the electronic energy $\mathscr{E}_{i,\text{el}}$ only in quantum terms. The canonical partition function (9.18) is then

$$Q(\Theta,V) = \frac{1}{N!}\frac{1}{h^{fN}}\prod_{i=1}^{N}\int_{\omega_{i,\text{kin CM}}} \exp\left(-\frac{\mathscr{E}_{i,\text{kin-CM}}}{\Theta}\right) d\omega_{i,\text{kin CM}} \cdots$$

$$\cdots \int_{\omega_{i,\text{boundary}}} \exp\left(-\frac{\Phi_{\text{boundary}}}{\Theta}\right) d\omega_{i,\text{boundary}} \cdots$$

$$\cdots \int_{\omega_{i,\text{vib}}} \exp\left(-\frac{\mathscr{E}_{i,\text{kin vib}} + \phi_{i,\text{internal}}}{\Theta}\right) d\omega_{i,\text{vib}} \cdots$$

$$\cdots \int_{\omega_{i,\text{kin rot}}} \exp\left(-\frac{\mathscr{E}_{i,\text{kin rot}}}{\Theta}\right) d\omega_{i,\text{kin rot}} \cdots$$

$$\cdots \sum_{\text{all states}} \exp\left(-\frac{\mathscr{E}_{i,\text{el}}}{\Theta}\right).$$

(9.90)

Each of the integrals in (9.90) is over the single particle (μ - space) coordinates for the ith molecule. That is $d\omega_{i,\text{kin CM}}$ contains only differentials of momenta of the CM, $d\omega_{i,\text{boundary}}$ only differentials of the coordinates locating the CM, $d\omega_{i,\text{kin rot}}$ only differentials of angular momenta about the three principle axes of rotation about the CM, and $d\omega_{i,\text{vib}}$ only differentials of momenta and coordinates along the axes of vibration. The summation in the electronic contribution is over all electronic states of the single molecule. Since the molecules are identical, the integrals, and the summation, in (9.90) are identical for each of the molecules. Therefore (9.90) becomes

$$Q(\Theta,V) = \frac{1}{N!}q^{N}(\Theta,V),$$

(9.91)

where

$$q(\Theta,V) = q_{\text{tr}}q_{\text{vib}}q_{\text{rot}}q_{\text{el}}.$$

(9.92)

is the molecular partition function.

In (9.92) the individual terms are

$$q_{\text{tr}} = \frac{1}{h^3}\int_{\omega_{\text{boundary}}} \exp\left(-\frac{\Phi_{\text{boundary}}}{\Theta}\right) d\omega_{\text{boundary}} \cdots$$

$$\cdots \int_{\omega_{\text{kin CM}}} \exp\left(-\frac{\mathscr{E}_{\text{kin-CM}}}{\Theta}\right) d\omega_{\text{kin CM}},$$

(9.93)

$$q_{vib} = \int_{\omega_{vib}} \exp\left(-\frac{\mathscr{E}_{kin\ vib} + \phi_{internal}}{\Theta}\right) d\omega_{vib}, \tag{9.94}$$

$$q_{rot} = \int_{\omega_{kin\ rot}} \exp\left(-\frac{\mathscr{E}_{kin\ rot}}{\Theta}\right) d\omega_{kin\ rot}, \tag{9.95}$$

and

$$q_{el} = \sum_{\text{all states}} \exp\left(-\frac{\mathscr{E}_{el}}{\Theta}\right). \tag{9.96}$$

are the translational, vibrational, rotational, and electronic partition functions respectively. In (9.93), (9.94), (9.95) and (9.96) we have dropped the subscript i, since all molecules are identical.

Translational Partition Function. For a boundary interaction potential appropriate for smooth walls $\Phi_{boundary}$ vanishes within the system and is infinite outside. This choice of $\Phi_{boundary}$ prevents molecules from existing outside of the system boundary because there $\exp\left(-\Phi_{boundary}/\Theta\right) = 0$. Then

$$\exp\left(-\frac{\Phi_{boundary}}{\Theta}\right) = \begin{cases} 1 & \text{inside the system} \\ 0 & \text{on the boundary and outside the system} \end{cases},$$

and the final integral contribution to (9.90) from interactions with the boundary is simply

$$\int_{\omega_{boundary}} \exp\left(-\frac{\Phi_{boundary}}{\Theta}\right) d\omega_{boundary} = V,$$

which is the volume of the container. Then with $\mathscr{E}_{kin\text{-}CM} = \left(p_x^2 + p_y^2 + p_z^2\right)/2m$ and $d\omega_{kin\ CM} = dp_x dp_y dp_z$, the translational partition function is

$$q_{tr} = \frac{1}{h^3} V \int_{-\infty}^{+\infty} \int_{-\infty}^{+\infty} \int_{-\infty}^{+\infty} \exp\left(-\frac{p_x^2 + p_y^2 + p_z^2}{2m\Theta}\right) dp_x dp_y dp_z$$

$$= V \left(\frac{2\pi m k_B T}{h^2}\right)^{\frac{3}{2}}, \tag{9.97}$$

where $\Theta = k_B T$.

Vibrational Partition Function. There may be a number of bonds between the atoms making up a molecule. Each has a potential energy, which is a function of the distance between the atomic nuclei. The force between the atoms depends upon this interatomic distance and the equilibrium distance or bond length between the respective atoms corresponds to a minimum in the potential energy. If the vibrations are small the potential energy may be approximated as a quadratic function of the distance between nuclei (see Sect. 9.10). If there are f_{vib} bonds among nuclei in the molecule there will be f_{vib} possible internal vibrational modes for the molecule. Each of these may be modeled as the motion of an effective mass in a quadratic potential [cf. [59], Chap. 3]. That is we model each as a simple harmonic oscillator.

The simple harmonic motion associated with the λth bond has a momentum p_λ^{vib} and a coordinate q_λ^{vib}. The energy $\mathscr{E}_{\text{kin vib}} + \phi_{\text{internal}}$ in (9.90) is a sum of terms of the form $\left(p_\lambda^{\text{vib}}\right)^2 / 2m_\lambda + k_\lambda \left(q_\lambda^{\text{vib}}\right)^2 / 2$, where m_λ is a reduced mass and $k_\lambda \left(q_\lambda^{\text{vib}}\right)^2 / 2$ is the quadratic approximation to the potential energy of the λth bond. With this understanding the vibrational partition function is

$$
q_{\text{vib}} = \frac{1}{h^{f_{\text{vib}}}} \int dq_1^{\text{vib}} dq_2^{\text{vib}} \cdots dp_1^{\text{vib}} dp_2^{\text{vib}} \cdots
$$

$$
\cdots \exp\left(-\frac{\sum_\lambda^{f_{\text{vib}}} \left(p_\lambda^{\text{vib}}\right)^2 / 2m_\lambda + k_\lambda \left(q_\lambda^{\text{vib}}\right)^2 / 2}{\Theta}\right), \tag{9.98}
$$

where the integration is over all vibrational coordinates and momenta.

Rotational Partition Function. There are generally three principal axes of rotation [cf. [59], Chaps. 4 and 5]. We shall designate the angular momenta about these axes as $\left(p_\vartheta, p_\varphi, p_\psi\right)$ and the moments of inertia as $\left(I_\vartheta, I_\varphi, I_\psi\right)$. The rotational partition function is then

$$
q_{\text{rot}} = \frac{1}{h^3} A \int_{-\infty}^{+\infty} \int_{-\infty}^{+\infty} \int_{-\infty}^{+\infty} \exp\left(-\frac{\sum_\lambda \frac{p_\lambda^2}{2I_\lambda}}{\Theta}\right) dp_\vartheta dp_\varphi dp_\psi, \tag{9.99}
$$

where the summation in the exponential is over $\lambda = \vartheta, \varphi, \psi$ and A is the result of the angular integration, which, in the field free case, is $(2\pi)^3$.

Electronic Partition Function. Finally there is a contribution to the molecular partition function from the energies of the electrons. We shall always consider the electron energies to be quantum energies and simply write the electronic partition function as (9.96).

9.13 Spinless Gases

In this section we introduce the quantum energies and degeneracies for the rotational, vibrational, and the electronic states to calculate the individual contributions to the molecular partition function for spinless molecules. As we show in the next chapter, this is a completely justifiable approach at moderate to high temperatures in gases.

We could include a quantum mechanical treatment of the translational energies as well by considering that the molecules are quantum particles moving independently of one another in a large box. However, the final result for the translational partition function is identical to our classical result in (9.97). We, therefore, stay with the classical treatment for translation in the present chapter.

For simplicity we shall limit our treatment to diatomic molecules. Because this is an introductory text, our treatment here will be general. We shall specifically ignore multiplicities associated with nuclear spin, such as in ortho- and para-hydrogen.

9.13.1 Vibration

The diatomic molecule has only one atomic bond. There is, therefore, only a single quantum oscillator in the molecule. The energy levels of the quantum oscillator are

$$\varepsilon_{i_v} = \left(i_v + \frac{1}{2}\right) h v \tag{9.100}$$

in which i_v is the vibrational quantum number and v is the natural frequency of the oscillator.

When $i_v = 0$ the energy of the oscillator is

$$\varepsilon_0 = \frac{1}{2} h v, \tag{9.101}$$

which is called the zero point energy. This zero point energy has no effect on the statistical mechanics of the molecules and we shall neglect it here.

The vibrational partition function is then

$$q_{vib} = \sum_{i_v} \exp\left(-\frac{i_v h v}{k_B T}\right). \tag{9.102}$$

Because

$$\frac{1}{1-x} = 1 + x + x^2 + x^3 \cdots = \sum_{i_v} x^{i_v},$$

we may write (9.102) as

$$q_{vib} = \left[1 - \exp\left(-\frac{\Theta_{vib}}{T}\right)\right]^{-1}, \tag{9.103}$$

in which

$$\Theta_{vib} = \frac{h v}{k_B} \tag{9.104}$$

is the vibrational temperature of the molecule. For spectroscopic measurements the convenient quantity is the wave number $\tilde{v}_{vib} = v/c = 1/\lambda$ where λ is the the wavelength and c is the speed of light in vacuum. In terms of $\tilde{v}_{vib}$ the vibrational energy takes the form

Table 9.1 Vibrational temperature Θ_{vib} calculated from spectroscopic data $(\tilde{v})$ for selected diatomic molecules

Molecule	$\tilde{v}_{vib}/cm^{-1}$	Θ_{vib}/K
N_2	2358.07	3393
NO	1904	2740
O_2	1580.361	2274
F_2	801.8	1154
$^{35}Cl_2$	559.71	805.3
CO	2169.8233	3122
H_2	4400.39	6331
HD	3811.924	5485
D_2	3118.46	4487
HF	4138.32	5954
HI	2308.09	3321

Source: D.E. Gray (ed.): American Institute of Physics Handbook, 3d ed. (McGraw-Hill, New York, 1972) reproduced with the permission of the McGraw-Hill Companies.

$$\varepsilon_{i_v} = \left(i_v + \frac{1}{2}\right) hc\tilde{v}_{vib} \tag{9.105}$$

and the vibrational temperature is

$$\Theta_{vib} = \frac{hc}{k_B}\tilde{v}_{vib}, \tag{9.106}$$

where hc/k_B has the numerical value $1.4388\,cm\,K$.

Vibrational temperatures are very high. The vibrational temperature for a select number of diatomic molecules is provided in Table 9.1.

9.13.2 Rotation

Diatomic molecules have a single moment of inertia and may be modeled as a dumbbell. This is the classical analog of the rigid quantum rotator with moment of inertia I. The energies available to the rigid quantum rotator are

$$\varepsilon_\ell = \frac{\hbar^2}{2I}\ell(\ell+1). \tag{9.107}$$

where ℓ is a positive integer, i.e. $\ell = 0,1,2,\ldots$ and $\hbar$ is Planck's constant (see Sect. 9.4.5) of action h divided by 2π.

The projection of the angular momentum onto an axis has the values $m_\ell\hbar$ where $m_\ell = -\ell, -\ell+1, \ldots, \ell-1, \ell$. In the spatially homogeneous (field free) case these all have the same energy. Each rotational energy state then contains a number

$$g_\ell = 2\ell + 1 \tag{9.108}$$

of states each with the same energy. This is termed degeneracy and we must include this g_ℓ as a factor in the rotational partition function, which then becomes

$$q_{rot} = \sum_{\ell=0}^{\infty} (2\ell+1)\exp\left(-\frac{\ell(\ell+1)\hbar^2}{2Ik_BT}\right). \tag{9.109}$$

We define the rotational temperature Θ_{rot} as

$$\Theta_{rot} = \frac{\hbar^2}{2Ik_B}. \tag{9.110}$$

Then (9.109) may be written as

$$q_{rot} = \sum_{\ell=0}^{\infty} (2\ell+1)\exp\left(-\ell(\ell+1)\frac{\Theta_{rot}}{T}\right). \tag{9.111}$$

For spectroscopic measurements it is convenient to write the rotational energy in the same form as (9.105). That is

$$\varepsilon_\ell = \ell(\ell+1)hcB \tag{9.112}$$

where $B = h^2/8\pi^2 cI$ is the rotational wave number analogous to $\tilde{v}_{vib}$. For diatomic molecules Θ_{rot} is very small because the moment of inertia I is large, unless the atoms are light, such as hydrogen. The rotational temperatures for a select number of diatomic molecules is provided in Table 9.2.

The energies are very closely spaced for temperatures normally considered. The energy spectrum is then almost a continuum and we can replace the sum in (9.111) by an integral. With $\xi = \ell(\ell+1)$, (9.111) is

$$q_{rot} = \int_0^{\infty} \exp\left(-\xi\frac{\Theta_{rot}}{T}\right)d\xi = \frac{T}{\Theta_{rot}}. \tag{9.113}$$

Table 9.2 Rotational temperature Θ_{rot} calculated from spectroscopic data (B) for selected diatomic molecules

Molecule	B/cm^{-1}	Θ_{rot}/K
N_2	1.9987	2.8757
NO	1.7048	2.4529
O_2	1.4457	2.0801
F_2	0.8828	1.2702
$^{35}Cl_2$	0.24407	0.35117
CO	1.9313	2.7788
H_2	60.864	87.571
HD	45.6378	65.664
D_2	30.442	43.800
HF	20.9557	30.151
HI	6.5108	9.3677

Source: D.E. Gray (ed.): American Institute of Physics Handbook, 3d ed., (McGraw-Hill, New York, 1972) reproduced with the permission of the McGraw-Hill Companies.

For symmetry reasons a correct analysis for homonuclear molecules such as O_2 and N_2 we must include a factor of $1/2$ in the rotational molecule partition function. To indicate this q_{rot} is written as

$$q_{rot} = \frac{T}{\sigma \Theta_{rot}}.$$ (9.114)

where $\sigma = 1$ for heteronuclear diatomic molecules and $\sigma = 2$ for homonuclear diatomic molecules.

A more accurate result at high temperature ($T \gg \Theta_{rot}$) can be obtained for q_{rot} if we use the Euler-MacLaurin sum formula [cf. [67, 162]. The result is

$$q_{rot} = \frac{T}{\sigma \Theta_{rot}} \left[1 + \frac{\Theta_{rot}}{3T} + \frac{1}{15} \left(\frac{\Theta_{rot}}{T} \right)^2 + \frac{4}{315} \left(\frac{\Theta_{rot}}{T} \right)^3 + \cdots \right]$$ (9.115)

In practical terms we gain, however, little from expansion techniques. At this time the sum in (9.111), and subsequent differentiations, can be easily handled on desk top computers.

9.13.3   Electronic

We may find the electronic partition function in the same way as we have found the rotation and vibration partition functions. A discussion of the electronic states in an atom can be found in any textbook on quantum mechanics.

Data for spectra can be found in sets of tables such as *National Standard Reference Data Series (NSRDS) 3 Sec. 1–10,1965–1983, Selected Tables on Atomic Spectra* [118] which can be accessed online at the *National Institute of Standards and Technology* (NIST). In most circumstances only the first few levels need be considered.

We choose atomic oxygen as an example,. The spectrum of atomic oxygen is found in *NSRDS 3 Sec. 7*. The first three levels above the ground state, referenced to the ground state, are presented in Table 9.3.

Table 9.3 shows that when ordering the energy levels only the outer or valence electrons ($2p^4$) need be considered. This is the first of what are referred to

Table 9.3 Atomic oxygen spectrum

Configuration	Designation	J	Level $\left[cm^{-1} \right]$
$2s^2 2p^4$	$2p^4\,^3P_2$	2	0.000
	$2p^4\,^3P_1$	1	158.265
	$2p^4\,^3P_0$	0	226.977
$2s^2 2p^4$	$2p^4\,^1D_2$	2	15867.862

Source: NSRDS 3 Sect. 7.

as Hund's[14] Rules (1925), which are particular cases in the molecular spin-orbit coupling theory. Here J is the total orbital angular momentum quantum number and the data are given in wave numbers with units of cm^{-1}.

In quantum mechanics angular momentum is defined in terms of the commutation rules obeyed by the angular momentum operators. These include quantum mechanical spin. The total angular momentum is a combination of the orbital and total spin angular momentum. We can speak of total angular momentum states but cannot speak of the angular momentum state of individual atomic electrons. The spectroscopic notation reflects this. Atomic states are designated as $^{(2S+1)}T_J$ in which S is the total spin of the atomic electrons, J is the total angular momentum, and T is the orbital identification letter: S, P, D, F, G, ... The identification letter refers to the value of the orbital angular momentum quantum number L. For example the orbital designation for $L = 0$ is S and for $L = 1$ is P. The number $(2S + 1)$ at the upper left is the multiplicity of the atomic state. If the atom has filled electronic shells the total spin S is zero and the multiplicity is equal to 1. The total angular momentum quantum number takes on the values $J = L + S, L + S - 1, \ldots, |L - S|$. For a particular orbital angular momentum quantum number L, and corresponding identification letter, there may be a number of values of total angular momentum quantum number J. Specifically in the part of the atomic oxygen spectrum in Table 9.3 the ^{3}P states have $J = 2, 1, 0$.

The Planck-Einstein formula for the energy of a photon with frequency $\nu = c/\lambda$ is

$$\varepsilon = h\nu = h\frac{c}{\lambda}.$$

The terms appearing in the electronic partition function then have the general form

$$g_J \exp\left(-\frac{\varepsilon_J}{k_B T}\right) = g_J \exp\left(-\frac{hc/\lambda_J}{k_B T}\right),$$

in which ε_J is the energy of the Jth electronic energy state relative to the ground state and g_J $(= 2J + 1)$ is the multiplicity of the quantum state ε_J with total orbital angular momentum quantum number J. This is the number of total orbital angular momentum states with the energy ε_J.

For the electronic state J we shall designate

$$\Theta_J = \frac{hc/\lambda_J}{k_B}$$

as a characteristic temperature for excitation. This is the electronic analog of the vibrational and rotational temperatures. For the ground state and the first three levels in atomic oxygen we have

---

[14] Friedrich Hund (1896–1997) was a German physicist known for his work on atoms and molecules. He was Max Born's assistant, working with quantum interpretation of band spectra of diatomic molecules, and later Professor at the Universities of Rostock, Leipzig, Jena, Frankfurt am Main, and Göttingen.

$$\Theta_0 = 0, \text{ with } g_0 = 2*2+1 = 5$$

$$\Theta_1 = \frac{158.265\,\mathrm{cm}^{-1}}{\frac{0.695\,06}{\mathrm{cm\,K}}} = 227.7\,\mathrm{K}, \text{ with } g_1 = 2*1+1 = 3$$

$$\Theta_2 = \frac{226.977\,\mathrm{cm}^{-1}}{\frac{0.695\,06}{\mathrm{cm\,K}}} = 326.56\,\mathrm{K}, \text{ with } g_2 = 2*0+1 = 1$$

$$\Theta_3 = \frac{15867.862\,\mathrm{cm}^{-1}}{\frac{0.695\,06}{\mathrm{cm\,K}}} = 22829.\,\mathrm{K}, \text{ with } g_3 = 2*2+1 = 5.$$

The approximate form of the electronic partition function for atomic oxygen is then

$$q_{\mathrm{el}} = 5 + 3\exp\left(-\frac{227.7}{T}\right) + \exp\left(-\frac{326.56}{T}\right) + 5\exp\left(-\frac{22829}{T}\right). \quad (9.116)$$

From (9.116) we see that for even high temperature gas dynamic considerations we need only the first three terms in the electronic partition function for atomic oxygen. This, however, will not be true for astrophysical considerations.

The corresponding data for atomic nitrogen are

$$\Theta_0 = 0\,\mathrm{K}, \text{ with } g_0 = 2*\frac{3}{2}+1 = 6$$

$$\Theta_1 = \frac{19224.464\,\mathrm{cm}^{-1}}{\frac{0.695\,06}{\mathrm{cm\,K}}} = 27659.\,\mathrm{K}, \text{ with } g_1 = 2*\frac{5}{2}+1 = 6$$

so that

$$q_{\mathrm{el}} = 4 + 6\exp\left(-\frac{27659}{T}\right). \quad (9.117)$$

For even high temperature gas dynamic considerations we may then equate q_{el} to 4 for atomic nitrogen.

The temperatures above which the atomic forms of these gases are observed are already fairly high (approximately $3,000\,\mathrm{K}$ for oxygen). At this temperature the first two exponential factors in (9.116) are unity, and the last terms in (9.116) and (9.117) are still negligible. We may in general assume, except for very high temperatures, that for atomic oxygen and nitrogen

$$q_{\mathrm{el}} = \text{constant}. \quad (9.118)$$

Similar results hold for other species with low characteristic temperatures of electronic excitation.

9.13.4 Molecular Partition

Combining (9.97), (9.103), (9.114), and (9.118) in (9.92) we have

$$q = V \left(\frac{2\pi m k_B T}{h^2} \right)^{\frac{3}{2}} \left[1 - \exp\left(-\frac{\Theta_{vib}}{T} \right) \right]^{-1} q_{rot} q_{el}. \tag{9.119}$$

Using (9.91) and Stirling's approximation, (9.47), for $N!$, we have the canonical partition function as

$$Q = \frac{1}{\sqrt{2\pi N}} e^N \left[\frac{V}{N} \left(\frac{2\pi m k_B T}{h^2} \right)^{\frac{3}{2}} \left[1 - \exp\left(-\frac{\Theta_{vib}}{T} \right) \right]^{-1} q_{rot} q_{el} \right]^N \tag{9.120}$$

and from (9.27) the Helmholtz energy for the spinless gas is

$$F = -N k_B T \left\{ \frac{3}{2} \ln(T) + \ln\frac{V}{N} - \ln\left[1 - \exp\left(-\frac{\Theta_{vib}}{T} \right) \right] \right.$$
$$\left. + \frac{3}{2} \ln\left(\frac{2\pi m k_B}{h^2} \right) + \ln q_{rot} + \ln q_{el} + 1 \right\}, \tag{9.121}$$

or, when $T \gg \Theta_{rot}$

$$F = -N k_B T \left\{ \frac{3}{2} \ln(T) + \ln\frac{V}{N} - \ln\left[1 - \exp\left(-\frac{\Theta_{vib}}{T} \right) \right] \right.$$
$$\left. + \frac{3}{2} \ln\left(\frac{2\pi m k_B}{h^2} \right) + \ln\left(\frac{T}{\sigma \Theta_{rot}} \right) + \ln q_{el} + 1 \right\}, \tag{9.122}$$

neglecting $\ln \sqrt{2\pi N} \ll N$, since N is very large.

Because we have used the high temperature approximation (9.114) for the rotational partition in (9.119), (9.120), and (9.122), these equations are only valid for high temperatures. In the next section we relax this approximation and perform a numerical calculation for the rotational partition function.

9.13.5 Applications

For the gas with the Helmholtz energy (9.122) the thermal equation of state is

$$P = -\left(\frac{\partial F}{\partial V} \right)_T = \frac{N k_B T}{V}, \tag{9.123}$$

which is that for the ideal gas. The rotation and vibration have no effect on the thermal equation of state.

We can obtain the entropy from the Helmholtz energy as

$$S = -\left(\frac{\partial F}{\partial T}\right)_V$$

$$= Nk_B \left\{ \frac{3}{2}\ln(T) + \ln\frac{V}{N} + \frac{\Theta_{vib}}{T}\left[\exp\left(\frac{\Theta_{vib}}{T}\right) - 1\right]^{-1} \right.$$

$$\left. -\ln\left[1 - \exp\left(-\frac{\Theta_{vib}}{T}\right)\right] + \frac{3}{2}\ln\left(\frac{2\pi mk_B}{h^2}\right) \right. \tag{9.124}$$

$$\left. + \frac{T}{q_{rot}}\left(\frac{\partial q_{rot}}{\partial T}\right)_V + \ln q_{rot} + \ln q_{el} + \frac{5}{2} \right\} \tag{9.125}$$

If we compare (9.124) with the Sackur-Tetrode equation (9.82) we see differences in the temperature dependence and the (constant) $\ln q_{el}$. The new temperature dependent terms result from the rotational and vibrational motion of the molecule. These are the rotational and vibrational entropies

$$S_{rot} = Nk_B \left[\frac{T}{q_{rot}}\left(\frac{\partial q_{rot}}{\partial T}\right)_V + \ln q_{rot} \right], \tag{9.126}$$

which for $T \gg \Theta_{rot}$ is

$$S_{rot} = Nk_B \left[\ln\left(\frac{T}{\sigma\Theta_{rot}}\right) + 1 \right], \tag{9.127}$$

and

$$S_{vib} = Nk_B \left\{ -\ln\left[1 - \exp\left(-\frac{\Theta_{vib}}{T}\right)\right] + \frac{\Theta_{vib}}{T}\left[\exp\left(\frac{\Theta_{vib}}{T}\right) - 1\right]^{-1} \right\}, \tag{9.128}$$

which, for $T \gg \Theta_{vib}$ is

$$S_{vib} = Nk_B \left[\ln\left(\frac{T}{\Theta_{vib}}\right) + 1 \right]. \tag{9.129}$$

The rotational and vibrational entropies then have the same mathematical form, but different normalized temperature scales, since $\Theta_{vib} \gg \Theta_{rot}$.

From (9.21) or from $\langle H \rangle = F + TS$ the average internal energy is

$$\langle H \rangle = Nk_B T \left[\frac{3}{2} + T\left(\frac{\partial \ln q_{rot}}{\partial T}\right)_V + \frac{\Theta_{vib}}{T}\left[\exp\left(\frac{\Theta_{vib}}{T}\right) - 1\right]^{-1} \right], \tag{9.130}$$

which for $T \gg \Theta_{rot}$ is

$$\langle H \rangle = Nk_B \left(\frac{5}{2}T + \Theta_{vib}\left[\exp\left(\frac{\Theta_{vib}}{T}\right) - 1\right]^{-1} \right). \tag{9.131}$$

Fig. 9.10 Rotational contribution to specific heat C_V/R as a function of T/Θ_r

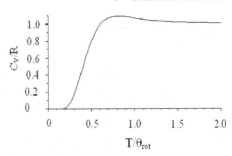

Classically there are seven quadratic energy terms in the Hamiltonian. According to the equipartition principle the average energy should then be $(7/2)Nk_BT$. Equation (9.131) produces only $(5/2)k_BT$ unless $T \gg \Theta_{vib}$, in which case $\langle H \rangle$ becomes the classical $(7/2)k_BT$.

In Fig. 9.10 we have plotted the low temperature contribution to the specific heat C_V from rotational energies for $0 \leq T \leq 2\Theta_{rot}$. Here we evaluated q_{rot} from a numerical summation in (9.111) and calculated the rotational contribution to the specific heat from $C_V = -T\left(\partial^2 F/\partial T^2\right)_V$ with (9.27). This rotational contribution to the specific heat vanishes as $T \to 0$ in agreement with the third law, and becomes equal to the classically expected value of R for diatomic molecules at high temperatures. We included 5,000 terms in the summation in (9.111). For the range considered in Fig. 9.10, however, we find no visible difference between these results and those obtained when only four terms were included in the summation.

It was this dependence of the specific heat of hydrogen for temperatures below 80 K which, in 1911, led Sommerfeld to propose that molecular degrees of freedom should be weighted, not counted (see Sect. 9.10).

The vibrational contribution to the internal energy in (9.131) appears in closed form. The vibrational contribution to the specific heat may then be evaluated in closed form as well. From (9.131) the specific heat of a collisionless gas of diatomic molecules is

$$C_V = \left(\frac{\partial \langle H \rangle}{\partial T}\right)_V$$
$$= \frac{5}{2}Nk_B + Nk_B\left(\frac{\Theta_{vib}}{T}\right)^2 \exp\left(\frac{\Theta_{vib}}{T}\right)\left[\exp\left(\frac{\Theta_{vib}}{T}\right) - 1\right]^{-2}. \quad (9.132)$$

The vibrational contribution to the specific heat is the second term on the right hand side in (9.132). In Fig. 9.11 we have plotted predictions from the theoretical specific heat in (9.132) and the NIST results for diatomic nitrogen (N_2) as an example. The curves in Fig. 9.11 indicate essential agreement between theoretical predictions and observations for the specific heat as a function of temperature. The quantum treatment of the vibrational energies has then removed the cloud which Lord Kelvin referred was over the dynamical theory of heat in Baltimore in 1884.

There is a small discrepancy at high temperatures, which results from anharmonicity in the potential as the vibrational energy increases (see Fig. 9.5). At higher

Fig. 9.11 Specific heat C_V/R for a diatomic gas. NIST results for N_2 (*solid curve*) [102]. Theoretical result from (9.132) (*dashed curve*)

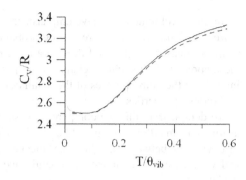

Fig. 9.12 Rotational and vibrational entropies for heteronuclear dipole molecule, S_{rot}/R as a function T/Θ_{rot} (**a**), S_{vib} as a function of T/Θ_{vib} (**b**)

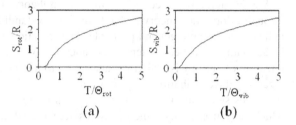

vibrational energies the moment of inertia is also affected so that the vibrational and rotational energies are no longer completely independent. The result is an additional contribution to the specific heat of

$$C_{V,\,anh} = C_m R \frac{\Theta_{rot}}{\Theta_{vib}^2} T,$$

where C_m is a constant specific to the molecule [[158], pp. 136–138], [[111], pp. 160–166], [99], p. 145].

In Fig. 9.12 we have plotted the contributions to the entropy of a heteronuclear diatomic molecule arising from the rotational and vibrational states. The calculation of the rotational entropy (Fig. 9.12(a)) is based on the summation in (9.111) (5,000 terms) and the calculation of the vibrational entropy (Fig. 9.12(b)) is based on (9.128). The temperature Θ_{rot} is very low and Θ_{vib} is very high. The rotational entropy then contributes to the entropy of the gas for all temperatures of general interest and vanishes as $T \to 0$. The vibrational entropy only contributes to the entropy of the gas at high temperatures.

9.14   Summary

Our goal in this chapter has been to provide a brief introduction to statistical mechanics, while remaining as mathematically rigorous as practicable for the level of this text. The reader wishing to consider more advanced aspects of the subject will be prepared to engage those without modification of what has been presented.

In our development we have emphasized that statistical mechanics does not require additional hypotheses regarding probabilities. It is based only on our ability to make measurements. And those are determined by the practical restrictions of the laboratory. Statistical mechanics then provides a framework in which we can investigate the consequences of the details of our microscopic picture of matter on measurable properties.

To demonstrate this we compared our statistical mechanical results with those available from NIST in Sects. 9.11 and 9.13.5. In each case we considered the causes of deviations between the statistical mechanical predictions and the NIST results and were able to obtain specific quantitative insight into the causes of the noted deviations.

We devoted a brief section to information theory. The primary insight obtained through information theory is that Gibbs' understanding of statistical mechanics is indicative of a more general principle. The statistical mechanical entropy of Gibbs is the mathematical equivalent of the uncertainty in information theory. The maximization of entropy is, therefore, analogous to the maximization of our uncertainty regarding the distribution of system states.

Jaynes suggested that what is lacking is a fundamental statistical theory from which the entropy would emerge as a result of a constrained maximization procedure applied to a statistical expression. This, he contends, would replace the confusion about order and disorder associated with the second law (see Sect. 9.7.2) [80].

Modern applications of statistical mechanics to nonequilibrium situations are based on the BBGKY hierarchy noted in Sect. 9.6.

Exercises

9.1. (a) Plot the phase space trajectory for a freely falling body.

(b) A general one dimensional, conservative system has the Hamiltonian $H = p^2/(2m) + V(x) = E$ show the difference in phase space plots for $E \gtrless V(x)$.

9.2. Consider a system of N linear, one dimensional harmonic oscillators with the single particle Hamiltonian (9.1). Find (a) the canonical partition function, (b) the Helmholtz energy, (c) the internal energy, (d) the specific heat C_V, and (e) the entropy for this system.

9.3. Magnetism in matter results from unpaired electrons in molecules. It is more energetically favorable for electrons to first individually fill the available levels. This is one of Hund's rules (see Sect. 10.2). Highly magnetic materials then have net spin angular momenta resulting from the unpaired electrons. Each molecule may then be treated as a magnetic moment.

Consider a system of magnetic moments and an externally applied magnetic field of intensity $\mathscr{H}$. The energy of the jth magnetic moment is $-\mu \cos \vartheta_j B$. For notational simplicity define $\mu_j = \mu \cos \vartheta_j$. In each system a particular magnetic moment may have a different orientation. For the jth magnetic moment in the σth system we write $\mu_j^\sigma = \mu \cos \vartheta_j^\sigma$. The system Hamiltonian for the σth system is then $H_\sigma = - \sum_j^N \mu_j^\sigma \mathcal{H}$ and the microscopic, single system magnetization is $M_\sigma = \sum_j^N \mu_j^\sigma$.

The magnetic susceptibility is defined by $\chi_m = (\partial \langle M \rangle / \partial \mathcal{H})_\Theta$ that is $\langle M \rangle = \chi_m \mathcal{H}$.

(a) Show that $\langle M \rangle = \Theta \frac{1}{Q} (\partial \ln Q / \partial \mathcal{H})_\Theta$.

(b) Then show that the correlation function $\langle \delta M^2 \rangle = \Theta^2 (\partial^2 \ln Q / \partial \mathcal{H}^2)_\Theta = \Theta \chi_m$, where $\delta M = M - \langle M \rangle$ is the fluctuation in the magnetization. {Hint: show that $(\partial^2 \ln Q / \partial \mathcal{H}^2)_\Theta = \frac{1}{Q} (\partial^2 Q / \partial \mathcal{H}^2)_\Theta - \frac{1}{Q^2} (\partial Q / \partial \mathcal{H})_\Theta^2$.]

9.4. The canonical ensemble relaxes the requirement of constant energy of the microcanonical ensemble. But we have shown that the results for the microcanonical and canonical ensemble are the same (in the thermodynamic limit). That implies that the energy fluctuations $\delta H = H - \langle H \rangle$, where H is the Hamiltonian, must be small in the canonical ensemble.

(a) Show first that the correlation function is generally given by $\langle \delta H^2 \rangle = \left(\partial^2 \ln Q / \partial (1/\Theta)^2 \right)_V$.

(b) Then using the partition function for the ideal gas $Q = Q = (1/N!) (2\pi m k_B / h^2)^{3N/2} V^N$ show that $\langle \delta H^2 \rangle / \langle H \rangle^2 \propto 1/N$, which, is very small for the expected values of N.

What do you then conclude?

9.5. In a two level system each particle may exist in one of two states: $\varepsilon_1 = -\Delta\varepsilon/2$ and $\varepsilon_2 = \Delta\varepsilon/2$. Consider a two level system of N indistinguishable particles. Find (a) the Helmholtz energy, (b) the internal energy, and (c) the specific heat for this system. Does C_V obey the third law?

9.6. Consider the magnetic system from another point of view. Assume that we know the total energy of the magnetic system and want to know the coefficient of probability P_σ of the σth system in an ensemble. The energy that we claim to know is $\langle H \rangle = - \sum_\sigma P_\sigma M_\sigma \mathcal{H}$ where the magnetization of the σth system is $M_\sigma = -\sum_j^N \mu_j^\sigma$. We assume the validity of the Gibbs entropy $S = -k_B \sum_\sigma P_\sigma \ln P_\sigma$. The coefficient of probability must satisfy $\sum_\sigma P_\sigma = 1$. We elect to maximize the entropy subject to the two constraints.

Show that the result is $P_\sigma = (1/Q) \exp (M_\sigma \mathcal{H} / \Theta)$. (This is an example of the MAXENT principle).

9.7. We can treat the magnetization problem classically, as was originally done. We consider a system of N identical magnetic moments μ oriented at various angles ϑ_j to the external magnetic field $\mathcal{H}$ as in Exercise 9.3. For magnetic moments fixed in space and freely rotating about the direction of the magnetic field, the classical partition function is then

$$Q = (1/N!h^{2N})(2\pi)^N \left\{ \int_0^\pi d\vartheta \sin\vartheta \exp[\mu H \cos\vartheta/(k_B T)] \right\}^N$$

(a) Obtain Q by integration.
(b) Find the magnetic susceptibility.
(c) Perform a Taylor series expansion of your result for high temperature to show that $\chi_m \approx N\mu^2/(3k_B T)$. This is the Curie Law.

9.8. The result for the susceptibility in Exercise 9.4 can be improved. We assume (Weiss) that the magnetic moments do not respond only to the external magnetic field, but also to the field resulting from the other magnetic moments in the immediate vicinity. That is we replace the external field by $\mathcal{H}_{eff} = \mathcal{H} + w\langle M\rangle$ where w is a constant. We note that the system is spatially homogeneous.

(a) Show that now the magnetic susceptibility is $\chi_m = N\mu^2(\coth\tilde{u} - 1/\tilde{u})$ where $\tilde{u} = [\mu/(k_B T)](\mathcal{H} + w\langle M\rangle)$.
(b) Show that in the high temperature limit the magnetization is now $\langle M\rangle = [C_C/(T - \Theta_C)]H$ where $C_C = N\mu^2/(3k_B)$ is the Curie constant and $\Theta_C = wC_C$ is the Curie temperature.

9.9. We can improve on our treatment of magnetism by introducing the quantum levels of a magnetic moment in an external field. For a molecule with total spin S the spin angular momentum is $S\hbar$. The component of total spin along any axis can only take on the values $(S, S-1,\ldots,-S+1,-S)\hbar$. The energy levels of the molecule in the presence of an external magnetic field $\mathcal{H}$ are then $-(S, S-1,\ldots,-S+1,-S)2\mu_B\mathcal{H}$ where μ_B is the Bohr magneton $\mu_B = q_e\hbar/(2m_e)$, where q_e and m_e are the charge and mass of the electron. We can then write a magnetic partition function (see Sect. 9.12) as $q = \sum_{m=-S}^{m=+S}\exp(m\alpha)$ where $\alpha = 2\mu_B\mathcal{H}/(k_B T)$.
Show that the magnetic partition function is

$$q = \exp(-S\alpha)\frac{\exp[(2S+1)\alpha/2] - 1}{\exp\alpha - 1}$$

$$= \frac{\sinh[(2S+1)\alpha/2]}{\sinh(\alpha/2)}.$$

[Hint: use the geometrical series ($\frac{1}{1-x} = 1 + x + x^2 + x^3 + \cdots$)]

9.10. The magnetic moment may be obtained from the magnetic partition function q of Exercise 9.9 in the same way as from Q in Exercise 9.3.

(a) Find the magnetization for the partition function q.
(b) Show that for small values of $\mu_B H/(k_B T)$ the susceptibility becomes $\chi_m = 4N\mu_B^2 S(S+1)/(3Vk_B T)$.

9.11. The kinetic theory of Maxwell, Boltzmann, and Clausius was based on an equilibrium distribution function for the number of particles per unit volume having a specific velocity, to within the limits of an infinitesimal volume in velocity space $d\mathbf{v} = dv_x dv_y dv_z = v^2 \sin \vartheta \, dv \, d\vartheta \, d\varphi$ (in spherical velocity space with $v = $ magnitude of $\mathbf{v}$). This number is $f(\mathbf{v})d\mathbf{v} = n [m/(2\pi k_B T)]^{3/2}$ $\exp \left[-mv^2/(2k_B T)\right] d\mathbf{v}$, where n_p is the density of the gas in particles per unit volume.

Kinetic theory is still the basis for modern work in theoretical plasma physics. There, however, we seek a time dependence of the distribution function.

(a) Show that for the velocity distribution given here is normalized such that $\int_{\mathbf{v}} f(\mathbf{v})d\mathbf{v} = n_p$, where the integral is over all velocity space.

(b) Show that the average velocity of a particle vanishes (i.e. the gas is stationary).

9.12. It is often more convenient to deal with a speed distribution defined in such a manner that the number of particles with a speed $v = |\mathbf{v}|$ within the limits on particle speed of dv is $f_S(v)\,dv$.

(a) Show that $f_S(v) = 4\pi n_p [m/(2\pi k_B T)]^{3/2} v^2 \exp\left[-mv^2/(2k_B T)\right]$.

(b) What is the most probable speed v_{mp} of a particle?

9.13. Find (a) the average speed $\bar{v}$ and (b) the root mean square (rms) v_{rms} speed of a particle in a distribution $f_S(v)$.

9.14. Calculate the number of particles striking a wall per unit area per unit time. Construct a cylinder (in velocity space) whose base is the area of interest and whose slant height is $v dt$ along the velocity vector. Particles striking the wall are those with nonzero components of velocity perpendicular to the wall. [Note that when using $f(\mathbf{v})$ particles are coming from both sides of an area in space.]

9.15. Calculate the pressure resulting from the rate of momentum transfer from the gas particles to the wall. Assume that the particles undergo completely elastic collisions with the wall, which is smooth. This calculation is very similar to that carried out in Exercise 9.14. The difference is that now you are interested in the momentum carried by the particles.

9.16. Calculate the internal energy density of particles in the gas. This will be all kinetic.

9.17. Compute the Boltzmann entropy $S = -Nk_B H$ with $H = \int f(\mathbf{v}) \ln(f(\mathbf{v})) d\mathbf{v}$. Notice that you have already calculated the energy density and the normalization of the distribution function. Compare your result with the Sackur-Tetrode result, realizing that Boltzmann's result does not include reference to the Heisenberg principle.

9.18. The grand canonical ensemble considers only the ensemble averaged number of particles known. Do a MAXENT process (see Exercise 9.6) with the constraints $\sum_\sigma P_\sigma = 1, \sum_\sigma P_\sigma H_\sigma = \langle H \rangle, \sum_\sigma P_\sigma N_\sigma = \langle N \rangle$ to obtain the coefficient of probability for the grand canonical ensemble.

Chapter 10
Quantum Statistical Mechanics

10.1 Introduction

In this chapter we will present a brief introduction to the statistical mechanics of quantum particles. This is the most advanced chapter in the text. We have included this chapter as a natural step in an introduction to modern thermodynamics.

We have made no attempt to introduce principles of the quantum theory. The applications to Bose-Einstein Condensation (BEC) and to degenerate Fermi systems require an understanding of the quantum description of a free particle in a container. Beyond this, no fundamental understanding of the quantum theory is required. The reader must, however, be familiar with the consequences of the Pauli principle, which requires that the state of a quantum system must be symmetric or antisymmetric upon exchange of particles, depending upon whether the particles have integer or half-integer spin. The Pauli principle may be thought of as the dividing line between the quasi-classical applications of the preceding chapter and those of the present chapter [125].

10.2 Particles with Spin

Real quantum particles have spin. Spin is a quantum phenomenon with no classical analog. But it has macroscopic consequences. For example spin is the cause of permanent magnetism in matter. Electrons have a spin of one half and a magnetic moment of $-928.476 \times 10^{-26}\,\mathrm{J\,T^{-1}}$. Protons and neutrons also have spin of one half and magnetic moments of $1.410 \times 10^{-26}\,\mathrm{J\,T^{-1}}$ (proton) and $-0.966 \times 10^{-26}\,\mathrm{J\,T^{-1}}$ (neutron). The quantum (non-classical) nature of the spin is particularly revealed by the neutron magnetic moment, which is anti-aligned with the spin. No classical argument involving spinning charge densities can explain this.

The magnetic moments of the nucleons provide the basis for nuclear magnetic resonance (NMR) and magnetic resonance imaging (MRI), which are important diagnostic tools in chemistry and medicine. Electron paramagnetic resonance (EPR)

C.S. Helrich, *Modern Thermodynamics with Statistical Mechanics*,
DOI 10.1007/978-3-540-85418-0_10, © Springer-Verlag Berlin Heidelberg 2009

is the diagnostic tool that uses electron spins rather than nuclear spins. EPR is an important tool in solid state and biophysical studies.

The quantum state of a complex atom is specified in terms of total angular momentum and total spin angular momentum (see Sect. 9.13.3). The atomic state with maximum multiplicity (maximum value of the total spin quantum number S) has the lowest energy. This is the second of Hund's Rules. We may think of this as an energetic requirement that the electrons first fill the valence states with spins aligned rather than coupled. The magnetic moment of an atom is determined by the total (electronic) spin quantum number. The quantum mechanical interaction between the spins of these atoms, known as spin exchange, causes the alignment of the net spins in ferromagnetism or the antialignment in antiferromagnetism.

Until this point we have been able to limit the consequences of a quantum treatment solely to quantization of energy levels. To treat systems of real quantum particles we must consider more general consequences of the quantum theory. The particles (atoms, molecules, or individual electrons in a solid) can no longer be treated as classical particles with internal quantum structures. We must now consider the entire system as a quantum system and speak of the quantum state of the entire system.

If we choose to represent the quantum state in spatial coordinates we have what is called the wave function[1] for the system. The wave function is a mathematical quantity which contains all the information we can obtain about the system, but has no physical meaning itself. Specifically the wave function is ordinarily a complex valued function, i.e. it has, in mathematical terms, real and imaginary parts. To extract physical information from the wave function we use quantum mechanical operators, such as the Hamiltonian operator for the system energy. The square of the modulus of the wave function is the probability density function for the location of particles in the system [13].

Each particle exists in a quantum state and the entire system is in a quantum state made up of the quantum states of the individual particles. Wolfgang Pauli recognized that the quantum state of a collection of indistinguishable particles must be either symmetric or antisymmetric upon exchange of particles between states. Symmetric system states are those for which the exchange of identical particles between two quantum states has no effect on the system wave function. Antisymmetric system states are those for which the exchange of identical particles results in a change in the algebraic sign of the system wave function. The symmetry properties of the quantum system state (wave function) have a profound effect on the behavior of matter [125].

Particles with half-integer spin $\left(\frac{1}{2}, \frac{3}{2}, \cdots\right)$ are called fermions after Enrico Fermi. The wave function of a system of fermions is antisymmetric on interchange of any two particles. The statistics obeyed by fermions are termed Fermi-Dirac (Paul A. M. Dirac) statistics. The quantum state cannot be antisymmetric if there is more than a single particle in each quantum state, since then interchange of particles would produce a quantum state identical to the first. Fermi-Dirac statistics, therefore, require

---

[1] The quantum state may be represented in other than spatial terms. Representation as a matrix in terms of energy states is often simpler.

that each single particle state in the system be either unoccupied or occupied by only one particle. The occupation numbers are then 0 or 1. The Fermi-Dirac statistics were discovered in 1926 [43]. For a personal reflection see [45].

Particles with integer spin $(1, 2, \cdots)$ are called bosons after Satyendra Nath Bose who discovered the statistics they obey. The quantum state (wave function) of a system of bosons is symmetric on interchange of particles. The statistics are called Bose-Einstein statistics because of Einstein's role in getting the paper by Bose, *Planck's Law and the Hypothesis of Light Quanta*, published in *Zeitschrift für Physik* [14]. Bose-Einstein statistics allow any number of particles per state. The occupation numbers are then $(0, 1, 2, \cdots)$.

The story of the introduction of the Bose-Einstein statistics and Einstein's role is told in detail by Abraham Pais in his book *Subtle is the Lord* [126]. Einstein extended the ideas of Bose to atoms and translated the article by Bose into German for the *Zeitschrift für Physik*. It was certainly partially Einstein's influence that provided the publication of Bose's article. But Einstein would not have done this as a favor had the physics not been correct and revealing of new ideas.

10.3 General Treatment

For simplicity we shall consider a system of non-interacting quantum particles. In a general quantum system of non-interacting particles, the energy states available to a single particle form a set $\{\varepsilon_\alpha\} = \varepsilon_0, \varepsilon_1, \varepsilon_2, \cdots$. These quantum states depend on external parameters such as the system volume and possible fields imposed on the system. The set of energy states $\{\varepsilon_\alpha\}$ is then a constant for all systems in the ensemble. We designate the number of particles in the αth quantum state by n_α.[2] Then the canonical partition function for our quantum system has the form

$$Q = \sum_{n_1, n_2, \cdots} \exp\left(-\frac{\sum_\alpha n_\alpha \varepsilon_\alpha}{k_B T}\right) \tag{10.1}$$

10.3.1 Grand Canonical Ensemble

Our treatments of both bosons and fermions will be simplified by using a device that Richard Feynman used [53]. Feynman pointed out that we can conveniently include the ensemble average of the number of particles in the system $\langle N \rangle = \langle \sum_\alpha n_\alpha \rangle$ if we consider

$$Q^{(\mu)} = \sum_{n_1, n_2, \cdots} \exp\left(-\frac{\sum_\alpha n_\alpha (\varepsilon_\alpha - \mu)}{k_B T}\right) \tag{10.2}$$

---

[2] This is the analog of the occupation number in a μ-space cell N_i. We use the lower case for the quantum occupation numbers.

$$= \sum_{n_1,n_2,\cdots} \prod_{\alpha} \exp\left(-\frac{(\varepsilon_\alpha - \mu)}{k_B T}\right)^{n_\alpha} \tag{10.3}$$

instead of (10.1) and drop the constraint that the number of particles in the system is known. The numbers in the set $\{n_\alpha\}$ are then independent of one another. The resulting simplification comes from the fact that we can now rewrite (10.3) as

$$Q^{(\mu)} = \prod_{\alpha} \sum_{n_\alpha} \exp\left(-\frac{(\varepsilon_\alpha - \mu)}{k_B T}\right)^{n_\alpha}, \tag{10.4}$$

which is a product of sums, rather than a sum of products. Because our primary interest is in the logarithm of $Q^{(\mu)}$ this represents a considerable mathematical simplification.

The function $Q^{(\mu)}$ is not a canonical partition function, since we have relaxed the constraint on the number of particles. It is, however, a partition function. We can obtain ensemble averaged (thermodynamic) quantities from $Q^{(\mu)}$ by partial differentiation of $\ln Q^{(\mu)}$ with respect to the parameters on which $Q^{(\mu)}$ depends. If we partially differentiate $k_B T \ln Q^{(\mu)}$ with respect to μ, we obtain.

$$k_B T \frac{\partial}{\partial \mu} \ln Q^{(\mu)} = \frac{1}{Q^{(\mu)}} \sum_{n_1,n_2,\cdots} \left(\sum_{\alpha} n_\alpha\right) \exp\left(-\frac{\sum_{\alpha} n_\alpha (\varepsilon_\alpha - \mu)}{k_B T}\right)$$

$$= \left\langle \sum_{\alpha} n_\alpha \right\rangle. \tag{10.5}$$

Since the number of particles in a particular system is $N = \sum_\alpha n_\alpha$, i.e. the sum of the number of particles in each quantum state, (10.5) is the ensemble average $\langle N \rangle$ of the number of particles in the system. That is, by introducing the parameter μ, and relaxing the limitation on the number of particles for each system in the ensemble, we have retained the ensemble average of the number of particles as a property of the ensemble. We have then, as Feynman claimed, conveniently included the ensemble average of the number of particles in the system as a property of the ensemble defined by the partition function $Q^{(\mu)}$.

Although the set of energy states is constant for all systems, the number of particles in a quantum state differs for each system in the ensemble. We may obtain the ensemble average of the number of particles in each quantum state as

$$\langle n_\alpha \rangle = -k_B T \frac{\partial}{\partial \varepsilon_\alpha} \ln Q^{(\mu)}. \tag{10.6}$$

From a comparison of (10.5) and (10.6) we see that $\langle N \rangle = \sum_\alpha \langle n_\alpha \rangle$.

The parameter μ is a constant over the systems in the ensemble. It is then a thermodynamic function of state, as is the temperature. We note that because the argument of the exponential in (10.2) contains the factor $(\varepsilon_\alpha - \mu)$ the parameter μ is then a thermodynamic energy per particle.

Equation (10.5) indicates that $\langle N \rangle$ is a function of μ. In principle this relationship can be inverted to obtain a dependence of μ on $\langle N \rangle$. In (10.2) $N\mu$ appears as a reference level for the system energy. And the energy of a system in the ensemble is a microscopic quantity dependent on the potential energies of interaction between the particles, the particle momenta, and the volume of the system. Among these only the volume is an ensemble (macroscopic) parameter. The parameter μ is then a function of $(V, \langle N \rangle)$.

The ensemble for which the temperature T and the ensemble average number of particles $\langle N \rangle$ are held constant is the grand canonical ensemble of Gibbs. Feynman's mathematical trick has then delivered the grand canonical ensemble into our hands with no detailed discussions. This is no less rigorous than the original treatment of Gibbs. And it is more direct. For this reason we stay with Feynman's approach here.

In Sect. 9.7.2 we simplified our work by assuming that the Gibbs formulation of the entropy (9.29) was valid for the microcanonical ensemble as well as for the canonical. Here we shall begin with the same assumption. The microscopic state of a system in a grand canonical ensemble is the set of occupation numbers $\{n_\alpha\}$. From (10.2), before the summation on the sets of occupation numbers, the probability of a set of occupation numbers $P_{\{n_\alpha\}}$ is

$$P_{\{n_\alpha\}} = \frac{1}{Q^{(\mu)}} \exp\left(-\frac{\sum_\alpha n_\alpha (\varepsilon_\alpha - \mu)}{k_B T} \right). \tag{10.7}$$

The Gibbs entropy for the grand canonical ensemble is then

$$S = -k_B \sum_{\{n_\alpha\}} P_{\{n_\alpha\}} \ln P_{\{n_\alpha\}}. \tag{10.8}$$

Using (10.7) and (10.2) the entropy in (10.8) becomes

$$S = k_B \ln Q^{(\mu)} + \frac{1}{T} \left(\langle H \rangle - \langle N \rangle \mu \right) \tag{10.9}$$

Because $\langle H \rangle = U$, the thermodynamic internal energy of the system, we obtain the Helmholtz energy for the grand canonical ensemble from (10.9) as

$$F(T, V, \langle N \rangle) = -k_B T \ln Q^{(\mu)} + \langle N \rangle \mu. \tag{10.10}$$

We then have the entropy of the system as

$$S = -\left(\frac{\partial F(T, V, \langle N \rangle)}{\partial T} \right)_{V, \langle N \rangle}. \tag{10.11}$$

and the thermodynamic pressure as

$$P = -\left(\frac{\partial F(T, V, \langle N \rangle)}{\partial V} \right)_{T, \langle N \rangle}, \tag{10.12}$$

which we see from (10.10) is identical to the mechanical definition of pressure,

$$P = \left\langle -\frac{\partial}{\partial V} \sum_\alpha n_\alpha \varepsilon_\alpha \right\rangle. \tag{10.13}$$

Finally we see from (10.10) that

$$\mu = \left(\frac{\partial F\left(T, V, \langle N \rangle\right)}{\partial \langle N \rangle} \right)_{T,V} = \frac{1}{N_A} \left(\frac{\partial F\left(T, V, \langle N \rangle\right)}{\partial n} \right)_{T,V} \tag{10.14}$$

where n is the number of mols present in the system. Therefore $N_A \mu$ is the thermo-dynamic chemical potential of the system. The designation of μ as the parameter in (10.2) is standard. The reader must simply remember that μ in the grand canonical ensemble is numerically the chemical potential per particle.

Using (10.2) and (10.7) the system internal energy $\langle H \rangle$ may generally be obtained from

$$\langle H \rangle = -\frac{\partial}{\partial\left(1/k_B T\right)} \ln Q^{(\mu)} + \langle N \rangle \mu. \tag{10.15}$$

In the subsequent sections we shall treat the Bose-Einstein and Fermi-Dirac systems separately and systematically using the results of this general development. We shall specialize each case to noninteracting, structureless gases. These specialized treatments include the systems used in treating Bose-Einstein condensation and the degenerate Fermi gas. We shall designate the quantities for the Bose-Einstein system with a subscript BE and those for the Fermi-Dirac system with FD.

10.3.2 Bose Gas

Because in the Bose-Einstein, or simply Bose system any number of particles may be in the same quantum state, the numbers n_α appearing in the sums in (10.4) take on all integer values, $n_\alpha = 0, 1, 2, \cdots$. Each sum appearing in Eq. (10.4) then has the form

$$\sum_{n_\alpha=0} \exp\left(-\frac{\varepsilon_\alpha - \mu_{BE}}{k_B T}\right)^{n_\alpha} = 1 + \exp\left(-\frac{\varepsilon_\alpha - \mu_{BE}}{k_B T}\right)$$
$$+ \left[\exp\left(-\frac{\varepsilon_\alpha - \mu_{BE}}{k_B T}\right)\right]^2 + \cdots,$$

which is

$$\sum_{n_\alpha=0} \left[\exp\left(-\frac{\varepsilon_\alpha - \mu_{BE}}{k_B T}\right)\right]^{n_\alpha} = \frac{1}{1 - \exp\left(-\frac{\varepsilon_\alpha - \mu_{BE}}{k_B T}\right)},$$

since

$$\frac{1}{1-x} = 1 + x + x^2 + x^3 + \cdots.$$

For the Bose system we then have the partition function

$$Q_{BE}^{(\mu)} = \prod_\alpha \frac{1}{1 - \exp\left(-\frac{\varepsilon_\alpha - \mu_{BE}}{k_B T}\right)}, \tag{10.16}$$

(see (10.3)) and the Helmholtz energy

$$F_{BE}(T, V, \langle N \rangle_{BE}) = k_B T \sum_\alpha \ln\left[1 - \exp\left(-\frac{\varepsilon_\alpha - \mu_{BE}}{k_B T}\right)\right]$$

$$+ \langle N \rangle_{BE}\, \mu_{BE}. \tag{10.17}$$

(see (10.10)).

From (10.6) we obtain the ensemble average number of particles in a particular state as

$$\langle n_\alpha \rangle_{BE} = \frac{1}{\exp\left(\frac{\varepsilon_\alpha - \mu_{BE}}{k_B T}\right) - 1}, \tag{10.18}$$

and from (10.5) we have the ensemble averaged number density for the system as

$$\langle N \rangle_{BE} = \sum_\alpha \frac{1}{\exp\left(\frac{\varepsilon_\alpha - \mu_{BE}}{k_B T}\right) - 1}, \tag{10.19}$$

which is the summation over α of (10.18). Because no state can ever be occupied by a negative number of particles, we see from (10.18) that $\exp\left((\varepsilon_\alpha - \mu_{BE})/k_B T\right) > 1$ or $\exp(\varepsilon_\alpha/k_B T) > \exp(\mu_{BE}/k_B T)$. The lowest value of ε_α is the particle ground state energy. The chemical potential μ_{BE} for the Bose gas must then be less then ε_0 and may be negative.

We obtain the thermal equation of state for the Bose system from (10.12) as

$$P_{BE} = -\left(\frac{\partial F_{BE}(T, V, \langle N \rangle_{BE})}{\partial V}\right)_{T, \langle N \rangle_{BE}}$$

$$= \left\langle \sum_\alpha -n_\alpha \frac{\partial}{\partial V} \varepsilon_\alpha \right\rangle_{BE}. \tag{10.20}$$

In (10.20) we have used (10.18) and the fact that the set of energy states $\{\varepsilon_\alpha\}$ and the volume V are constants over the ensemble. We then retrieve the expected relationship between the thermodynamic pressure and the energies of the quantum states $\{\varepsilon_\alpha\}$ (see (10.13)).

The average energy of the systems in the ensemble is

$$\langle H \rangle_{BE} = \left\langle \sum_\alpha n_\alpha \varepsilon_\alpha \right\rangle_{BE}, \tag{10.21}$$

as we see from (10.15) and (10.18), and the fact that $\{\varepsilon_\alpha\}$ is constant over the ensemble.

10.3.3 Structureless Bose Gas

We now limit our treatment to free, structureless quantum particles with integer spin. The components $\lambda \; (=x,y,z)$ of the linear momentum of free quantum particles are $p_\lambda = \hbar k_\lambda$ and k_λ takes on the discrete values $2\pi n_\lambda / L_\lambda$, where n_λ is restricted to be a positive integer [[113], p. 66] and L_λ is the dimension of the containment volume in the direction λ. The quantum energy states for particles moving freely in a volume with sides L_λ are given by

$$\varepsilon_\alpha = \frac{\hbar^2}{2m}\left(k_x^2 + k_y^2 + k_z^2\right) = \frac{2\hbar^2\pi^2}{mL^2}\left(n_x^2 + n_y^2 + n_z^2\right), \tag{10.22}$$

where we have used the subscript α to designate the set of numbers $\left(n_x, n_y, n_z\right)$.

Each quantum energy state may then be represented by a point (k_x, k_y, k_z) at the center of a cube (in k−space) with sides equal to $2\pi/L$. Each state then occupies a volume of $(2\pi/L)^3$ in k− space. If the volume (in Cartesian space (x,y,z)) is large, i.e. macroscopic, the allowed values of k_λ will be very closely spaced and may be considered to form a continuum. The energy spectrum is then a continuum and we may replace our sums over energy states by integrals over energy differentials corresponding to intervals $\left(\Delta n_x, \Delta n_y, \Delta n_z\right)$ in the quantum numbers.

For convenience we shall define the dimensionless quantity

$$\nu_{\mathrm{BE}} = \exp\left(\frac{\mu_{\mathrm{BE}}}{k_B T}\right), \tag{10.23}$$

which is the fugacity (see Sect. 13.4). Because, as we discovered in the preceding section, $\mu_{\mathrm{BE}} \leq \varepsilon_0$, for the Bose gas the fugacity has the limits $0 \leq \nu_{\mathrm{BE}} \leq 1$, with upper limit attained if we choose the energy scale such that $\varepsilon_0 = 0$.

The spin angular momentum S has a set of projection states m_S. For material particles, those with non-zero rest mass, the values of S and m_S are related by $-S \leq m_S \leq S$. A material particle of spin $S = 1$ then has three spin states with $m_S = -1, 0, +1$. For each linear momentum state of the material particle p_α with spin $S = 1$ there are then three spin states.

A photon is a quantum of the electromagnetic field, which has spin of 1, but has no spin angular momentum along the direction of propagation ([113], p. 569). Therefore only the spin states ± 1 are included for each linear momentum state.

We shall account for each of these situations by including a factor s which takes on the value 3 for material particles with spin $S = 1$ and the value 2 for photons.

We begin our treatment of the structureless Bose gas with a calculation of the logarithm of the partition function in (10.16). For a continuous spectrum $\{\varepsilon_\alpha\}$ we can then replace the summation by an integral,

$$\begin{aligned} \ln Q_{\mathrm{BE}}^{(\mu)} &= -\sum_\alpha \ln\left[1 - \exp\left(-\frac{(\varepsilon_\alpha - \mu_{\mathrm{BE}})}{k_B T}\right)\right] \\ &= -\frac{4\pi}{h^3} V s \left(2mk_B T\right)^{\frac{3}{2}} \int_0^\infty \ln\left[1 - \nu_{\mathrm{BE}}\exp\left(-x^2\right)\right] x^2 dx, \end{aligned} \tag{10.24}$$

in which $x^2 = p^2/2mk_BT$, corresponding to ε_α/k_BT. Integrating by parts we obtain

$$\ln Q_{BE}^{(\mu)} = Vs \left(\frac{2\pi mk_BT}{h^2}\right)^{\frac{3}{2}} \zeta_{\frac{5}{2}}(v_{BE}) \tag{10.25}$$

where

$$\zeta_{m/2}(v_{BE}) = \sum_{n=1}^{\infty} \frac{v_{BE}^n}{n^{m/2}}. \tag{10.26}$$

The sum in (10.26) converges only if $v_{BE} \leq 1$, which we have already discovered is guaranteed for the Bose Gas. The maximum value of (10.26) is attained when $v_{BE} = 1$. We note that when $v_{BE} = 1$ the function $\zeta_{m/2}(v_{BE})$ becomes the Riemann $\zeta-$ function of $m/2$, which is defined by [61]

$$\zeta\left(\frac{m}{2}\right) = \sum_{n=1}^{\infty} \frac{1}{n^{m/2}}. \tag{10.27}$$

We shall use the conventional designation for the Riemann $\zeta-$ function, which is ζ with no subscript.

Using (10.10) and (10.25) the Helmholtz energy for the structureless Bose gas is

$$F_{BE}(T,V,\langle N\rangle_{BE}) = -k_BTVs \left(\frac{2\pi mk_BT}{h^2}\right)^{\frac{3}{2}} \zeta_{5/2}(v_{BE})$$
$$+ \langle N\rangle_{BE}\mu_{BE}. \tag{10.28}$$

Noting that

$$\frac{\partial}{\partial\mu_{BE}}\zeta_{5/2}(v_{BE}) = \frac{1}{k_BT}\zeta_{3/2}(v_{BE}), \tag{10.29}$$

and using (10.5) and (10.25), the ensemble average of the number of particles in the structureless Bose gas is

$$\langle N\rangle_{BE} = Vs \left(\frac{2\pi mk_BT}{h^2}\right)^{\frac{3}{2}} \zeta_{3/2}(v_{BE}), \tag{10.30}$$

for $\langle N\rangle_{BE}$.

From the Helmholtz energy in (10.28) we obtain the thermodynamic pressure as

$$P_{BE} = -\left(\frac{\partial}{\partial V}F_{BE}(T,V,\langle N\rangle_{BE})\right)_{T,\langle N\rangle_{BE}}$$
$$= \langle N\rangle_{BE}\frac{k_BT}{V}\frac{\zeta_{5/2}(v_{BE})}{\zeta_{3/2}(v_{BE})}, \tag{10.31}$$

and the entropy as

$$S_{BE} = -\frac{\partial}{\partial T} F_{BE}(T, V, \langle N \rangle_{BE})$$

$$= k_B \langle N \rangle_{BE} \left[\frac{5}{2} \left(\frac{\zeta_{5/2}(v_{BE})}{\zeta_{3/2}(v_{BE})} \right) - \frac{\mu_{BE}}{k_B T} \right]. \tag{10.32}$$

In the differentiation to obtain the entropy we used the relationship

$$\left(\frac{\partial}{\partial T} \frac{\mu_{BE}}{k_B T} \right)_{V, \langle N \rangle_{BE}} = -\frac{3}{2} \frac{1}{T} \frac{\zeta_{3/2}(v_{BE})}{\zeta_{1/2}(v_{BE})}, \tag{10.33}$$

which is obtained by differentiation of Eq. (10.30) and noting that $\rho_{BE} = \langle N \rangle_{BE} / V$ is a constant.

Since

$$\frac{\zeta_{\frac{5}{2}}(v_{BE})}{\zeta_{\frac{3}{2}}(v_{BE})} = 1.0 - 0.17678 v_{BE} - 0.06580 v_{BE}^2 - 0.03647 v_{BE}^3 - \cdots, \tag{10.34}$$

we see that for small v_{BE}, which results from high thermodynamic temperatures, the thermodynamic pressure in Eq. (10.31) becomes that of an ideal gas made up of point particles, and the entropy in Eq. (10.32) becomes

$$S_{BE} = k_B \langle N \rangle_{BE} \left[\frac{5}{2} + \ln \frac{V}{\langle N \rangle_{BE}} + \frac{3}{2} \ln T + \ln s \left(\frac{2\pi m k_B}{h^2} \right)^{\frac{3}{2}} \right], \tag{10.35}$$

which is the Sackur-Tetrode equation for the entropy of the ideal gas made up of point particles (see (9.82)).

Because of the simple form of (10.25), the ensemble average system energy for the structureless Bose gas is most easily calculated directly from (10.15). This results in

$$\langle H \rangle_{BE} = \frac{3}{2} k_B T \langle N \rangle_{BE} \frac{\zeta_{5/2}(v_{BE})}{\zeta_{3/2}(v_{BE})}. \tag{10.36}$$

Using (10.34) we see that for small values of v_{BE} (10.36) reduces to the internal energy of the ideal gas of point particles.

Comparing (10.31) and (10.36) we see that

$$P_{BE} V = \frac{2}{3} \langle H \rangle_{BE}. \tag{10.37}$$

This is a general thermodynamic relationship for an ideal gas made up of point particles. It was first obtained by Bernoulli (see Sect. 8.2.1, (8.7)).

The thermodynamic functions for the structureless Bose gas are then identical to those we obtained for the classical ideal gas made up of point particles provided the thermodynamic temperature is high enough. This is not a drastic assumption and applies to most ordinary gases at temperatures of usual interest. It does not, however, apply to light gases at low temperatures, such as, particularly, Ne, He, and H_2 [38]. The practical consequence is that we may consider a Bose gas to be well represented by a classical ideal gas at temperatures of interest in normal applications. Although the transition from quantum to classical mechanics is subtle [125], for many applications we may treat the motion of the center of mass of molecules as classical point particles and the internal motion quantum mechanically as done in Sect. 9.12.

10.3.4 Bose-Einstein Condensation

Bose-Einstein condensation (BEC) is a phenomenon originally noticed as a theoretical possibility by Einstein resulting from an application of Bose's statistics to atomic gases. It seemed possible that the atoms could all occupy the ground state. Einstein wondered about this possibility in December of 1924 in a letter to Paul Ehrenfest[3] [126]. The condensation that seemed to be predicted occurred without attractive forces. That is this condensation may be expected for a noninteracting gas. As Einstein wrote, "The theory is pretty, but is there also some truth to it?" In 1924 the quantum theory had not been fully developed and particularly the Pauli exclusion principle had not been proposed.

BEC results in superfluidity, which was discovered in liquid helium by Pyotr Leonidovich Kapitsa,[4] John F. Allen, and Don Misener in 1937. Helium IV, the most abundant helium isotope undergoes a transition to a superfluid, in which it has no viscosity, at temperatures below 2.17 K. This transition is referred to as a lambda-transition because of the particular form of the specific heat as a function of temperature, which resembles the Greek letter lambda (λ). However, because the transition to the superfluid state in helium IV is in a liquid rather than in a gas, superfluid helium IV is not commonly referred to as a BE condensate. In a liquid there are relatively strong interaction forces among the molecules and the transition to the lowest state is not strictly quantum based.

The first real BE condensate was produced by Eric Cornell and Carl Wieman of the *Joint Institute for Laboratory Astrophysics* (JILA) at the *National Institute of Standards and Technology* (NIST) and *University of Colorado at Boulder* in 1995. They cooled a gas of rubidium atoms to 170 nK (nanokelvins). The cooling is done by what is called laser cooling followed by collecting the atoms in a magneto optical trap (MOT) and a final loss of energy by boiling off a few atoms. The result is that the great majority of the atoms is in the ground state of the gas.

We will consider the details of the experiment before conducting an analysis [32].

---

[3] Paul Ehrenfest (1880–1933) was an Austrian/Dutch theoretical Physicist. Ehrenfest was a student of Boltzmann's, who made contributions to statistical physics and quantum mechanics [46]. He was professor of physics at Leiden.

[4] Pyotr Leonidovich Kapitsa (1894–1984) was a Soviet/Russian physicist and Nobel laureate (1978). The Nobel Prize was awarded for Kapitsa's work in superfluidity.

Experiment

In the first demonstration of BEC a gas of rubidium-87 ($^{87}_{37}$Rb) atoms was used. Because in BEC the particles all settle into the ground state as a result of the requirements of the quantum theory not because of the complex forces present in the liquid state, it is necessary to preserve a metastable gaseous state below the conditions for condensation. We recall that Maxwell observed that such a metastable state would result if there were no nucleation sites in the gas (see Sect. 8.3.3). In a very pure gas confined in a smooth-walled container a nucleation site can be produced at an atomic level by a three body collision among the gas atoms. To form a molecule two atoms must collide and stick. This is possible if some of the collisional energy is carried away by a third atom. That is three body collisions produce molecules, which can be sites for nucleation and the transition to a liquid. To avoid three body collisions the density of the gas must be kept very low.

The experiment was conducted in a small glass box. To keep the atoms from colliding with the walls of the box, from which they would gain energy, laser cooling was used. This pushed the atoms toward the center of the box as they were being cooled.

Laser cooling is based on absorption of coherent light in a particular direction and then re-emission in random directions. Antiparallel laser beams are tuned to a slightly lower frequency than an electronic transition of the atoms to be cooled. Then only atoms moving toward one of the beams will absorb a quantum (photon) of the Doppler shifted light. The momentum of the absorbed photon decreases the velocity of the atom. The atom the emits a photon at the natural frequency, which means a lower momentum, in the atom's frame, than the absorption. A net decrease in kinetic energy of the atom results.

After laser cooling to temperatures in the range of 10μ K the laser was turned off and a fairly weak spatially dependent magnetic field with a quadratic minimum $B(x)$, where x is the spatial coordinate, was turned on. The atoms then acquired an energy $\sim s \cdot B(x)$. Atoms with spins in one direction were attracted toward the minimum of the magnetic field and those with opposing spins were repelled from the minimum. The magnetic field was arranged to form an energy cup with a lip allowing atoms with sufficient energy to escape by evaporation, which is a very efficient mechanism for cooling. The magnetic field could be distorted and the height of the lip decreased to regulate the energy of the evaporating atoms.

Finally a radio frequency magnetic field was applied, which results in a spin flip of some of the atoms increasing their energy enough to allow then to escape over the energy lip. The radio frequency required for the spin flip is a function of the location of the atom in the magnetic field. So this frequency can be tuned to "shave off" only the highest energy atoms at the rim of the magnetic cup leaving the lowest energy atoms in the cup. As more high energy atoms are shaved off the average energy of the remaining atoms decreases until BEC is attained.

Theory

We shall treat the condensing gas as a collection of noninteracting particles contained in a cubical container and use the expressions developed in the preceding section. This is not the correct picture for the experiment we just discussed, in which the particles were trapped in a (quadratic) potential well resulting from the spin interaction with a magnetic field. There are some differences in the resulting thermodynamic expressions because the energy spectrum of a particle in a quadratic potential well (harmonic oscillator) differs from that used for the integration we performed in (10.24). The differences are, however, not extreme and the basic physics is revealed very well in picture we use here.

We begin with (10.18) for the ensemble average of the particles in a particular state. With the requirement that $\mu_{BE} \leq \varepsilon_0$ for the Bose gas the ground state is at least not empty. Because the quantum numbers for a free particle in a volume are positive integers, the ground state energy has a non-zero, although very small, value. In quantum mechanical terms the ground state is that for which the quantum numbers are $(n_x = 1, n_y = 1, n_z = 1)$ and the quantum the wave function is "nodeless."

Since the energy scale is arbitrary, we may choose the scale to be such that this energy is $\varepsilon_0 = 0$. We emphasize that this is done solely for mathematical convenience and that the ground state energy does not physically vanish. We shall also assume that there is no accidental degeneracy. Accidental degeneracy occurs if the energy of the first state is equal to that of the ground state, i.e. if $\varepsilon_1 = \varepsilon_0$. Then $\varepsilon_1 > \varepsilon_0$. BEC will take place if the number of particles in the ground state $\langle n_0 \rangle_{BE}$ is vastly greater than the number of particles in the first state. Setting $\varepsilon_0 = 0$ and assuming that $\langle n_0 \rangle_{BE}$ is large we can obtain an approximation for μ_{BE} from Eq. (10.18). This is

$$\mu_{BE} = -k_B T \ln\left(1 + \frac{1}{\langle n_0 \rangle_{BE}}\right) \approx -\frac{k_B T}{\langle n_0 \rangle_{BE}}. \qquad (10.38)$$

The chemical potential μ_{BE} then approaches the zero of our energy scale, i.e. approaches ε_0, as the occupation of the ground state $\langle n_0 \rangle_{BE}$ becomes very large, even though $T > 0$.

Using (10.38) in (10.18) we can obtain an approximation for the number of particles in the energy state ε_1 as

$$\langle n_1 \rangle_{BE} = \left[\exp\left(\frac{\varepsilon_1}{k_B T} + \frac{1}{\langle n_0 \rangle_{BE}}\right) - 1\right]^{-1}$$

$$\approx \left[\exp\left(\frac{\varepsilon_1}{k_B T}\right) - 1\right]^{-1}, \qquad (10.39)$$

since $1/\langle n_0 \rangle_{BE}$ is a very small number. Then in the limit as $T \to 0$ the number of particles in the first state $\varepsilon_1 > 0$ vanishes and only the ground state is populated.

We may obtain a more detailed look at the physics of the process resulting in a transition to the ground state by considering the difference between the ensemble averaged number of particles in the system and that in the ground state. This is

$$\langle N \rangle_{BE} - \langle n_0 \rangle_{BE} = \sum_{\alpha=1}^{\infty} \left[\exp \left(\frac{\varepsilon_\alpha}{k_B T} - \frac{\mu_{BE}}{k_B T} \right) - 1 \right]^{-1}, \qquad (10.40)$$

which is (10.19) with the ground state excluded. Because we have made the spacing between the energy levels arbitrarily small by increasing the volume, inclusion of the level ε_0 has no affect on the value of the integral, and we obtain (10.30) for the summation on the right hand side of (10.40). From (10.38) we see that $\mu_{BE} \to 0$ for $T > 0$ because of the presence of $\langle n_0 \rangle_{BE}$ in the denominator. Then at the limiting condition for which $\mu_{BE} \to 0$, (10.40) becomes

$$\langle N \rangle_{BE} - \langle n_0 \rangle_{BE} = V s \left(\frac{2\pi m k_B T}{h^2} \right)^{\frac{3}{2}} \zeta_{3/2}(1). \qquad (10.41)$$

The function $\zeta_{3/2}(1)$ is the Riemann ζ-function of $\frac{3}{2}$, which is $\zeta\left(\frac{3}{2}\right) \approx 2.612$. Then the difference between the ensemble averages of the total number of particles in the system and the number in the ground state vanishes as the temperature $T \to 0$ and all of the particles in the system are in the ground state.

If we define the temperature T_C as that for which $\nu_{BE} = 1$ we can obtain an expression for T_C in terms of the ensemble averaged particle density $\rho_{BE} = \langle N \rangle_{BE} / V$. Using (10.30) this temperature is

$$T_C = \frac{h^2}{2\pi m k_B} \left(\frac{\rho_{BE}}{s \zeta_{3/2}(1)} \right)^{\frac{2}{3}}. \qquad (10.42)$$

If we then use this in (10.41) we have, for $T < T_C$, the result

$$\langle n_0 \rangle_{BE} \approx \langle N \rangle_{BE} \left[1 - \left(\frac{T}{T_C} \right)^{\frac{3}{2}} \right]. \qquad (10.43)$$

So at $T = T_C$ the condensation into the ground state is not yet complete. The gas continues to condense as the thermodynamic temperature decreases below T_C. From (10.43) we can obtain a plot of $\langle n_0 \rangle_{BE} / \langle N \rangle_{BE}$. This is shown in Fig. 10.1. The

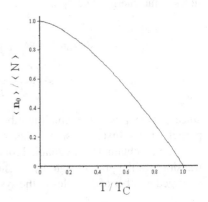

Fig. 10.1 Ensemble average ground state occupation below critical temperature for BEC

number $\eta = \langle n_0 \rangle_{\mathrm{BE}} / \langle N \rangle_{\mathrm{BE}}$ is the order parameter for the transition to the quantum ground state. Data of Ensher et al. follow the curve in Fig. 10.1 very closely, even though in their system the particles were contained in a magnetic trap and particles were evaporated as described above (Fig. 1 of [51]). These data provide experimental agreement with the approximate (10.43).

From (10.36) and (10.30) the total energy for the ranges $T > T_C$ and $T < T_C$ has the forms

$$\langle H \rangle_{\mathrm{BE}} = \begin{cases} \frac{3}{2} k_B T V s \left(2\pi m k_B T / h^2 \right)^{\frac{3}{2}} \zeta_{3/2} (v_{\mathrm{BE}}) & \text{for } T > T_C \\ \frac{3}{2} k_B T V s \left(2\pi m k_B T / h^2 \right)^{\frac{3}{2}} \zeta_{3/2} (1) & \text{for } T < T_C \end{cases}. \tag{10.44}$$

Plotting the energy $\langle H \rangle_{\mathrm{BE}}$ as a function of T/T_C in the region of the critical temperature we have the plot in Fig. 10.2.

From (10.44) we can obtain the specific heat at constant volume as $n C_{V,\mathrm{BE}} = (\partial \langle H \rangle_{\mathrm{BE}} / \partial T)_{V, \langle N \rangle_{\mathrm{BE}}}$. We have then two equations. One for $T > T_C$

$$nC_{V,\ \mathrm{BE}} = \frac{15}{4} k_B V s \left(\frac{2\pi m k_B T}{h^2} \right)^{\frac{3}{2}} \zeta_{5/2} (v_{\mathrm{BE}})$$

$$- \frac{9}{4} k_B V s \left(\frac{2\pi m k_B T}{h^2} \right)^{\frac{3}{2}} \frac{\zeta_{3/2}^2 (v_{\mathrm{BE}})}{\zeta_{1/2} (v_{\mathrm{BE}})}, \tag{10.45}$$

and one for $T < T_C$

$$nC_{V,\mathrm{BE}} = \frac{15}{4} k_B V s \left(\frac{2\pi m k_B T}{h^2} \right)^{\frac{3}{2}} \zeta_{5/2} (1) \quad \text{for } T < T_C, \tag{10.46}$$

where n is the number of mols present in the sample. In Fig. 10.3 we have plotted these two functional forms of $C_{V,\ \mathrm{BE}}$ over a temperature range that includes the critical temperature. This $C_{V,\mathrm{BE}}$ plot resembles the λ shape of the experimental result for the helium IV transition. The specific heat $C_{V,\mathrm{BE}}$ is, however, continuous at T_C in BEC. The discontinuity is in the derivative of $C_{V,\mathrm{BE}}$ at T_C. We recall from the expressions for C_V and C_P in (5.23) and (5.24) and the fundamental relationship between C_V and C_P in (5.32) that a discontinuity in the derivative of $C_{V,\mathrm{BE}}$ at T_C is

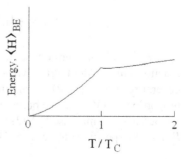

Fig. 10.2 Energy of a Bose gas $\langle H \rangle_{\mathrm{BE}}$ as a function of T/T_C in the region of the critical temperature

Fig. 10.3 Specific heat for a Bose gas $C_{V,\text{BE}}$ near the (BEC) critical temperature T_C. The form of $C_{V,\text{BE}}$ at T_C resembles the lamda (λ) form of the He IV transition to a superfluid

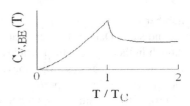

indication of a discontinuity in $\left(\partial^3 G/\partial T^3\right)_{N,P}$. This is what Ehrenfest defined as a third order phase transition. Common usage, however, only distinguishes between first-order and higher-order transitions, the latter being simply referred to as second order [cf. [77], p. 35]. For a BEC condensate confined in a quadratic potential, such as the magnetic cup described above, $C_{V,\text{BE}}$ is discontinuous at $T = T_C$. This is predicted theoretically and has been observed by Ensher et al. [51].

For the thermodynamic description of the system we turn to the general expression for the Helmholtz energy of the structureless Bose gas, (10.28), from which we obtained equations for the thermodynamic pressure and entropy in (10.31) and (10.32) respectively. These each have two forms depending upon whether the temperature T is greater than or less than the critical temperature T_C. Using (10.30) we obtain the pressure when $T > T_C$ as

$$P_{\text{BE}} = k_B T s \left(\frac{2\pi m k_B T}{h^2}\right)^{\frac{3}{2}} \zeta_{5/2}\left(\nu_{\text{BE}}\right) \tag{10.47}$$

and when $T < T_C$ as

$$P_{\text{BE}} = k_B T s \left(\frac{2\pi m k_B T}{h^2}\right)^{\frac{3}{2}} \zeta_{5/2}\left(1\right) \tag{10.48}$$

Similarly we obtain the entropy when $T > T_C$ as

$$S_{\text{BE}} = k_B V s \left(\frac{2\pi m k_B T}{h^2}\right)^{\frac{3}{2}} \zeta_{3/2}\left(\nu_{\text{BE}}\right) \left[\frac{5}{2}\frac{\zeta_{5/2}\left(\nu_{\text{BE}}\right)}{\zeta_{3/2}\left(\nu_{\text{BE}}\right)} - \frac{\mu_{\text{BE}}}{k_B T}\right] \tag{10.49}$$

and when $T < T_C$ as

$$S_{\text{BE}} = \frac{5}{2}k_B V s \left(\frac{2\pi m k_B T}{h^2}\right)^{\frac{3}{2}} \zeta_{5/2}\left(1\right) \tag{10.50}$$

For $T < T_C$ the thermodynamic pressure is independent of the volume and is a function only of the thermodynamic temperature. This may be deduced from the general relationship i(10.37) and the result in (10.44), which shows that $\langle H \rangle_{\text{BE}}$ is proportional to V. It is also born out in our result (10.47) and (10.48). At the critical temperature we may use (10.42) to obtain a general relationship between pressure P_C and the specific volume $\nu_C = \left(1/\rho_{\text{BE}}\right)]_{T=T_C}$. This is

Fig. 10.4 Pressure as a function of volume for BEC

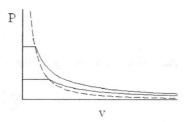

$$P_C = \frac{h^2}{2\pi m s^{\frac{2}{3}}} \frac{1}{v_C^{\frac{5}{3}}} \frac{\zeta_{5/2}(1)}{(\zeta_{3/2}(1))^{\frac{5}{3}}} \tag{10.51}$$

In Fig. 10.4 we plot (10.51) as a dashed line and two isotherms as solid lines. The isothermal coefficient of compressibility κ_T then becomes infinite in the region $T < T_C$ and the specific heats become equal (see Sect. 5.3.2, (5.32)), vanishing as $T \to 0$, which we see from (10.46). Because the thermodynamic pressure is proportional to $\langle H \rangle_{BE}/V$ it is also proportional to the energy of the ground state as $T \to 0$. For computational reasons we chose an energy scale on which the ground state energy was the zero point. Quantum mechanically we recall that neither the ground state energy of the free particle in a box nor the ground state energy of the oscillator (in the magnetic cup) is zero. Therefore neither the pressure nor the actual energy ever vanish as $T \to 0$.

As we see from (10.50) the entropy of the BEC vanishes as $T \to 0$. The specific heats and the entropy then vanish as $T \to 0$. This is consistent with the third law.

10.4 Fermi Gas

For the Fermi-Dirac or Fermi gas the occupation numbers n_α in (10.4) can only take on the values $n_\alpha = 0, 1$. The partition function in (10.4) is then

$$Q_{FD}^{(\mu)} = \prod_\alpha \sum_{n_\alpha = 0,1} \exp\left(-\frac{\varepsilon_\alpha - \mu_{FD}}{k_B T}\right)^{n_\alpha}$$

$$= \prod_\alpha \left[1 + \exp\left(-\frac{\varepsilon_\alpha - \mu_{FD}}{k_B T}\right)\right]. \tag{10.52}$$

With (10.10) and (10.52) we obtain the Helmholtz energy for the Fermi-Dirac gas. The result is

$$F_{FD}(T,V,\langle N \rangle_{FD}) = -k_B T \sum_\alpha \ln\left[1 + \exp\left(-\frac{\varepsilon_\alpha - \mu_{FD}}{k_B T}\right)\right]$$

$$+ \langle N \rangle_{FD} \mu_{FD}. \tag{10.53}$$

We may obtain the ensemble average of the number of particles in the αth state

$$\langle n_\alpha \rangle_{FD} = \left[\exp\left(\frac{\varepsilon_\alpha - \mu_{FD}}{k_B T} \right) + 1 \right]^{-1} \qquad (10.54)$$

and the ensemble average of the total number of particles in the system

$$\langle N \rangle_{FD} = \sum_\alpha \left[\exp\left(\frac{(\varepsilon_\alpha - \mu_{FD})}{k_B T} \right) + 1 \right]^{-1} \qquad (10.55)$$

using (10.6) and (10.5). From (10.15) and (10.53) the ensemble averaged energy is

$$\langle H \rangle_{FD} = \left\langle \sum_\alpha n_\alpha \varepsilon_\alpha \right\rangle_{FD}, \qquad (10.56)$$

where we have used (10.54) and the fact that $\{\varepsilon_\alpha\}$ is constant over the ensemble.

10.4.1 Structureless Fermi Gas

We now consider a gas of free structureless fermions. Electrons are examples of structureless fermions, with spin of one half, as are the neutrons and protons in a neutron star. A collection of free and noninteracting electrons contained within a specific volume would then be an example of such a gas.

This is, however, a fairly crude model of electron behavior. Electron interactions cannot be neglected in, for example, a high temperature thermonuclear plasma. And the effects of crystal symmetries on the quantum mechanical behavior of electrons in solids cannot be neglected. We will, however, learn much about the general behavior of electrons in our study here.

The appearance of the theoretical equations we obtain for the structureless Fermi gas differ little in form from those of the structureless Bose gas. But the difference in the behaviors of the two systems is dramatic. Rather than a condensation, such as we found in the Bose gas, we encounter a degeneracy of the Fermi gas. This degenerate state has practical importance for the behavior of electrons in solids, and of neutrons in neutron stars.

The basic quantum description of free structureless fermions is the same as that for free structureless bosons. The energy spectrum may again be considered continuous and sums replaced by integrals. We shall again introduce the notation $\nu_{FD} = \exp(\mu_{FD}/k_B T)$. For high temperatures $\nu_{FD} \ll 1$.

Just as we did for the case of the structureless Bose gas, we begin our treatment of the structureless Fermi gas with a calculation of the logarithm of the partition function from (10.4). From (10.52) we have

$$
\begin{aligned}
\ln Q_{FD}^{(\mu)} &= \sum_\alpha \ln\left(1 + \exp\left(-\frac{(\varepsilon_\alpha - \mu_{FD})}{k_B T} \right) \right) \\
&= \frac{4\pi}{h^3} V (2mk_B T)^{\frac{3}{2}} \int_0^\infty \ln\left[1 + \nu_{FD} \exp\left(-x^2 \right) \right] x^2 dx \qquad (10.57)
\end{aligned}
$$

where again $x^2 = p^2/2mk_B T$. An integration by parts results in

$$\ln Q_{FD}^{(\mu)} = V \left(\frac{2\pi m k_B T}{h^2} \right)^{\frac{3}{2}} Z_{5/2} (v_{FD}), \qquad (10.58)$$

where we now define

$$Z_{m/2} (v_{FD}) = \sum_{n=1}^{\infty} (-1)^{n+1} \frac{v_{FD}^n}{n^{m/2}}. \qquad (10.59)$$

Using (10.58) in (10.10), the Helmholtz energy of the structureless Fermi gas is

$$F_{FD} (T, V, \langle N \rangle_{FD}) = -k_B T V \left(\frac{2\pi m k_B T}{h^2} \right)^{\frac{3}{2}} Z_{5/2} (v_{FD})$$
$$+ \langle N \rangle_{FD} \mu_{FD}. \qquad (10.60)$$

Noting that

$$\frac{\partial}{\partial \mu_{FD}} Z_{5/2} (v_{FD}) = \frac{1}{k_B T} Z_{3/2} (v_{FD}), \qquad (10.61)$$

we obtain the ensemble average of the number of particles

$$\langle N \rangle_{FD} = V \left(\frac{2\pi m k_B T}{h^2} \right)^{\frac{3}{2}} Z_{3/2} (v_{FD}), \qquad (10.62)$$

from (10.5), with the help of (10.58). Then using (10.15) and (10.58) with (10.62) we obtain

$$\langle H \rangle_{FD} = \frac{3}{2} k_B T V \left(\frac{2\pi m k_B T}{h^2} \right)^{\frac{3}{2}} Z_{5/2} (v_{FD})$$
$$= \frac{3}{2} k_B T \langle N \rangle_{FD} \frac{Z_{5/2} (v_{FD})}{Z_{3/2} (v_{FD})}. \qquad (10.63)$$

for the ensemble averaged energy for the structureless Fermi gas. We may expand the ratio $Z_{\frac{5}{2}} (v_{FD}) / Z_{\frac{3}{2}} (v_{FD})$ in powers of v_{FD}. The result is

$$\frac{Z_{\frac{5}{2}} (v_{FD})}{Z_{\frac{3}{2}} (v_{FD})} = 1 + 0.176777 v_{FD} - 0.065800 v_{FD}^2 + 0.036465 v_{FD}^3 + \cdots . \qquad (10.64)$$

For high temperature, i.e. very small v_{FD} then (10.63) reduces to

$$\langle H \rangle_{FD} = \frac{3}{2} \langle N \rangle_{FD} k_B T, \qquad (10.65)$$

which is the classical result for a gas of structureless particles, i.e. hard spheres.

We obtain the thermodynamic pressure from the Helmholtz energy in (10.60) as

$$P = -\left(\frac{\partial}{\partial V} F_{FD}\left(T, V, \langle N\rangle_{FD}\right)\right)_{T, \langle N\rangle_{FD}}$$

$$= k_B T \left(\frac{2\pi m k_B T}{h^2}\right)^{\frac{3}{2}} Z_{5/2}\left(v_{FD}\right)$$

$$= \frac{1}{V} \langle N\rangle_{FD} k_B T \frac{Z_{5/2}\left(v_{FD}\right)}{Z_{3/2}\left(v_{FD}\right)}. \tag{10.66}$$

We used (10.62) to obtain the last line of (10.66). For high temperatures v_{FD} is small and the expansion in (10.64) is approximately equal to unity. Therefore, for high temperatures the thermal equation of state of a Fermi gas (10.66) becomes

$$P_{FD} V \approx \langle N\rangle_{FD} k_B T,$$

which is the ideal gas equation of state.

We can obtain the entropy from the Helmholtz energy in (10.60) as

$$S_{FD} = -\left(\frac{\partial}{\partial T} F_{FD}\left(T, V, \langle N\rangle_{FD}\right)\right)_{V, \langle N\rangle_{FD}}$$

$$= \frac{5}{2} k_B V \left(\frac{2\pi m k_B T}{h^2}\right)^{\frac{3}{2}} Z_{5/2} - \langle N\rangle_{FD} \frac{\mu_{FD}}{T} \tag{10.67}$$

$$= \frac{5}{2} k_B \langle N\rangle_{FD} \frac{Z_{5/2}}{Z_{3/2}} - \langle N\rangle_{FD} \frac{\mu_{FD}}{T} \tag{10.68}$$

In the high temperature limit, for which we can neglect $v_{FD}^2 \ll v_{FD}$, (10.68) becomes

$$S_{FD} = \langle N\rangle_{FD} k_B \left[\ln\left(\frac{V}{\langle N\rangle_{FD}}\right) + \frac{3}{2}\ln\left(T\right)\right.$$

$$\left. + \ln\left(\frac{2m k_B}{h^2}\right)^{\frac{3}{2}} + \frac{5}{2}\right], \tag{10.69}$$

which is the Sackur-Tetrode equation for the entropy of the classical gas of structureless particles (see Sect. 9.11.1).

Under the same conditions the Helmholtz energy in (10.60) becomes

$$F_{FD}\left(T, V, \langle N\rangle_{FD}\right) = \langle N\rangle_{FD} k_B \left[\ln\left(\frac{V}{\langle N\rangle_{FD}}\right) + \frac{3}{2}\ln\left(T\right)\right.$$

$$\left. + \ln\left(\frac{2m k_B}{h^2}\right)^{\frac{3}{2}} + 1\right], \tag{10.70}$$

which is that for the ideal gas (see Sect. 9.11.1).

As in the case of the structureless Bose gas, the expressions we have obtained for the thermal equation of state in (10.66), the ensemble averaged energy in (10.63), the entropy in (10.69), and finally for the Helmholtz energy in (10.70) are identical to those we obtained for the classical ideal gas made up of point particles if the temperature is sufficiently high.

At sufficiently high temperatures, then, both the structureless Bose and Fermi gases behave like ideal gases. Therefore, although the atoms or molecules of ordinary gases have a net spin of integer or half integer, at reasonable to high temperatures we may ignore the net spin and use the results of the preceding chapter.

10.4.2 Degenerate Fermi Gas

As the temperature drops the particles in a Fermi gas begin occupying the lowest possible energy levels, just as was the case for the Bose gas. The difference is that the Pauli principle requires that only one fermion can occupy each quantum state.

There are two spin states associated with each of the quantum energy states in (10.22). So each quantum energy state contains two fermions. Each of the volumes in $k-$ space containing an energy state is infinitesimal because of the large value of L_λ, so we may think of the energy states as a densely packed lattice of points in $k-$ space. Because (10.22) is a sphere in $k-$ space, at $T = 0$ the fermions occupy all of the densely packed lattice of points in a spherical volume with a radius determined by the number of fermions present in the system. This sphere is called the Fermi sphere. The $k-$ space radius of this sphere is the Fermi radius k_F and the energy corresponding to this value of k is called the Fermi energy $E_F^0 = \hbar^2 k_F^2/2m$.

The Fermi sphere has a volume equal to $4\pi k_F^3/3$. There are then $4\pi k_F^3/3 \, (L/2\pi)^3$ states in the Fermi sphere. Since each state contains two fermions, the number of fermions in the Fermi sphere is then $N_F = L^3 k_F^3/3\pi^2$ and the number density of the fermions is $n_F = N_F/L^3 = k_F^3/3\pi^2$.

We can then calculate the Fermi energy in terms of the number density of fermions. The result is

$$E_F^0 = \left(\hbar^2/2m\right)\left(3\pi^2 n_F\right)^{\frac{2}{3}}. \tag{10.71}$$

From the Fermi energy we define the Fermi temperature T_F as $T_F = E_F^0/k_B$ and from the Fermi radius we define the Fermi velocity as $v_F = \hbar k_F/m$.

Copper (Cu) and silver (Ag) have single valence electrons, which may, as an approximation, be considered free fermions. We may then calculate the density of free electrons n_F for Cu and Ag from the measured densities of the materials. Using (10.71) we have the Fermi energies and finally the Fermi temperatures for Cu and Ag. The results are given in Table 10.1.

To provide a more complete picture of the Fermi gas at low temperatures we must return to the statistical mechanical treatment.

The ensemble average of the number particles in each state is given by (10.54). The thermodynamic function μ_{FD} depends upon the temperature. As $T \to 0\,K$ the

Table 10.1 Fermion properties of Cu and Ag

Metal	$n_F/10^{22}\,\mathrm{cm}^{-3}$	$v_F/10^8\,\mathrm{cm\,s}^{-1}$	E_F^0/eV	$T_F/10^4\,\mathrm{K}$
Cu	8.47	1.57	7.03	8.16
Ag	5.86	1.39	5.50	6.38

value of μ_{FD} becomes μ_0, which is numerically equal to the Fermi energy E_F^0. Then

$$\lim_{T\to 0}\langle n_\alpha\rangle_{FD} = \lim_{T\to 0}\left[\exp\left(\frac{(\varepsilon_\alpha - E_F^0)}{k_B T}\right) + 1\right]^{-1}$$

$$= \begin{cases} 0 \ \text{if } \varepsilon_\alpha > E_F^0 \\ 1 \ \text{if } \varepsilon_\alpha < E_F^0 \end{cases}.$$

That is, all the energy levels are filled up to the value E_F^0, and all occupied energy states are within the Fermi sphere. In Fig. 10.5 we have plotted the Fermi distribution

$$f_{FD}(\varepsilon, T) = \left[\exp\left(\frac{\varepsilon - \mu_{FD}}{k_B T}\right) + 1\right]^{-1} \tag{10.72}$$

(see (10.54)) for the Fermi gas near $T = 0$. The rectangle is the density at $T = 0\,\mathrm{K}$. Densities for temperatures of $0.01T_F$, $0.05T_F$, $0.1T_F$, and $0.2T_F$ are shown by dashed lines for comparison.

For temperatures near $T = 0$ the distribution of particles among the energy states $\langle n_\alpha\rangle_{FD}$ is such that most of the particles fill the lowest states and only a few with energies near E_F^0 have a distribution which is temperature dependent. We can find the distribution of these particles from (10.54). For energies $\varepsilon_\alpha > \mu_0$ and $0 < T \ll T_F$ we have $\exp\left((\varepsilon_\alpha - \mu_0)/k_B T\right) \gg 1$. Then (10.72) becomes

$$f_{FD}(\varepsilon, T) \approx \exp\left(-\frac{\varepsilon_\alpha - \mu_0}{k_B T}\right), \tag{10.73}$$

and $\mu_0 \approx E_F^0$. The particles are then distributed among the states near E_F^0 according to a Maxwell-Boltzmann distribution.

The values of T_F are of the order of $10^4\,\mathrm{K}$ (see Table 10.1), which is considerably greater than the temperatures ($\sim 300\,\mathrm{K}$) of interest for metallic conductors. Therefore, in the approximation that the valence electrons are free particles, we may

Fig. 10.5 Fermi distribution $f_{FD}(\varepsilon, T)$ near $T = 0$. At $T = 0$ distribution is rectangular (*solid line*). Distributions for $T = 0.01T_F, 0.05T_F, 0.1T_F,$ and $0.2T_F$ are shown (*dashed lines*)

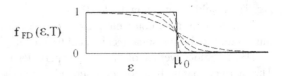

use the $T = 0\,\mathrm{K}$ distribution of Fig. 10.5 as the occupation number for most of the states available to the valence electrons. In the case of Cu, for example, $300\,\mathrm{K}$ is approximately $0.04T_F$ and the distribution lies between the first and second dashed curves in Fig. 10.5.

For $T = 0$, with all the energy states in the Fermi sphere occupied, the number of fermions per unit volume with energies in the interval $d\varepsilon$ may be equated to the product of a density of states $D_F^0(\varepsilon)$ and the energy differential $d\varepsilon$. The density of states can be found from (10.22), which is a relationship between the energy and the radius $k = \sqrt{k_x^2 + k_y^2 + k_z^2}$ of a shell of thickness dk in $k-$ space (see exercises). Using (10.22) the number of energy states (per unit volume) in the energy shell $d\varepsilon$ corresponding to dk is $4\pi k^2 dk/(2\pi/L)^3 = (2m)^{\frac{3}{2}} \varepsilon^{\frac{1}{2}} d\varepsilon/4\pi^2\hbar^3$. Each energy state contains two fermions. Therefore the density of occupied states in the Fermi sphere is

$$D_F^0(\varepsilon)\,d\varepsilon = \frac{1}{2\pi^2}\left(\frac{2m}{\hbar^2}\right)^{\frac{3}{2}}\varepsilon^{\frac{1}{2}}\,d\varepsilon. \tag{10.74}$$

Using (10.74) we can obtain the ensemble averaged energy (per unit volume) as

$$u_{FD}^0 = \frac{\langle H\rangle_{FD}^0}{V} = \int_0^{E_F^0} \varepsilon D_F^0(\varepsilon)\,d\varepsilon = \frac{3}{5}n_F E_F^0. \tag{10.75}$$

This value is orders of magnitude above that of the classical gas at $300\,\mathrm{K}$.

For $T > 0$ the density of states $D_{FD}(\varepsilon,T)$ is a product of the density $D_F^0(\varepsilon)$ in (10.74) with $f_{FD}(\varepsilon,T)$, which is

$$D_{FD}(\varepsilon,T) = D_F^0(\varepsilon)\,f_{FD}(\varepsilon,T)$$
$$= \frac{1}{2\pi^2}\left(\frac{2m}{\hbar^2}\right)^{\frac{3}{2}}\frac{\varepsilon^{\frac{1}{2}}}{\exp\left(\frac{(\varepsilon-\mu_{FD})}{k_B T}\right)+1} \tag{10.76}$$

We have plotted $D_{FD}(\varepsilon,T)$ in Fig. 10.6. The temperatures are the same as those in Fig. 10.5 except that we have not plotted the density of states for $T = 0$ in Fig. 10.6.

Using the density of states in e(10.76) for $T > 0$, the particle density in the degenerate Fermi gas is

Fig. 10.6 Density of states $D_{FD}(\varepsilon,T)$ for free fermions near $T = 0$. Plots of $D_{FD}(\varepsilon,T)$ for $T = 0.01T_F, 0.05T_F, 0.1T_F$, and $0.2T_F$ are shown (*dashed lines*)

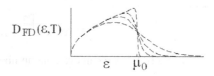

$D_{FD}(\varepsilon,T)$

ε μ_0

$$\rho_{FD} = \frac{\langle N \rangle_{FD}}{V} = \int_0^\infty D_F(\varepsilon, T)\, d\varepsilon$$

$$= a\,(k_B T)^{\frac{3}{2}} \int_0^\infty \frac{y^{\frac{1}{2}}\, dy}{\exp(y - \lambda) + 1} \tag{10.77}$$

and the energy density is

$$u_{FD} = \frac{\langle H \rangle_{FD}}{V} = \int_0^\infty \varepsilon D_F(\varepsilon, T)\, d\varepsilon \tag{10.78}$$

$$= a\,(k_B T)^{\frac{5}{2}} \int_0^\infty \frac{y^{\frac{3}{2}}\, dy}{\exp(y - \lambda) + 1}.$$

where $y = \varepsilon / k_B T$ and $\lambda = \mu_{FD} / k_B T$ and

$$a = \frac{1}{2\pi^2} \left(\frac{2m}{\hbar^2} \right)^{\frac{3}{2}}.$$

The number and energy densities are of the general form

$$\rho_{FD} = \int_0^\infty \frac{\phi_\rho(y)}{\exp(y - \lambda) + 1}\, dy \tag{10.79}$$

and

$$u_{FD} = \int_0^\infty \frac{\phi_u(y)}{\exp(y - \lambda) + 1}\, dy \tag{10.80}$$

with $\phi_\rho(y) = a\,(k_B T)^{\frac{3}{2}} y^{\frac{1}{2}}$ and $\phi_u(y) = a\,(k_B T)^{\frac{5}{2}} y^{\frac{3}{2}}$. In general

$$J = \int_0^\infty \frac{\phi(y)}{\exp(y - \lambda) + 1}\, dy$$

$$\approx \int_0^\lambda \phi(y)\, dy + \frac{\pi^2}{6} \phi'(\lambda), \tag{10.81}$$

where the prime indicates the first derivative. Using (10.81) we then have for the particle and energy densities

$$\rho_{FD} \approx a \left[\frac{2}{3} \mu_{FD}^{\frac{3}{2}} + \frac{\pi^2}{12} (k_B T)^2 \mu_{FD}^{-\frac{1}{2}} \right] \tag{10.82}$$

and

$$u_{FD} \approx a \left[\frac{2}{5} \mu_{FD}^{\frac{5}{2}} + \frac{\pi^2}{4} \mu_{FD}^{\frac{1}{2}} (k_B T)^2 \right] \tag{10.83}$$

From (10.82) we can evaluate number density at $T = 0$ as

$$\rho_{FD}^0 = \frac{2a}{3} \mu_0^{\frac{3}{2}}. \tag{10.84}$$

Since ρ_{FD} is a constant over the ensemble, we can equate (10.82) and (10.84) to obtain

$$\mu_{FD} \approx \mu_0 \left[1 - \frac{\pi^2}{12} \left(\frac{k_B T}{\mu_0} \right)^2 \right] \tag{10.85}$$

and generally

$$\mu_{FD}^{m/2} \approx \mu_0^{m/2} \left[1 - \frac{m}{3} \frac{\pi^2}{8} \left(\frac{k_B T}{\mu_0} \right)^2 \right]$$

to second order in $k_B T / \mu_0$. Using (10.85) we obtain an expression for the energy density of the degenerate Fermi gas very close to $T = 0$.

$$\lim_{T \to 0} u_{FD} \approx \frac{2a}{5} \mu_0^{5/2} + \frac{\pi^2 a}{6} \mu_0^{1/2} (k_B T)^2$$

$$\approx \rho_{FD}^0 \left(\frac{3}{5} \mu_0 + \frac{\pi^2}{4\mu_0} (k_B T)^2 \right) \tag{10.86}$$

The energy of the degenerate Fermi Gas has a quadratic dependence on T near $T = 0$ and has the limit of Eq. (10.75) at $T = 0$.

From (10.86) we can obtain the molar specific heat at constant volume near $T = 0$. as

$$\lim_{T \to 0} C_{V,FD} = \frac{N_A}{V} \frac{\pi^2}{2\mu_0} k_B^2 T. \tag{10.87}$$

The molar specific heat of the degenerate Fermi Gas then vanishes at $T = 0$ consistent with the third law.

If we are considering our fermions to be electrons in a solid they are not the sole constituents of the solid [79, 154]. Quantum mechanically the behavior of electrons in a solid is determined by the symmetries of the crystal lattice of the solid. And the vibrations of the atoms in the crystal lattice contribute to the specific heat of the solid, which is proportional to T^3 at temperatures close to $T = 0$. At very low temperatures the linear dependence on T of the electronic contribution to the specific heat dominates the T^3 dependence of the contribution from the lattice vibrations. And at high temperatures the linear electron dependence may also be observed, since there the lattice contribution becomes constant as T increases.

Neutron stars are formed as remnants of supernovae. The electrons are captured by the protons in the dense residue forming neutrons and neutrinos (cf. [159]). The resulting matter resembles a single nucleus confined by a gravitational force. Some protons are present due to the neutron/proton+electron interchange.

The measured masses of neutron stars are approximately 1.4 solar masses packed into a sphere with a radius of approximately 10–20 km. The actual conditions inside a neutron star have not (yet) been sampled, but we may assume that the matter in neutron stars may be considered to be a Fermi gas. For a neutron star of 1.4 solar masses and a radius 10–15 km the density is $1.97 - 6.64 \times 10^{17}$ kg m^{-3}. If we choose an average of these, then, with the neutron mass as 1.67×10^{-27} kg, the neutron density is 1.37×10^{44} m^{-3} and the Fermi temperature is $T_F = 6.13 \times 10^{11}$ K.

The temperature from the spectrum is $T \sim 10^8$ K. We may then consider that the neutron star is a degenerate Fermi gas.

10.5 Summary

In this chapter we presented the basis of quantum statistical mechanics in sufficient detail to provide the reader with the background necessary to understand some of the interesting applications. The quantum theory post dates Gibbs' development of the statistical mechanics by a quarter century. Nevertheless, because of the generality of the statistical mechanics, the basic ensemble theory embraces both classical and quantum descriptions of matter.

We began our treatment with a general formulation that could be applied to any quantum particles. We introduced the grand canonical ensemble following the development of Feynman, which is more physically straightforward than the original of Gibbs. This development particularly revealed that the grand canonical ensemble is the natural ensemble for quantum statistical mechanics.

We then specialized to the two types of quantum particles distinguished by the value of the spin: bosons and fermions. For each of these we used the ideal gas approximation. This enabled us to consider some applications that are a part of present research into the behavior of matter. These applications included Bose-Einstein condensation and degenerate Fermi gases. At the time of this writing BEC and the state of matter in neutron stars are active areas of research. The behavior of electrons in solids is well established as the basis of modern applications in electronics, which is itself an active area of work.

Exercises

10.1 Consider a closed cavity the walls of which are at a temperature T. According to Einstein's heuristic picture of light this cavity is filled with a photon gas. For simplicity we shall neglect polarization. The energy of the jth state available to the photon is ε_j and the number of photons in this level in n_j^{σ} in the σth system. Photons are bosons. The ground state energy does not vanish, but is very small if the cavity is large. Find the (a) partition function, (b) the ensemble average number of photons in the jth state, and the (c) entropy of the photon gas. Does the entropy obey the third law?

10.2. Show that the logarithm of the Bose-Einstein partition function is given by (10.25). This involves primarily the evaluation of the integral $\int_0^\infty x^2 \ln\left[1 - \alpha \exp\left(-x^2\right)\right] dx$, which is a necessary detail in an understanding of the quantum results.

10.3 Show that $\langle N \rangle_{\mathrm{BE}}$ is given by (10.30). This requires the relationship in (10.29), which must be derived.

10.4. Show that $\langle E \rangle_{\text{BE}}$ is correctly given by (10.36).

10.5. Show that the entropy of the Bose gas satisfies the requirements of the third law.

10.6. What is the energy of a Bose gas at absolute zero?

10.7. What is the energy density of a Fermi gas at absolute zero?

Chapter 11
Irreversibility

11.1 Introduction

Until this point we have considered systems in equilibrium. We have followed this course because we can only define thermodynamic functions at equilibrium. This has been true of our considerations of statistical mechanics, although there we have pointed to modern interests in nonequilibrium. We have even mentioned some of the original studies of conditions that involved kinetics. But our actual treatment has not strayed from equilibrium.

In modern applications of thermodynamics we are interested in irreversibility. We cannot understand chemical reactions or biological processes if we consider only equilibrium systems. And engineering applications almost always deal with irreversible processes.

In this chapter we will consider nonequilibrium, irreversible processes. Our approach will be based on the second law and specifically on the concept of entropy production. We will try to keep the development general although we will concentrate on specific examples in an effort to keep the concepts transparent.

This will brings us back to the original concept of Clausius regarding the reality of the entropy. We recall that Clausius chose the Greek $\eta\tau\rho o\pi\eta$, which means transformation, to identify this property (see Sect. 1.3.2). Entropy is not a conserved quantity. Entropy is produced in a process.

11.2 Entropy Production

11.2.1 General

The Clausius inequality requires that there will be an increase in entropy resulting from any irreversible (spontaneous) process occurring anywhere within a closed system(see Sect. 2.3.3, (2.22)). This is the basic physical law on which we must base any discussion of irreversibility. However, The Clausius inequality is not completely adequate for the study of the details of the irreversible process that may have

C.S. Helrich, *Modern Thermodynamics with Statistical Mechanics*,
DOI 10.1007/978-3-540-85418-0_11, © Springer-Verlag Berlin Heidelberg 2009

occurred in a small region within the system. If we wish to study the details of irreversible processes we must begin with the consequences of the second law applied to infinitesimal processes. The differential statement of the second law is part of Carathéodory's principle.

For the interested reader we have provided a geometrically based outline of Carathéodory's principle in the appendix. It is, however, not necessary to understand Carathéodory's principle in order to follow the development in this chapter. It is only necessary to understand the second law as a requirement that the differential of the entropy $dS \geq 0$, with $dS = 0$ for reversible quasistatic adiabatic processes. This is the infinitesimal equivalent of Clausius inequality in (2.22).

Carathéodory's principle is based on closed adiabatic systems as is the Clausius inequality. So we must begin with a closed adiabatic system and extend our discussion to interacting, non-adiabatic systems.

We consider a closed adiabatic system composed of a number of closed subsystems that are not individually adiabatic. We assume that the subsystems interact with one another through work and heat transfer. An infinitesimal interaction among these subsystems will result in a change in the entropy of the μth subsystem that we shall designate as $d_e^{(\mu)} S$. This entropy change is given by the Gibbs equation, which we write as

$$d_e^{(\mu)} S = \frac{1}{T_\mu} dU_\mu + \frac{P_\mu}{T_\mu} dV_\mu. \tag{11.1}$$

In addition to this change in entropy we consider that an infinitesimal irreversible process takes place within the μth subsystem. The result is an increase in entropy independent of external interaction. This internal change in entropy is designated as $d_i^{(\mu)} S$. In a linear approximation the total change in the entropy of the μth subsystem is

$$d^{(\mu)} S = d_e^{(\mu)} S + d_i^{(\mu)} S. \tag{11.2}$$

The entropy change for the closed adiabatic system is then

$$dS = \sum_\mu \left[d_e^{(\mu)} S + d_i^{(\mu)} S \right] > 0.$$

If the walls of the original adiabatic system are fixed there is no external entropy change for the system as a whole. Then

$$\sum_\mu d_e^{(\mu)} S = 0.$$

and

$$\sum_\mu d_i^{(\mu)} S > 0.$$

This result must hold even if there is an internal process occurring in only one of the subsystems. Therefore, the entropy change resulting from any infinitesimal irreversible process occurring in any closed system must be positive definite and the terms $d_i^{(\mu)}S$ are sources of entropy These are referred to as entropy production terms. That is the second law requires that

$$d_iS \geq 0 \tag{11.3}$$

for any irreversible process with the equality holding only when the irreversible process ceases. We have dropped the superscript in (11.3) as superfluous.

We may then conclude that for any general closed system the change in entropy is given by

$$dS = d_eS + d_iS. \tag{11.4}$$

In (11.4) the entropy change resulting from interactions with the exterior is given by the Gibbs equation

$$d_eS = \frac{1}{T}dU + \frac{P}{T}dV. \tag{11.5}$$

Equation (11.4) with the definition (11.5) and the requirement (11.3) will be the basis of our discussion in this chapter.

11.2.2   Forces and Flows

Irreversible processes are driven by generalized thermodynamic forces. When treating nonequilibrium such forces include chemical affinities and gradients of quantities such as temperature and concentration. These result in chemical reaction rates and flows of quantities such as heat or mass. The flow of a quantity, which we shall identify by a subscript, is defined as the amount of that quantity transported across a unit area per unit time. Specifically if dX_μ, is the amount of the quantity μ transported across a unit area in the infinitesimal time interval dt, then the flow of μ is given by

$$J_\mu = \frac{dX_\mu}{dt}. \tag{11.6}$$

We must now relate thermodynamic forces and flows to the production of entropy.

Entropy production is generally a localized phenomenon. The flow J_μ may then be non-zero only in a volume ΔV which is much less than the system volume. The entropy production rate per unit volume, which we shall designate as σ, is then the quantity of primary interest. We shall define the generalized thermodynamic force, F_μ, that drives the flow J_μ by the requirement that the product of this force and the resultant flow is the rate of entropy production per unit volume from the process driven by F_μ. That is

$$\sigma_\mu = \left[\frac{1}{\Delta V}\frac{d_iS}{dt} \right]_\mu = F_\mu J_\mu \tag{11.7}$$

for the process producing the flow J_μ.

Fig. 11.1 Transport of heat through a wall

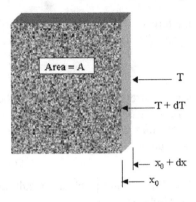

Area = A

T

T + dT

x_0 + dx

x_0

Example 11.2.1. As an example we consider entropy production from the flow of heat through a diathermal wall of area A separating two systems with uniform temperatures $T+dT$ and T respectively. We designate the direction perpendicular to the wall as x. The wall extends from the plane at $x = x_0$ to that at $x = x_0+dx$ as shown in Fig. 11.1.

The infinitesimal temperature difference results in the flow of an infinitesimal quantity of heat, δQ, from x_0 to x_0+dx in a time dt. In this time interval the system in the region $x < x_0$ loses entropy $d_e S(x_0) = \delta Q/(T+dT)$ due to a flow of heat out of the system. The system in the region $x > x_0+dx$ gains entropy $d_e S(x_0 + dx) = \delta Q/T$ due to a flow of heat into the system.

If the temperatures remain uniform in the systems on either side of the wall there is no entropy production in either of these systems. The volume, ΔV, in which entropy production occurs is then the volume of the wall $A dx$. The rate at which entropy enters the wall at x_0 and the rate at which entropy leaves the wall at x_0+dx due to the heat flow rate $\dot{Q} = \delta Q/dt$. The entropy production rate per unit volume in the diathermal wall is then

$$\sigma_q = \left[\frac{1}{\Delta V}\frac{d_i S}{dt}\right]_q = \frac{1}{A dx}\left[\frac{\dot{Q}}{T+dT} - \frac{\dot{Q}}{T}\right] = \left[\frac{d}{dx}\left(\frac{1}{T}\right)\right]\frac{\dot{Q}}{A}.$$

The two terms J_q and F_q can be identified from this expression. The flow of heat, J_q, is

$$J_q = \frac{\dot{Q}}{A}. \tag{11.8}$$

The thermodynamic force producing this heat flow is

$$F_q = \frac{d}{dx}\left(\frac{1}{T}\right). \tag{11.9}$$

11.2.3 Phenomenological Laws

Phenomenological laws generally have their origin in experiment. A phenomenological law is an empirical (experimentally obtained) mathematical relationship between a flow and what is measured as the driving force. The form of the empirical equation is a result of the intuition or objectives of the experimentalist. No hypothesis is made regarding atomic or molecular structure and often the observed phenomenon was well known and studied before a theoretical understanding of any form was available. An example is Fick's law of diffusion.

In 1855 Adolf Fick[1] published an empirical equation relating the rate of aqueous diffusion of a substance a to the gradient in the concentration of that substance dc_a/dx,

$$J_a = -D_a \frac{dc_a}{dx}, \qquad (11.10)$$

in which m D_a is the diffusion coefficient for the substance in the aqueous medium [[72], p. 313].

A companion law is Ohm's[2] circuit law relating the current (density) J_q through a resistance to the voltage across the resistance, which we write as negative the gradient of the potential φ, as [124]

$$J_e = -R \frac{d\varphi}{dx}. \qquad (11.11)$$

In a linear approximation we may expect the motion of a charged substance in an aqueous medium then to obey a mathematical equation of the form

$$J_\lambda = -A_c \frac{dc_\lambda}{dx} - A_\varphi \frac{d\varphi}{dx}. \qquad (11.12)$$

We have left the details of the respective coefficients unspecified for simplicity (see Sect. 11.3.2)

In a linear approximation we may generally relate flows to thermodynamic forces by phenomenological equations of the form

$$J_\mu = \sum_v L_{\mu v} F_v. \qquad (11.13)$$

where $L_{\mu v}$ are phenomenological coefficients. Individual empirical laws such as expressed in Eqs. (11.10) and (11.11) often do not fit the form in (11.13), but they can be brought into this form. The following example shows how this is done for Fourier's law [149] (Jean Baptiste Joseph Fourier (1768–1830)).

Example 11.2.2. For a temperature gradient in the $x-$ direction, Fourier's law of heat transfer provides the heat flow as

---

[1] Adolf Eugen Fick (1829–1901) was a German physiologist.
[2] Georg Simon Ohm (1789–1854) was a German physicist.

$$J_q = -\kappa \frac{dT}{dx}, \tag{11.14}$$

where κ is the thermal conductivity of the substance in which the heat flows. Introducing the thermodynamic force in (11.9), Fourier's law becomes

$$J_q = \kappa T^2 F_q. \tag{11.15}$$

Equation (11.15) has the general form of (11.13). We may then identify the phenomenological coefficient, L_{qq}, relating the flow, J_q, to the force, F_q, is

$$L_{qq} = \kappa T^2. \tag{11.16}$$

Equations (11.7) and (11.13) can be combined to give

$$\sigma_\mu = \sum_\nu F_\mu L_{\mu\nu} F_\nu \tag{11.17}$$

for each force μ. From (11.17) the requirement of the second law is then that

$$\sigma = \sum_\mu \sigma_\mu = \sum_{\mu,\nu} F_\mu L_{\mu\nu} F_\nu \geq 0 \tag{11.18}$$

for an arbitrary number of processes, provided the conditions in the system are such that we may expect a linear approximation to $d_i S$ and linear phenomenological equations to be valid. The equality only holds in a state of unconstrained stable equilibrium.

The system may be constrained by imposing boundary conditions. For example different parts of the system boundary may be maintained at different temperatures. If these boundary conditions are fixed in time the conditions in the system will not vary with time. But under these conditions σ will not be zero.

11.2.4 Onsager Symmetry

In 1931 Lars Onsager[3] established that the phenomenological coefficients $L_{\mu\nu}$ appearing in (11.13) and (11.17) are symmetric. That is $L_{\mu\nu} = L_{\nu\mu}$ and the coupling of the νth force to the μth flow is the same as the coupling of the μth force to the νth flow ([96], p.354).

In his proof Onsager used the fact that Newtonian mechanics is symmetric under time reversal and assumed that the average time decay of fluctuations about an equilibrium state is determined by the same laws that govern the decay of macroscopic variations about equilibrium (cf. [96], pp. 355–358)

---

[3] Lars Onsager (1903–1976) was a Norwegian born American physical chemist and theoretical physicist. He held the Gibbs Professorship of Theoretical Chemistry at Yale University and was winner of the 1968 Nobel Prize in Chemistry.

In addition to the specific requirements of the second law expressed in (11.18) we shall consider the Onsager symmetry relations to be primary requirements imposed on the phenomenological coefficients. The second law in the presence of irreversible phenomena then has the mathematical form

$$\sigma = \sum_{\mu,\nu} F_\mu L_{\mu\nu} F_\nu \geq 0 \text{ with } L_{\mu\nu} = L_{\nu\mu}. \tag{11.19}$$

11.3 Sources of Entropy Production

We have already considered entropy production from the transfer of heat in Example 11.2.1. We found there that the entropy production was confined to the wall in which there was a gradient in the temperature. Generally any transport driven by the gradient of a thermodynamic property will produce entropy. We shall investigate this in detail as we consider membrane transport. There we shall find that the entropy production is confined to the membrane itself, which in biological systems may be quite thin. Perhaps the most important entropy production source is a chemical reaction. In this section we consider three important sources of entropy production. These are the mixing of fluid systems, the transport of substances across membranes, and chemical reactions. In the treatment of chemical reactions we shall consider that the reactions occur uniformly throughout the system resulting in a non-local entropy production rate.

We shall begin with a treatment of entropy production from mixing.

11.3.1 Mixing

We treated the problem of mixing in Sect. 4.7.2. There we obtained the basic equations for ideal gas mixing entropy and the Pfaffians for the potentials for mixtures. Here we shall seek the rate of entropy production arising from mixing.

To isolate the entropy production resulting from mixing alone we shall assume a system contained in a rigid adiabatic vessel so that no work is done and no heat transferred. To obtain the rate of entropy production from mixing we combine (4.83) and (2.4). Dropping the terms δW_s and δQ the result is

$$dS = \frac{1}{T} d\Phi - \frac{1}{T} \sum_\lambda \mu_\lambda dn_\lambda. \tag{11.20}$$

The term $d\Phi$ in (11.20) is the change in energy of the system resulting from the transport of matter across the boundaries. This term is not necessarily negligible. If there is mixing at all we must have matter crossing a boundary. This term is not, however, the mixing term we seek. The mixing term is the last term in (11.20). That is

$$d_i S_{mix} = -\frac{1}{T} \sum_\lambda \mu_\lambda \, dn_\lambda. \qquad (11.21)$$

The chemical potential μ_λ for a mixture of ideal gases is given in (4.80), which we repeat here for continuity.

$$\mu_\lambda (T,P,\{n_\lambda\}) = \mu_{pure,\lambda} (T,P) + RT \ln \frac{n_\lambda}{n}. \qquad (11.22)$$

In (11.22) the chemical potential of the pure gas λ is

$$\mu_{pure,\lambda} (T,P) = C_{P,\lambda} \left[T - T \ln \frac{T}{T_{ref}} \right] + RT \ln \frac{P}{P_{ref}}$$
$$- C_{P,\lambda} T_{ref} + h_{ref,\lambda} - T s_{ref, \lambda}.$$

Similar results are obtained for ideal solutions (see Sect. 15.8). The rate of entropy production then has the form

$$\frac{d_i S_{mix}}{dt} = -\frac{1}{T} \sum_\lambda \left(\mu_{pure,\lambda} (T,P) + RT \ln \frac{n_\lambda}{n} \right) \frac{dn_\lambda}{dt}. \qquad (11.23)$$

The difficulty in any possible application of (11.23) results from the fact that entropy production is process dependent. The process by which the mixing takes place, even within a rigid adiabatic container, determines the form of the terms n_λ and dn_λ/dt.

If we choose to mix the gases as we did in Sect. 15.8, by removing partitions, there will be considerable spatial and temporal variations in n_λ and dn_λ/dt. It is even probable that there will be spatial and temporal variations in temperature and pressure during the mixing process. The rate of entropy production will be very high for a short span of time.

Or we may choose to mix the gases quasistatically using permeable partitions for the gases. We could contrive to remove shields from the outside leaving these permeable partitions through which the gases could pass very slowly. We could then remove the permeable partitions after the mixing is complete. Depending on the thickness of the partition barriers, we can make the spatial and time dependence of the n_λ and dn_λ/dt as small as we wish. The entropy production rate would then be very small and the total mixing time very large.

The total entropy change due to mixing is the same in either case.

11.3.2 Membrane Transport

A membrane is a thin boundary which may be permeable to substances including ions in solution. For example, the membrane most commonly considered in biological applications is the phospholipid bilayer, which is a double layer of molecules with polar groups forming the membrane surfaces and long non-polar hydrocarbon tails forming the interior of the membrane. The phospholipid bilayer is almost impermeable because of the dense packing of the lipids and the fact that the interior of the membrane is hydrophobic, i.e. it repels water. Channels formed by membrane bound proteins are generally selective for specific molecules or ions. At the level of a thermodynamic description the bilayer with ion channels present is a selectively permeable membrane.

We shall consider a closed system consisting of two subsystems, a and b, separated by a permeable membrane. The subsystems differ in temperature, molecular composition, and electrical potential. We assume that the subsystems contain ionic solutions and that the membrane is permeable to all ions in either direction. If transport between the two subsystems is slow each subsystem remains in thermodynamic equilibrium and thermodynamic properties may be defined for each subsystem.

Equation (2.4) holds for each open subsystem. The differential of the internal energy for open systems is given by (4.83). Combining these and setting $\delta W_s = 0$, we have

$$dS = \frac{1}{T}\delta Q + \frac{1}{T}d\Phi - \frac{1}{T}\sum_\lambda \mu_\lambda dn_\lambda \qquad (11.24)$$

for the entropy change in each of the open subsystems. Since the total system is closed but not isolated, the heat transfer, δQ, may include an exchange of heat with the exterior, $\delta_e Q$, and an internal heat transfer through the membrane, $\delta_i Q$. The heat transfer into the systems a and b from either of these sources will be designated by superscripts. Energy conservation requires that $\delta_i Q^{(b)} = -\delta_i Q^{(a)}$. The contribution to the entropy change due to a change in system composition, $\sum_\lambda \mu_\lambda dn_\lambda / T$, may involve the transport of matter between the two subsystems and changes in composition resulting from chemical reactions and mixing that occur in each subsystem. We shall assume, at this point, that no chemical reactions occur and designate the change in composition due to transport of matter through the membrane by terms $d_e n_\lambda^{(a,b)}$ for the subsystems a and b. Mass conservation requires that $d_e n_\lambda^{(b)} = -d_e n_\lambda^{(a)}$. The chemical potentials in the subsystems will be designated as $\mu_\lambda^{(a,b)}$ for each substance λ. Equation (11.24) then becomes

$$
\begin{aligned}
dS = {} & \left(\frac{1}{T_a}\delta_e Q^{(a)} + \frac{1}{T_b}\delta_e Q^{(b)}\right) + \left(\frac{1}{T_a} - \frac{1}{T_b}\right)\delta_i Q^{(a)} \\
& + \left(\frac{1}{T_a}d\Phi^{(a)} + \frac{1}{T_b}d\Phi^{(b)}\right) - \sum_\lambda \left(\frac{\mu_\lambda^{(a)}}{T_a} - \frac{\mu_\lambda^{(b)}}{T_b}\right) d_e n_\lambda^{(a)} \qquad (11.25)
\end{aligned}
$$

The only contribution we have not yet taken into account is that resulting from differences in the electrical potentials of the subsystems. A difference in electrical potential provides a generalized thermodynamic force resulting in the flow of ions. The contribution of ionic flow to the change in subsystem energies results in the terms $d\Phi^{(a,b)}$ in (11.25). If z_λ is the ionic charge on substance λ, then the charge per mol of this substance is $z_\lambda \mathfrak{F}$ where $\mathfrak{F}$ is the Faraday constant ($\mathfrak{F} = 0.9649 \times 10^5\,C\,mol^{-1}$). The transport of $d_e n_\lambda$ mols of substance λ into a region in which the electrical potential is φ results in a contribution

$$z_\lambda \mathfrak{F} \varphi \mathrm{d_e} n_\lambda$$

to the change in internal energy. Because it is proportional to $\mathrm{d_e} n_\lambda$, this change in the internal energy is most easily introduced into (11.25) by defining an electrochemical potential as

$$\tilde{\mu}_\lambda = \mu_\lambda + z_\lambda \mathfrak{F} \varphi. \tag{11.26}$$

For the subsystems a and b we have electrochemical potentials

$$\tilde{\mu}_\lambda^{(a,b)} = \mu_\lambda^{(a,b)} + z_\lambda \mathfrak{F} \varphi^{(a,b)},$$

since the subsystems are at different electrical potentials. Then (11.25) becomes

$$\mathrm{d}S = \left(\frac{1}{T_a} \delta_e Q^{(a)} + \frac{1}{T_b} \delta_e Q^{(b)} \right) + \left(\frac{1}{T_a} - \frac{1}{T_b} \right) \delta_i Q^{(a)} - \sum_\lambda \left(\frac{\tilde{\mu}_\lambda^{(a)}}{T_a} - \frac{\tilde{\mu}_\lambda^{(b)}}{T_b} \right) \mathrm{d_e} n_\lambda^{(a)}. \tag{11.27}$$

In (11.27) the entropy change from interactions with the exterior, $\mathrm{d_e}S$, is that resulting from the heat transfer terms $\delta_e Q$. This is

$$\mathrm{d_e}S = \left(\frac{1}{T_a} \delta_e Q^{(a)} + \frac{1}{T_b} \delta_e Q^{(b)} \right).$$

The entropy production, $\mathrm{d_i}S$, is then

$$\mathrm{d_i}S = \left(\frac{1}{T_a} - \frac{1}{T_b} \right) \delta Q_i - \sum_\lambda \left(\frac{\tilde{\mu}_\lambda^{(a)}}{T_a} - \frac{\tilde{\mu}_\lambda^{(b)}}{T_b} \right) \mathrm{d_e} n_\lambda^{(a)}.$$

Because the subsystems are in thermodynamic equilibrium this entropy production occurs only in the membrane separating the subsystems. In carefully designed experiments the membrane may be planar and of macroscopic dimensions. This will not be the case, however, in living cells, which are systems of potential interest.[4] We, therefore, consider an infinitesimal section of the membrane with area ΔA_M and thickness Δx. The entropy production rate per unit volume in this section of the membrane is

$$\sigma = \frac{1}{\Delta x} \left(\frac{1}{T_a} - \frac{1}{T_b} \right) J_q - \sum_\lambda \frac{1}{\Delta x} \left(\frac{\tilde{\mu}_\lambda^{(a)}}{T_a} - \frac{\tilde{\mu}_\lambda^{(b)}}{T_b} \right) J_{n_\lambda}. \tag{11.28}$$

Using (11.8), the heat flow, J_q, through the area ΔA_M is

$$J_q = \frac{1}{\Delta A_M} \frac{\delta_i Q}{\mathrm{d}t} = \frac{\dot{Q}_i}{\Delta A_M}$$

---

[4] In living cells membrane transport is often coupled to chemical reactions.

and from (11.6) the molar rate of flow of substance λ per unit area is

$$J_{n_\lambda} = \frac{1}{\Delta A_M} \frac{d_e n_\lambda}{dt} \tag{11.29}$$

In the limit as $\Delta x \to 0$ (11.28) becomes

$$\sigma = \frac{d}{dx}\left(\frac{1}{T}\right) J_q - \sum_\lambda \frac{d}{dx}\left(\frac{\tilde{\mu}_\lambda}{T}\right) J_{n_\lambda}. \tag{11.30}$$

In (11.30) the thermodynamic forces are readily identified as

$$F_q = \frac{d}{dx}\left(\frac{1}{T}\right),$$

(see (11.9)), and

$$F_{n_\lambda} = -\frac{d}{dx}\left(\frac{\tilde{\mu}_\lambda}{T}\right)$$

The entropy production rate per unit volume in (11.30) is then of the general form (11.7).

If the temperature of the system is uniform, $F_q = 0$. The entropy production then results only from the thermodynamic force

$$\begin{aligned} F_{n_\lambda} &= -\frac{1}{T}\frac{d}{dx}(\mu_\lambda + z_\lambda \mathfrak{F}\varphi) \\ &= -\frac{1}{T}\left[\frac{d\mu_\lambda}{dx} + z_\lambda \mathfrak{F}\frac{d\varphi}{dx}\right] \end{aligned} \tag{11.31}$$

driving the flow of matter through the membrane. It is then the gradient in electrochemical potential that produces a flow of matter. Fick's and Ohm's laws are the corresponding phenomenological laws (see Sect. 11.2.3). The first term on the right hand side of (11.31) is the generalization of Fick's and the second term is the generalization of Ohm's law, which appeared in (11.12).

In Sect. 11.3.1 we pointed out that the form of the chemical potential in (11.22) is of the same form as that for ideal solutions. Using (11.22) we have

$$\frac{d\mu_\lambda}{dx} = \frac{RT}{c_\lambda}\frac{dc_\lambda}{dx},$$

where $c_\lambda = n_\lambda/V$ is the concentration of the substance λ. Then (11.31) becomes

$$F_{n_\lambda} = -\frac{R}{c_\lambda}\left[\frac{dc_\lambda}{dx} + \frac{z_\lambda \mathfrak{F}c_\lambda}{RT}\frac{d\varphi}{dx}\right]. \tag{11.32}$$

For motion of ions in a fluid the mobility u_λ is introduced and the flow J_λ of ions resulting from an electric field is written as

$$J_\lambda = -z_\lambda \mathfrak{F} u_\lambda c_\lambda \frac{\mathrm{d}\varphi}{\mathrm{d}x}. \tag{11.33}$$

For biophysical studies the units commonly used for mobility are $\mathrm{cm^2\,V^{-1}\,s^{-1}}$.
The Nernst-Einstein relationship between mobility and the diffusion coefficient
D_λ is [119], [48]

$$D_\lambda = \frac{RT}{z_\lambda \mathfrak{F}} u_\lambda. \tag{11.34}$$

Combining (11.33) and (11.34) we have

$$J_\lambda = -z_\lambda \mathfrak{F} D_\lambda \frac{z_\lambda \mathfrak{F}}{RT} c_\lambda \frac{\mathrm{d}\varphi}{\mathrm{d}x}. \tag{11.35}$$

For ions Fick's law has the form

$$J_\lambda = -z_\lambda \mathfrak{F} D_\lambda \frac{\mathrm{d}c_\lambda}{\mathrm{d}x}. \tag{11.36}$$

Adding (11.35) and (11.36) we obtain the linear phenomenological law for ion
transport in a solution (see (11.13))

$$J_\lambda = L_{\lambda\lambda} F_\lambda$$
$$= -z_\lambda \mathfrak{F} D_\lambda \left(\frac{\mathrm{d}c_\lambda}{\mathrm{d}x} + \frac{z_\lambda \mathfrak{F} c_\lambda}{RT} \frac{\mathrm{d}\varphi}{\mathrm{d}x} \right). \tag{11.37}$$

Equation (11.37) is the Nernst-Planck equation for ion transport in a solution
(electrodiffusion).

The Nernst-Planck equation can be integrated using an integrating factor (see
Sect. A.7). That is (11.37) can be written as

$$J_\lambda \left[\exp\left(\frac{z_\lambda \mathfrak{F} \varphi}{RT} \right) / D_\lambda \right] = -z_\lambda \mathfrak{F} D_\lambda \left[\exp\left(\frac{z_\lambda \mathfrak{F} \varphi}{RT} \right) \right] \left(\frac{\mathrm{d}c_\lambda}{\mathrm{d}x} + \frac{z_\lambda \mathfrak{F} c_\lambda}{RT} \frac{\mathrm{d}\varphi}{\mathrm{d}x} \right)$$
$$= -z_\lambda \mathfrak{F} \frac{\mathrm{d}}{\mathrm{d}x} \left[c_\lambda \exp\left(\frac{z_\lambda \mathfrak{F} \varphi}{RT} \right) \right]. \tag{11.38}$$

Then

$$J_\lambda = -z_\lambda \mathfrak{F} \beta_\lambda \frac{c_{\mathrm{m},\lambda} \exp\left(\frac{z_\lambda \mathfrak{F} \varphi_\mathrm{m}}{RT} \right) - c_{0,\lambda}}{\int_{x=0}^{d_\mathrm{m}} \left[\exp\left(\frac{z_\lambda \mathfrak{F} \varphi}{RT} \right) / D_\lambda \right] \mathrm{d}x}. \tag{11.39}$$

Here $c_{\mathrm{m},\lambda}$ is the concentration c_λ at $x = d_\mathrm{m}$ and $c_{0,\lambda}$ is the concentration at $x = 0$
and we have introduced the partitioning coefficient β_λ to account for the fact that
the concentrations just inside the channel differ from those outside of the channel. If
we assume that the diffusion coefficient is a constant through the membrane and that
the electrical potential varies linearly (the electric field is constant), we can easily
integrate the term in the denominator in (11.39) to obtain

$$J_\lambda = \frac{z_\lambda^2 \mathfrak{F}^2 \varphi_m}{RT} D_\lambda \beta_\lambda \frac{c_{m,\lambda} - c_{0,\lambda} \exp\left(-\frac{z_\lambda \mathfrak{F} \varphi_m}{RT}\right)}{1 - \exp\left(-\frac{z_\lambda \mathfrak{F} \varphi_m}{RT}\right)}. \tag{11.40}$$

Here φ_m is the electrical potential at $x = d_m$ and the potential at $x = 0$ is taken to be zero. Equation (11.40) is the Goldman-Hodgkin-Katz (GHK) equation for ion transport through a membrane [58, 74, 86].

If we introduce a permeability coefficient $P_\lambda = D_\lambda \beta_\lambda$ ([72], p. 444). This combination of the partitioning coefficient and the diffusivity accounts for the fact that the biological ion channel responds to the chemical identity of the ion seeking to pass through the channel. Then (11.40) becomes

$$J_\lambda = \frac{z_\lambda^2 \mathfrak{F}^2 \varphi_m}{RT} P_\lambda \frac{c_{m,\lambda} - c_{0,\lambda} \exp\left(-\frac{z_\lambda \mathfrak{F} \varphi_m}{RT}\right)}{1 - \exp\left(-\frac{z_\lambda \mathfrak{F} \varphi_m}{RT}\right)}. \tag{11.41}$$

The total current passing through the channel is the sum over the individual ion currents $J = \sum_\lambda J_\lambda$.

The assumptions of constant diffusion coefficient and constant field are poor for diffusion through a biological membrane. There the flow is through protein ion channels, which have considerable structure. The GHK equation is, however, the only reasonable algebraic equation available for studies of transport in biological membranes. Many protein ion channels do, in fact, exhibit GHK characteristics.

11.3.3   Chemical Reactions

For the sake of simplicity we shall consider spatially uniform systems. We begin with mass and energy conservation for reactions before proceeding to the requirements of the second law. In each case we shall consider closed and open systems. Based on the second law we will be able to establish conditions that must be satisfied for reactions to proceed.

11.3.3.1   Mass Conservation

Closed Systems. A general chemical reaction is represented by a stoichiometric equation of the form

$$\alpha_A A + \alpha_B B + \cdots \rightarrow \alpha_D D + \alpha_E E + \cdots . \tag{11.42}$$

Here A,B,... are reactants and D,E,... are products. The α_K are the stoichiometric coefficients. If we introduce the coefficients ν_K that have the magnitude of the corresponding stoichiometric coefficients, but are positive for products and negative for reactants, we can write the equation expressing mass conservation for the reaction in Eq. (11.42) as

$$\sum_\lambda v_\lambda M_\lambda = 0 \tag{11.43}$$

in which M_λ is the molar mass of the constituent λ.

Using the coefficients v_K we can express the change in mass of the component λ as the reaction in (11.42) progresses from reactants to products by the equation

$$dm_\lambda = v_\lambda M_\lambda d\xi. \tag{11.44}$$

Here $d\xi$ is the extent of reaction introduced by Théophile De Donder[5] [39]. The extent of reaction is a function of time that equals zero at equilibrium. From (11.44) the change in mol number ($n_\lambda = m_\lambda / M_\lambda$) for the component λ is

$$dn_\lambda = v_\lambda d\xi \tag{11.45}$$

Example 11.3.1. As an example we consider the reaction

$$2H_2 + O_2 \rightarrow 2H_2O. \tag{11.46}$$

If in a particular time interval dt there are dn_{H_2O} mols of H_2O produced requiring dn_{H_2} mols of H_2 and dn_{O_2} mols of O_2 the extent of reaction for the reaction in (11.46) is

$$d\xi = \frac{dn_{H_2}}{2} = \frac{dn_{O_2}}{1} = \frac{dn_{H_2O}}{2}.$$

The extent of reaction per unit time is called the velocity of the reaction, $v = d\xi/dt$. In terms of the velocity of the reaction, the rate of change of the mass of component λ due to the chemical reaction is

$$\frac{dm_\lambda}{dt} = v_\lambda M_\lambda v. \tag{11.47}$$

The rate of change of the number of mols of the component λ is then

$$\frac{dn_\lambda}{dt} = M_\lambda v. \tag{11.48}$$

Summing (11.47) over all components and using (11.43) yields

$$\sum_\lambda \frac{dm_\lambda}{dt} = v \sum_\lambda (v_\lambda M_\lambda) = 0. \tag{11.49}$$

Mass conservation is guaranteed by the stoichiometric relationship for the reaction.

---

[5] Théophile De Donder (1872–1957) was a physical chemist who incorporated the uncompensated heat of Clausius into the formalism of the second law introducing the concept of chemical afinity.

We may extend these results to the situation in which there are a number of reactions occurring simultaneously. For each reaction, τ, there is a stoichiometric coefficient for the component λ, which we designate as $\alpha_\lambda^{(\tau)}$. This serves to define a corresponding term $v_\lambda^{(\tau)}$. Each reaction has an extent of reaction $d\xi^{(\tau)}$ and a corresponding velocity of reaction,

$$\mathfrak{v}^{(\tau)} = \frac{d\xi^{(\tau)}}{dt}. \tag{11.50}$$

The differential increases of mass and mol number of the component λ then include summation over all the reactions in which the component λ enters. That is

$$dm_\lambda = M_\lambda \sum_\tau v_\lambda^{(\tau)} d\xi^{(\tau)} \tag{11.51}$$

and

$$dn_\lambda = \sum_\tau v_\lambda^{(\tau)} d\xi^{(\tau)}. \tag{11.52}$$

The corresponding rates of change of mass and mol numbers are

$$\frac{dm_\lambda}{dt} = M_\lambda \sum_\tau v_\lambda^{(\tau)} \mathfrak{v}^{(\tau)} \tag{11.53}$$

and

$$\frac{dn_\lambda}{dt} = \sum_\tau v_\lambda^{(\tau)} \mathfrak{v}^{(\tau)}. \tag{11.54}$$

If we sum (11.53) over the substance index, λ, we obtain

$$\sum_\lambda \frac{dm_\lambda}{dt} = \sum_\tau \sum_\lambda v_\lambda^{(\tau)} M_\lambda \mathfrak{v}^{(\tau)}. \tag{11.55}$$

Stoichiometry for the τth reaction requires that

$$\sum_\lambda v_\lambda^{(\tau)} M_\lambda = 0. \tag{11.56}$$

Therefore (11.55) becomes

$$\sum_\lambda \frac{dm_\lambda}{dt} = \sum_\tau \sum_\lambda v_\lambda^{(\tau)} M_\lambda \mathfrak{v}^{(\tau)} = 0, \tag{11.57}$$

which is the same result as (11.49). That is the stoichiometric requirement for each reaction results in mass conservation for the entire closed system whether or not the reactions are coupled.

Open Systems. The system in which the chemical reactions are occurring may be open to exchange of mass with the exterior. The change in the mass of component λ is then written as the sum of a contribution from the exterior and a contribution from the chemical reactions occurring in the system. Designating the contribution from the internal chemical reactions by a subscript i and the contribution from mass crossing the system boundary by a subscript e, the total change in mass of the component λ is

$$\mathrm{d}m_\lambda = \mathrm{d}_e m_\lambda + \mathrm{d}_i m_\lambda. \tag{11.58}$$

With (11.51) this is

$$\mathrm{d}m_\lambda = \mathrm{d}_e m_\lambda + M_\lambda \sum_\tau v_\lambda^{(\tau)} \mathrm{d}\xi^{(\tau)}. \tag{11.59}$$

Summing (11.59) over λ and using (11.56) we obtain

$$\begin{aligned} \mathrm{d}m &= \sum_\lambda \mathrm{d}m_\lambda = \sum_\lambda \mathrm{d}_e m_\lambda + \sum_\tau \mathrm{d}\xi^{(\tau)} \sum_\lambda M_\lambda v_\lambda^{(\tau)} \\ &= \sum_\lambda \mathrm{d}_e m_\lambda. \end{aligned} \tag{11.60}$$

That is the total mass of the open system can only change as a result of matter crossing the system boundary.

11.3.3.2   Energy Conservation

The study of the consequences of energy conservation for a system in which there are chemical reactions is the subject of what is called thermochemistry.

Closed Systems. The energy conservation equation for closed systems may take either of the forms of the first law

$$\delta Q = \mathrm{d}U + P \mathrm{d}V \tag{11.61}$$

or

$$\delta Q = \mathrm{d}H - V \mathrm{d}P. \tag{11.62}$$

For systems in which chemical reactions occur we must include variations in composition in the expressions for $\mathrm{d}U$ and $\mathrm{d}H$. Because of the conditions encountered in the laboratory, it is convenient to choose $U = U(T,V,\{n_\lambda\})$ and $H = H(T,P,\{n_\lambda\})$. Equations (4.83) and (4.84) then take the form

$$\mathrm{d}U = \sum_\lambda n_\lambda C_{V,\lambda} \mathrm{d}T + \left(\frac{\partial U}{\partial V}\right)_{T,\{n_\lambda\}} \mathrm{d}V + \sum_\lambda \left(\frac{\partial U}{\partial n_\lambda}\right)_{n_\beta \neq n_\lambda, T, V} \mathrm{d}n_\lambda \tag{11.63}$$

and

$$dH = \sum_\lambda n_\lambda C_{P,\lambda} dT + \left(\frac{\partial H}{\partial P}\right)_{T,\{n_\lambda\}} dP + \sum_\lambda \left(\frac{\partial H}{\partial n_\lambda}\right)_{n_\beta \neq n_\lambda, T, P} dn_\lambda \qquad (11.64)$$

where $C_{V,\lambda}$ and $C_{P,\lambda}$ are the specific heats at constant volume and constant pressure for the substance λ.

Equation (11.63) is appropriate for experiments using a *constant volume* (bomb) *calorimeter*. In these experiments the reactants are sealed in a metal vessel that is immersed in a water bath and the reaction is initiated electrically. The heat transferred from the bomb is found from the rise in temperature of the water bath. For very accurate measurements the contribution from the first term on the right hand side of (11.63) must be taken into account, since this contribution may be of the order of a few hundredths of one percent to several tenths of a percent. We shall consider that these and the heat transferred to the wall of the bomb are taken into account as part of the experiment and shall not consider them explicitly here. Combining (11.61) and (11.63) the heat transferred from the contents of the constant volume bomb to the water bath is

$$\Delta Q_V = \sum_\lambda \int_{\{n_\lambda\}(\text{reactants})}^{\{n_\lambda\}(\text{products})} \left(\frac{\partial U}{\partial n_\lambda}\right)_{n_\beta \neq n_\lambda, T, V} dn_\lambda$$

$$= \sum_{\{n_\lambda\}(\text{products})} n_\lambda u_\lambda - \sum_{\{n_\lambda\}(\text{reactants})} n_\lambda u_\lambda$$

$$= \sum_\lambda \nu_\lambda u_\lambda, \qquad (11.65)$$

since $\nu_\lambda = +n_\lambda$ (products) $= -n_\lambda$ (reactants). The measurement of the heat transferred to the bath is then a direct measurement of internal energy change in the reaction. This is called the reaction energy and is designated as ΔU_{rxn} per mol of product,

$$\Delta U_{\text{rxn}} = \sum_\lambda \nu_\lambda u_\lambda. \qquad (11.66)$$

Equation (11.64) is appropriate for experiments conducted using a *steady-flow calorimeter*. The steady-flow calorimeter is an open system into which gaseous reactants are pumped and out of which gaseous products flow under steady conditions. The pressure is constant in the steady-flow calorimeter. Heat from the reaction is transferred to the coolant flowing around the calorimeter. Considerations of corrections are similar to those of the bomb calorimeter. Combining (11.62) and (11.64), the heat transferred from the calorimeter to the coolant is

$$\Delta Q_P = \sum_\lambda \int_{\{n_\lambda\}(\text{reactants})}^{\{n_\lambda\}(\text{products})} \left(\frac{\partial H}{\partial n_\lambda}\right)_{n_\beta \neq n_\lambda, T, V} dn_\lambda$$

$$= \sum_\lambda \nu_\lambda h_\lambda \qquad (11.67)$$

The measurement of the heat transferred to the coolant is then a direct measurement of enthalpy change in the reaction. This is called the reaction enthalpy and is designated as ΔH_{rxn} per mol of product.

$$\Delta H_{rxn} = \sum_{\lambda} v_{\lambda} h_{\lambda}. \tag{11.68}$$

Here h_{λ} is the specific molar enthalpy for the substance λ.

Most applications of thermochemistry are to reactions taking place under conditions of constant pressure and with initial and final temperatures equal, such as for reactions studied in the open laboratory. Accordingly reaction enthalpies are given in tables of thermochemical data. At the time of this writing NIST maintains *NIST Chemistry WebBook: NIST Standard Reference Database Number 69, June 2005 Release*, available at http://webbook.nist.gov/chemistry/. At the time of this writing the database may be accessed free of charge. NIST, however, reserves the right to charge for access to this database in the future.[6]

A standard state has been defined for the tabulation of reaction enthalpies. The standard conditions are $T^0 = 298.15$ K ($25\,^{\circ}$C) and $P^0 = 1$ bar $= 100$ kPa (≈ 1 atm). The natural state of a substance is its state (gas, liquid, or solid) at the standard conditions. The standard enthalpy of an element in its natural state is defined to be zero. The standard enthalpy of a compound, C, which is called the standard enthalpy of formation for the compound and designated by $\Delta H_f^0 [C]$,[7] is the enthalpy of the reaction required to form the compound from its constituent elements under conditions of standard temperature and pressure. In thermochemistry the physical state of all substances is explicitly indicated by a g (gas), l (liquid), or s (solid). If the compound C is a gas we write C(g).

Experiments may be conducted at other than standard temperature and pressure conditions. For example combustion measurements in a steady-flow calorimeter may be conducted at a pressure $P > 100$ kPa a temperature $T > 298.15$ K. The measured enthalpy change is appropriate for the elevated temperature and pressure, but is not the enthalpy of formation of the product at standard conditions. In thermodynamic terms we have measured the difference between the enthalpy of the products and the enthalpy of the reactants at (T,P). To obtain he enthalpy of the products at (T,P) we must know the enthalpies of the reactants at (T,P).

We may calculate the molar enthalpies of the reactants from a numerical integration of the Pfaffian for dh, i.e. (5.12) with (5.18), if we have thermodynamic data for C_P, β and the molar volume, v, as functions of (T,P). That is

---

[6] This site provides thermochemical, thermophysical, and ion energetics data compiled by NIST under the Standard Reference Data Program.

The National Institute of Standards and Technology (NIST) uses its best efforts to deliver a high quality copy of the Database and to verify that the data contained therein have been selected on the basis of sound scientific judgment. However, NIST makes no warranties to that effect, and NIST shall not be liable for any damage that may result from errors or omissions in the Database.

The NIST Chemistry WebBook was developed in part with funds from the Systems Integration for Manufacturing Applications (SIMA) program at NIST.

[7] This is also designated as $\Delta_f H^0$.

$$h(T,P) = h\left(T^0,P^0\right) + \int_{T=T^0,P=\text{constant}=P^0}^{T} dT C_P\left(T,P^0\right)$$

$$+ \int_{P^0,T=\text{constant}=T}^{P} dP\left[-T\beta\left(T,P\right)v\left(T,P\right)+v\left(T,P\right)\right]. \quad (11.69)$$

If we wish to use our data, gathered at (T,P), to obtain standard reaction enthalpies, then a similar integration must be performed requiring data for properties
of the products.

Data for specific heats at constant pressure, C_P, are often given in the form of
tabulated coefficients (a,b,c) for the empirical equation

$$C_P = a + bT + cT^2.$$

If we are dealing with substances for which the pressure dependence of the enthalpy is small, e.g. gases, and if the pressure, P, is not too high, the pressure integration in (11.69) may be negligible. Equation (11.69) then becomes

$$h(T,P) = h\left(T^0,P^0\right) + a\left(T - T^0\right) + \frac{1}{2}b\left(T - T^0\right)^2 + \frac{1}{3}c\left(T - T^0\right)^3.$$

The relationship between reaction energy and reaction enthalpy, at standard conditions, is

$$\Delta H_{\text{rxn}}^0 = \sum_\lambda v_\lambda\left(u_\lambda^0 + P^0 v_\lambda^0\right)$$

$$= \Delta U_{\text{rxn}}^0 + P^0 \Delta V_{\text{rxn}}^0,$$

where ΔV_{rxn}^0 is the difference in volumes between the products and reactants in the
formation of one mol of products. If the reactants and products are gases we may
use the ideal gas thermal equation of state as an approximation for the volumes.
Then

$$\Delta V_{\text{rxn}}^0 = RT^0 \sum_\lambda v_\lambda$$

$$= \Delta n_{\text{rxn}} RT^0,$$

where

$$\Delta n_{\text{rxn}} = \sum_\lambda v_\lambda$$

is the change in the number of mols in the reaction, and

$$\Delta H_{\text{rxn}}^0 = \Delta U_{\text{rxn}}^0 + \Delta n_{\text{rxn}} RT^0.$$

Example 11.3.2. As an example we shall assume that we have conducted a bomb
calorimeter experiment to determine the reaction energy for the combustion of carbon monoxide to produce carbon dioxide at 25 °C. Our results are

$$CO(g) + \frac{1}{2}O_2(g) \rightarrow CO_2(g) \quad \Delta U_{rxn}^0 = -281.71 \, kJ \, mol^{-1}.$$

For this reaction

$$\Delta n_{rxn}RT^0 = (1 - 1.5) \, mol \, (8.31441 \times 10^{-3} \, kJ \, mol^{-1} \, K^{-1}) \, (298.15 \, K)$$
$$= -1.2395 \, kJ \text{ per mol of } CO_2 \text{ produced.}$$

Then

$$\Delta H_{rxn}^0 = -(281.71 + 1.2395) \, kJ \, mol^{-1}$$
$$= -282.95 \, kJ \, mol^{-1}.$$

We note that the reaction transfers heat to the bath ($\Delta Q_V < 0$) and that if the reaction had been carried out at constant pressure P^0 and temperature T^0 the change in volume would have been

$$\Delta V = \frac{\Delta n_{rxn}RT^0}{P^0} = -12.3951 \, m^3 \text{ per mol of } Co_2 \text{ produced.}$$

For this example the work that would have been done by the surroundings on a reaction vessel maintained at the constant pressure P^0 and temperature T^0 is the difference between ΔU_{rxn}^0 and ΔH_{rxn}^0 and is small (0.4%) compared to the magnitudes of these quantities. Any corrections to the ideal gas thermal equation of state in calculating the difference between reaction energy and enthalpy is then unwarranted. This is generally true for gaseous reactions at standard conditions.

A reaction which requires a heat flow from the surroundings is termed endothermic. Reactions which give up heat to the surroundings, such as the one in the preceding example, are termed exothermic. Specifically if $\Delta H_{rxn}^0 < 0$ heat is removed from the system ($\Delta Q_P < 0$) and the reaction is termed exothermic. If $\Delta H_{rxn}^0 > 0$ heat is added to the system ($\Delta Q_P > 0$) and the reaction is termed endothermic. The definitions are based on reaction enthalpies rather than reaction energies.

Practical conditions limit the reactions for which reaction enthalpies or energies can be experimentally obtained. To be a candidate for laboratory study the reaction must (1) have no side reactions producing other products, (2) go to completion, and (3) take place rapidly.

The first of these conditions is the basic experimental requirement of greatest simplicity. We wish to isolate phenomena as much as practicable in our laboratory studies.

The second is related to the first. If a reaction does not go to completion we are faced with additional measurements of final composition, which introduce error, and calculations, which may be labor intensive.

Finally, if the reaction we are studying does not take place rapidly steady flow calorimeters cannot be used, and unavoidable errors are introduced into the measurements of the difference in bath temperature for the bomb calorimeter arising from heat losses to the surroundings.

Combustion reactions taking place in an excess of oxygen are those reactions satisfying these conditions and most thermochemical data are based ultimately on laboratory measurements conducted on combustion reactions.

Because enthalpy is a thermodynamic property differences in enthalpy are independent of the path we may choose for the calculation of that difference. We may, therefore, consider that a reaction, which takes a set of reactants in an initial thermodynamic state to a corresponding set of products in a final state, results from a series of steps involving specific simpler reactions.

In this we need not claim that the steps actually occur, or even approximately occur, in the reaction considered. We only require that the steps fit together such that the sum of the stoichiometric equations results in the reaction of interest. Mass conservation is then guaranteed (see (11.57)). The reaction enthalpy of interest is then the algebraic sum of the reaction enthalpies for the (imagined) sequence of reactions.

The fact that the enthalpy change in a reaction is independent of the steps by which it occurs is often considered to be one of the two main principles of thermochemistry. The second principle is that ΔH_{rxn} for the reverse of any reaction is the negative of ΔH_{rxn} for the reaction in the forward direction. Like the first principle, this is also based on the fact that the change in enthalpy is state and not process dependent.

We may also make these same two statements in terms of reaction energies for the same thermodynamic reason: the change in a thermodynamic property is dependent only on the initial and final thermodynamic states. The general principles of thermochemistry are then contained in the requirements of mass conservation and energy conservation expressed in terms of reaction enthalpies or energies.

We may use these general principles of thermochemistry to obtain reaction enthalpies (or energies) for reactions that cannot be actually studied in the laboratory because they do not satisfy our requirements for calorimetric measurements. As an example we consider the formation of a compound $AB_2C_3(g)$ at standard conditions. Here A, B, and C are arbitrary elements. We assume, for illustration, that the standard state of A is solid, of B is gas, and of C is liquid. The reaction by which the compound, $AB_2C_3(g)$, is formed from its elements is then

$$A(s) + B_2(g) + 3C(l) \rightarrow AB_2C_3(g) \qquad \Delta H_f^0 [AB_2C_3] \qquad (11.70)$$

Generally this reaction cannot be studied calorimetrically. However, we shall assume that we have thermochemical data for the reactions

$$A(l) + B_2(g) \rightarrow AB_2(g) \qquad \Delta H_{rxn}(1) \qquad (11.71)$$

and

$$AB_2(g) + 3C(l) \rightarrow AB_2C_3(g) \qquad \Delta H_{rxn}(2) \qquad (11.72)$$

The substance A(s) must first be melted to form A(l) for the reaction (11.71). This requires an amount of heat termed the latent heat of melting, which is the molar enthalpy for the "reaction"

$$A(s) \rightarrow A(l) \qquad \Delta H_{rxn}(3). \tag{11.73}$$

Adding the left and right hand sides of the reactions (11.71),(11.72), and (11.73) we have the reaction (11.70). The combination of the three reactions (11.71), (11.72), and (11.73) then satisfy mass conservation in the same way as the initial stoichiometric equation. The enthalpy of formation of AB_2C_3 (g) is then

$$\Delta H_f^0 [AB_2C_3] = \Delta H_{rxn}(3) + \Delta H_{rxn}(1) + \Delta H_{rxn}(2).$$

We may similarly define the Gibbs energy, or Gibbs free energy of formation as

$$\Delta G_f^0 = \sum_\lambda v_\lambda \mu_\lambda^0$$

11.3.3.3   Entropy Production

Closed Systems. For closed systems a combination of the first law in Eq. (2.2) and the differential energy change in internal energy with variable composition in Eq. (4.83), gives the entropy change in a system with chemical reactions as

$$dS = \frac{1}{T}\delta Q_{rev} - \sum_\lambda \frac{\mu_\lambda}{T} d_i n_\lambda. \tag{11.74}$$

In(11.74) the internal change in mol numbers, $d_i n_\lambda$, is due to chemical reactions. This is given by (11.52). Using (11.52), (11.74) becomes

$$dS = \frac{1}{T}\delta Q_{rev} - \frac{1}{T}\sum_\tau \sum_\lambda \mu_\lambda v_\lambda^{(\tau)} d\xi^{(\tau)} \tag{11.75}$$

The chemical affinity, $A^{(\tau)}$, for the reaction τ was introduced by De Donder as

$$A^{(\tau)} = -\sum_\lambda \mu_\lambda v_\lambda^{(\tau)}. \tag{11.76}$$

With this the entropy change for a system in which chemical reactions occur is

$$dS = \frac{1}{T}\delta Q_{rev} + \frac{1}{T}\sum_\tau A^{(\tau)} d\xi^{(\tau)}. \tag{11.77}$$

For a closed system the entropy change due to interaction with the exterior is associated with the heat transferred to the system,

$$d_e S = \frac{1}{T}\delta Q_{rev},$$

and the entropy production due to chemical reactions is the second term on the right hand side of (11.77),

$$d_i S = \frac{1}{T} \sum_\tau A^{(\tau)} d\xi^{(\tau)} > 0. \tag{11.78}$$

Eq. (11.78) applies to the case in which there are multiple coupled reactions occurring in the system. This may be simplified to single reactions by dropping the summation on τ and the superscripts. For a set of chemical reactions to actually occur the condition in (11.78) must be satisfied. It is possible that $A^{(\tau)} d\xi^{(\tau)} < 0$ for a single reaction in the set τ while the summation is still positive. This is the physical basis for catalysis.

Open Systems. For open systems we must add the effect of transport of matter across the system boundary to the change in composition. We shall assume that a certain subset of the substances present is transported across the boundary and designate the mol numbers of this subset by the index ℓ. The change in mol number for substance ℓ is then

$$dn_\ell = d_e n_\ell + d_i n_\ell$$

where $d_e n_\ell$ is the contribution from transport across the system boundary.

We shall designate the open system of interest as a and the exterior to a as system b. Transport of mass takes place between systems a and b. We assume that the system b serves only as a reservoir and that no chemical reactions occur in system b. We shall assume further that a and b are subsystems of a larger closed system, which we shall designate as c. This arrangement allows us to use our previous results. Specifically if we consider that the systems a and b are separated by a permeable membrane (11.27) gives the entropy change in these systems. To (11.27) we need only add the internal entropy production from the chemical reactions occurring in subsystem a. Using (11.78), the result for the entropy change in the system c is then

$$dS = \left(\frac{1}{T_a} \delta_e Q^{(a)} + \frac{1}{T_b} \delta_e Q^{(b)} \right) + \left(\frac{1}{T_a} - \frac{1}{T_b} \right) \delta_i Q^{(a)}$$
$$+ \frac{1}{T_a} \sum_\tau A^{(\tau)} d\xi^{(\tau)} - \sum_\lambda \left(\frac{\tilde{\mu}_\lambda^{(a)}}{T_a} - \frac{\tilde{\mu}_\lambda^{(b)}}{T_b} \right) d_e n_\lambda^{(a)}. \tag{11.79}$$

If we assume that there is no heat transfer between the subsystems a and b, the entropy production in the closed system c is

$$d_i S_c = \frac{1}{T_a} \sum_\tau A^{(\tau)} d\xi^{(\tau)} - \sum_\ell \left(\frac{\tilde{\mu}_\ell^{(a)}}{T_a} - \frac{\tilde{\mu}_\ell^{(b)}}{T_b} \right) d_e n_\ell^{(a)} > 0. \tag{11.80}$$

This expression is representative of the processes occurring in biology. In biological systems chemical reactions occurring in cells are linked to the transport across the membranes of the cell. We specifically included the electrochemical potential and the possibility that subsystems a and b are at different electrical potentials because that is the situation encountered in living cells.

11.4 Minimum Entropy Production

For a system in stable equilibrium $\sigma = 0$ throughout the system and there are no
spatial or temporal variations in any of the thermodynamic variables. We may con-
strain a system by imposing boundary conditions that do not vary with the time. In
such a constrained system there is no time variation in the properties at points within
the system. But the system is not in equilibrium. Because of the spatial variation in
the thermodynamic properties the entropy production in the system is not zero. Systems, constrained by time independent boundary conditions, are in nonequilibrium
stationary (time independent) states. These are also referred to as steady states and
are widely encountered in engineering and, to some degree, in biology.

Ilya Prigogine[8] has demonstrated that

*In the linear regime, the total entropy production in a system subject to the flow of
energy and matter,*

$$\frac{\mathrm{d_i}S}{\mathrm{d}t} = \int \sigma \mathrm{d}V, \tag{11.81}$$

attains a minimum value at the nonequilibrium stationary state.

With (11.18), (11.81) becomes

$$\frac{\mathrm{d_i}S}{\mathrm{d}t} = \int \sum_{\mu,\nu} F_\mu L_{\mu\nu} F_\nu \mathrm{d}V. \tag{11.82}$$

The numerical value of the total rate of entropy production in (11.82) is then determined by the spatial form of the thermodynamic forces, F_μ, which are dependent
on the spatial variations of thermodynamic properties within the system. In mathematical terms (11.82) is a functional of the thermodynamic properties. Prigogine's
principle requires that the spatial dependencies of the thermodynamic properties
within a system in a steady state condition are such that the numerical value of
$\mathrm{d_i}S/\mathrm{d}t$ in (11.82) is a minimum.

Variational calculus is the study of extrema of functionals such as that in (11.82)
[cf. [17], pp. 375–385]. The intention of the calculus of variations is to obtain a
set of differential equations for the functions, which in this case are the thermodynamic properties, based on the extremum condition for the functional in question. The set of differential equations resulting from the requirement that such a
functional is an extremum is often already known. For example the Euler-Lagrange
treatment of mechanics, based on a variational principle, produces Newton's laws. A
variational treatment can make the incorporation of constraint conditions more systematic. Variational principles can also provide the basis for numerical solutions.
Basically, however, a variational principle is a philosophical statement that provides
an insight into the structure of nature (cf. [164], pp. 162–197).

---

[8] Ilya Prigogine (1917–2003), winner of the 1977 Nobel Prize in Chemistry for his work in nonequilibrium thermodynamics.

The Prigogine principle above only holds for the linear approximation. No extremum principle of this nature exists for situations in which a nonlinear approximation must be used, which are situations far from equilibrium.

11.5 Summary

In this chapter we have presented the concept of entropy production in irreversible processes. We considered entropy production as a localized phenomenon in heat flow and membrane transport and as a systemic phenomenon in mixing and in chemical reactions. The choice to treat mixing and chemical reactions as systemic phenomena was because of the great difficulty in doing otherwise. Both of these can be local.

It is the rate of entropy production that is of importance. The entropy itself as a thermodynamic property was of secondary importance. The methods of thermodynamics are the only tools we have for understanding and dealing with complex systems, particularly those in biology, where the flow of matter is coupled to electrical and chemical processes. And, when dealing with irreversibility, there is no other path to follow than to search for the sources of entropy production. Establishing those sources in complex systems provides insight into the phenomena themselves and an understanding of their coupling.

There is a new branch of statistical mechanics that is considering some of the fundamental questions related to irreversibility, notably time irreversibility [cf. [66]].

Exercises

11.1. In principle methane gas $CH_4(g)$ can be formed from solid graphite $C(s)$ and hydrogen gas $H_2(g)$ according to the reaction

$$C(s) + 2H_2(g) \rightarrow CH_4(g). \tag{11.83}$$

But this reaction will not occur in practice. Therefore, laboratory data cannot be obtained directly for this reaction. But there are a number of reactions that will occur in the laboratory and can be considered as steps involved in the final formation of methane gas.

We can burn graphite to produce $CO_2(g)$ according to

$$C(s) + O_2(g) \rightarrow CO_2(g) \quad \Delta H_{rxn} = -393.5 \text{ kJ mol}^{-1} \text{ C}.$$

We can burn hydrogen to produce water vapor according to

$$H_2(g) + \frac{1}{2}O_2(g) \rightarrow H_2O(g) \quad \Delta H_{rxn} = -241.8 \text{ kJ mol}^{-1} \text{ H}_2.$$

We can convert vapor vapor to liquid according to

$$H_2O(g) \rightarrow H_2O(l) \quad \Delta H_{fusion} = -44.0\,kJ\,mol^{-1}\ H_2O.$$

We can burn methane to produce gaseous CO_2 and liquid water

$$CH_4(g) + 2O_2(g) \rightarrow CO_2(g) + 2H_2O(l) \quad \Delta H_{rxn} = -890.7\,kJ\,mol^{-1}\ CH_4.$$

What is the reaction enthalpy for the formation of methane gas?

11.2. A thermocouple is a temperature sensor. To produce a thermocouple two dis-similar wires are spot welded together at the ends. This end is the sensor and is at the temperature of the body being studied. Near the other end is at a reference temperature (an ice bath) One of the wires contains a voltmeter so that no current flows and the voltage difference between the two ends is measured by the voltmeter. This is shown in Fig. 11.2. Consider a differential volume of the top wire $dV = A dx$.

Fig. 11.2 Thermocouple

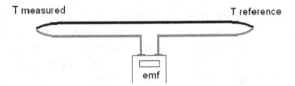

Show that the entropy production rate in this differential volume is $\sigma = (d/dx)(1/T)J_q + J_e(1/T)(d\varphi/dx)$, where J_e is the flow of charge per time area and φ is the electric potential. Write the linear phenomenological equations and integrate it along the top wire to obtain the emf $\Delta\varphi$ read by the voltmeter in terms of the temperature difference between the junctions.

11.3. The linear phenomenological equations for Exercise 11.2 are $J_q = L_{qe}F_e + L_{qq}F_q$ and $J_e = L_{ee}F_e + L_{eq}F_q$. Identify the thermodynamic forces in the result of Exercise 11.2 to write these in differential form. In the thermocouple $J_e = 0$. Use this condition to obtain a differential equation for $\varphi(T)$. Integrate this to obtain value of the measured temperature in terms of the potential difference $\Delta\varphi$. Assuming that the temperature above reference ΔT is much smaller than the reference temperature obtain an approximate result for $\Delta\varphi$ in terms of ΔT.

11.4. The selectivity of a biological ion channel can be investigated using the form of the Goldman-Hodgkin-Katz equation (11.41). Assume that you are studying a particular ion channel, which you have inserted into an artificial planar bilayer. You are able to vary the potential across the bilayer as well as the concentrations in the solutions on each side. You have three ions on each side with differing concentrations. For definiteness, as well as their biological importance, consider that the ions are Na, K, and Cl. The concentrations on each side are designated by the subscripts m and 0. You are able to measure the current through the ion channel and find that a certain value of the potential across the membrane E the current becomes zero. That is $0 = J_{Na} + J_K + J_{Cl}$. Show that E causing the value of the current to go to zero satisfies

$$\exp\left(\frac{\mathfrak{F}E}{RT}\right) = \frac{P_{Na}[Na]_0 + P_K[K]_0 + P_{Cl}[Cl]_m}{P_{Na}[Na]_m + P_K[K]_m + P_{Cl}[Cl]_0}.$$

Chapter 12
Stability

12.1 Introduction

Our treatment of irreversibility in the preceding chapter was based on the realization that irreversible processes produce entropy. At some point an isolated, inanimate system reaches a state of stable equilibrium and no further spontaneous process is possible. We now ask what characterizes that state. The conditions characterizing stable equilibrium under various conditions will provide the basis for our studies of chemical equilibrium and reaction kinetics, phase transitions, and solutions.

The second law deals with the direction of change in a system not yet in equilibrium. If no spontaneous process is possible in the system, then a state of stable equilibrium has been reached. Entropy production has then ceased and the entropy, as a thermodynamic state function for the system, will be a maximum. Once we have cast this basic physical fact into mathematical terms we shall have the basic principle of thermodynamic equilibrium with which we can begin.

Our first step will be to discover the consequences for the structure of the fundamental surface. Because all that we can know about the physical behavior of matter is contained in the properties of the fundamental surface we expect that the characteristics of thermodynamic stability will be reflected in those properties. We shall discover that the basic requirement that the entropy is a maximum under conditions of stable equilibrium will lead to the requirement that all the potentials are minima at stable equilibrium.

In Sect. 5.3.3 we related the curvatures of the potentials to basic laboratory quantities, which we claimed must be positive. The claim was correct, although not based on anything that can be considered scientific. In this chapter we shall show that the positive nature of the specific heats and compressibilities are consequences of the laws of physics expressed in the laws of thermodynamics.

The section of this chapter on Form and Stability (Sect. 12.4) is that which places the positive nature of the specific heats and compressibilities on a firm scientific basis. The argument has been adapted from A.H. Wilson [162], which I have tried to make visual while retaining the basic ideas. But the final steps still involve a Jacobian matrix. It is important that the reader realizes that we now have a base in the laws of physics. The details are not important.

C.S. Helrich, *Modern Thermodynamics with Statistical Mechanics*,
DOI 10.1007/978-3-540-85418-0_12, © Springer-Verlag Berlin Heidelberg 2009

12.2 Entropy

If a function $\Psi(\xi,\eta)$ has a maximum value at the point (ξ_P,η_P) then any deviation in any of the variables ξ and η from the values ξ_P and η_P will produce a value of $\Psi(\xi,\eta)$ which is less than the value $\Psi(\xi_P,\eta_P)$. That is

$$\Psi(\xi_P+\delta\xi,\eta_P+\delta\eta)-\Psi(\xi_P,\eta_P)=\delta\Psi<0. \tag{12.1}$$

Therefore the mathematical requirement of the second law is that variations in the system state about a condition of stable equilibrium must produce a variation in the entropy that is less than zero. For an isolated system composed of subsystems a and b the requirement for stable equilibrium is then that

$$\delta S=\delta S_a+\delta S_b<0, \tag{12.2}$$

where δS is the variation of the total system entropy and δS_a and δS_b are the variations of the subsystem entropies.

Since there is no entropy production for an isolated system in stable equilibrium the variations δS_a and δS_b result from mutual exchange of entropy between the subsystems. That is

$$\delta S_a<-\delta S_b \tag{12.3}$$

for any variations in subsystem entropies about a condition of stable equilibrium in an isolated system. If we assume that the subsystems are open to an exchange of matter the variation in entropy of each subsystem is given by (see (4.83))

$$T_i\delta S_i=\delta U_i+P_i\delta V_i-\sum_\lambda^r \mu_\lambda^{(i)}\delta n_\lambda^{(i)}$$

where i=a,b. The condition in (12.3) is then

$$\delta S_a<-\frac{\delta U_b+P_b\delta V_b-\sum_\lambda \mu_\lambda^{(b)}\delta n_\lambda^{(b)}}{T_b}. \tag{12.4}$$

The original system is isolated and in stable equilibrium. Therefore $T_b=T_a$, $\delta U_b=-\delta U_a$ and $\delta V_b=-\delta V_a$. Since the subsystems are open with respect to one another, $P_b=P_a$, $\mu_\lambda^{(b)}=\mu_\lambda^{(a)}$, and $\delta n_\lambda^{(b)}=-\delta n_\lambda^{(a)}$. The inequality in (12.4) then becomes

$$\delta S_a<\frac{\delta U_a+P_a\delta V_a-\sum_\lambda \mu_\lambda^{(a)}\delta n_\lambda^{(a)}}{T_a}. \tag{12.5}$$

This is the general condition that must be satisfied by an open subsystem of an isolated system in stable equilibrium. Without loss of generality we may consider that the subsystem b surrounds a and determines the boundary conditions imposed on a. For example b may be any part of a laboratory that may be considered isolated from the rest of the world.

The condition in (12.5) is also valid if we isolate the subsystem a from b by closing it and fixing the internal energy and volume. Then $\delta U_a = \delta V_a = \delta n_\lambda^{(a)} = 0$ and (12.5) becomes

$$\delta S]_{U,V,\{n_\lambda\}} < 0. \tag{12.6}$$

We then have the mathematical result that the *entropy* of a system, for which the internal *energy*, *volume* and *composition* are *constant*, is a *maximum for stable equilibrium*. The thermodynamic statement is not simply that entropy is a maximum!

12.3 The Potentials

Our primary interest is not in the entropy but in the potentials. We must then obtain the stability conditions for the potentials from the general stability condition in the inequality (12.5). We can accomplish this for the internal energy $U(S,V,\{n_\lambda\})$ by simply rearranging (12.5), since $U(S,V,\{n_\lambda\})$ and $S(U,V,\{n_\lambda\})$ are the same surface. The result is the inequality

$$\delta U_a > T_b \delta S_a - P_a \delta V_a + \sum_\lambda \mu_\lambda^{(a)} \delta n_\lambda^{(a)}. \tag{12.7}$$

For a system constrained so that the entropy, volume, and composition are constants, that is when $\delta S_a = \delta V_a = \delta n_\lambda^{(a)} = 0$, the inequality (12.7) is

$$\delta U_{\text{system}}]_{S,V,\{n_\lambda\}} > 0. \tag{12.8}$$

We then have the mathematical result that the internal energy of a system, for which the *entropy*, *volume* and *composition* are constant, is a *minimum* when the system is in a state of *stable equilibrium*.

With the definition of the enthalpy,

$$H = U + PV,$$

the inequality (12.7) becomes

$$\delta H_a > T_b \delta S_a + V_a \delta P_a + \sum_\lambda \mu_\lambda^{(a)} \delta n_\lambda^{(a)}. \tag{12.9}$$

For systems constrained so that entropy, pressure and composition are constants, $\delta S_a = \delta P_a = \delta n_\lambda^{(a)} = 0$, the inequality (12.9) is

$$\delta H_{\text{system}}]_{S,P,\{n_\lambda\}} > 0 \tag{12.10}$$

Therefore, in a state of stable equilibrium in a system for which *entropy*, *pressure* and *composition* are constant, the *enthalpy* is a *minimum*.

Similarly, from the definition of the Helmholtz energy,

$$F = U - TS,$$

we conclude that

$$\delta F_{\text{system}}]_{T,V,\{n_\lambda\}} > 0, \tag{12.11}$$

and from the definition of the Gibbs energy,

$$G = H - TS,$$

that

$$\delta G_{\text{system}}]_{T,P,\{n_\lambda\}} > 0 \tag{12.12}$$

Therefore, the Helmholtz and the Gibbs energies are minima at states of stable equilibrium under conditions for which their characteristic variables and the composition are constant.

We then conclude that the second law requires that at states of stable equilibrium the potentials are all minima under conditions for which their respective characteristic variables and the composition are held constant.

12.4 Form and Stability

We obtained the curvatures of the potentials in Sect. 5.3.3. And we were able to relate those curvatures to caloric and mechanical properties giving us the signs of those curvatures. What we do not yet have, however, is a general stability requirement involving also the cross partial derivatives of the potentials. That is we are lacking a full stability requirement for the form of the potentials.

12.4.1 Fundamental Relationships

Stability of real substances is reflected in the curvatures of the fundamental surface $U(S,V,\{n_\lambda\})$. These curvatures result in thermal and mechanical stability requirements for real substances reflected in limits on the values of the specific heats, C_P and C_V and the compressibilities. κ_T and κ_S. We obtained the curvatures for the fundamental surface in Sect. 5.3.3. For the internal energy these are given in (5.38), which we repeat here for continuity.

$$\left(\frac{\partial^2 U}{\partial S^2}\right)_V = \left(\frac{\partial T}{\partial S}\right)_V = \frac{T}{nC_V} > 0 \tag{12.13}$$

and

$$\left(\frac{\partial^2 U}{\partial V^2}\right)_S = -\left(\frac{\partial P}{\partial V}\right)_S = \frac{1}{V\kappa_S} > 0, \tag{12.14}$$

where we have used

$$T = (\partial U/\partial S)_V \tag{12.15}$$

and

$$P = -(\partial U/\partial V)_S \tag{12.16}$$

(see Sects. 4.2, (4.3) and (4.4)). That is the curvature conditions in (12.13) and (12.14) are the general stability requirements

$$\left(\frac{\partial T}{\partial S}\right)_V > 0 \tag{12.17}$$

and

$$-\left(\frac{\partial P}{\partial V}\right)_S > 0. \tag{12.18}$$

We now seek some more general stability requirements based on the basic form of the fundamental surface. Particularly we seek a relationship that will tie the curvatures to the crossed second partial derivative.

The fundamental surface curves upward along the $U-$ axis. This is the general requirement of stability resulting from the second law. If we choose a general state (a point) on the fundamental surface, which we designate as a, and construct a plane tangent to the surface at that point then all points on the fundamental surface will lie above the tangent plane at a. The situation is illustrated in Fig. 12.1. In Fig. 12.1 the surface we have selected is a convenient surface for the illustration and is not the fundamental surface of any particular substance. We have also chosen the point a, at which the tangent plane is located, to be at the base of the surface for ease of illustration only. It is important only to realize that for any such tangent plane, constructed at any point on such a surface with an upward curvature, the surface will always lie above the plane. That is a line parallel to the $U-$ axis from the surface $U(S,V)$ will always strike the tangent plane at a point below the surface.

We shall designate the points on the surface in Fig. 12.1 as $U(S,V)$ and the points on the tangent plane as $U'(S,V;S_a,V_a)$. The points on the plane $U'(S,V;S_a,V_a)$ are specified by the function

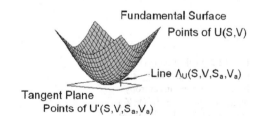

Fig. 12.1 Fundamental surface with points $U(S,V)$ and a tangent plane with points $U'(S,V,S_a,V_a)$

$$U'(S,V;S_a,V_a) = U_a + \left(\frac{\partial U'}{\partial S}\right)_{V\Big]_a}(S-S_a) + \left(\frac{\partial U'}{\partial V}\right)_{S\Big]_a}(V-V_a)$$
$$= U_a + T_a(S-S_a) - P_a(V-V_a). \tag{12.19}$$

We now construct a line from the tangent plane to the fundamental surface. This line, which we call $\Lambda_U(S,V;S_b,V_b)$, is shown in Fig. 12.1. Its length is the distance measured along the $U-$ axis between the point on the fundamental surface $U(S,V)$ and the point on the tangent plane $U'(S,V;S_a,V_a)$. The length is given by

$$\Lambda_U(S,V;S_a,V_a) = U(S,V) - U'(S,V;S_a,V_a)$$
$$= U(S,V) - U_a + T_a(S-S_a) - P_a(V-V_a). \tag{12.20}$$

For real substances this length is positive because the fundamental surface is above the tangent plane. That is $\Lambda_U(S,V;S_a,V_a) > 0$.

We also construct a second plane tangent to the fundamental surface at a point $(b) = (S_b,V_b)$ in the neighborhood of (a). This is not shown in Fig. 12.1. The points in this second plane we designate as $U''(S,V;S_b,V_b)$ where

$$U''(S,V;S_b,V_b) = U_b + T_b(S-S_b) - P_a(V-V_b). \tag{12.21}$$

We define a line between the fundamental surface and this plane just as we did for the tangent plane at the point a.

$$\Lambda_U(S,V;S_b,V_b) = U(S,V) - U'(S,V;S_b,V_b)$$
$$= U(S,V) - U_b + T_b(S-S_b) - P_b(V-V_b). \tag{12.22}$$

And $\Lambda_U(S,V;S_b,V_b) > 0$.

Because it is on the fundamental surface, the point (b) also lies above all points in the plane $U'(S,V;S_a,V_a)$. Likewise the point (a) lies above all points in the plane $U''(S,V;S_b,V_b)$. This situation is illustrated in Fig. 12.2 where we have plotted the projection of the fundamental surface and the planes onto a constant entropy plane. On this plot we have indicated the two neighboring states (a) and (b) and drawn the tangent lines at these points with slopes $-P_a$ and $-P_b$. Points on these tangent lines are $U'(V,V_a)$ and $U''(V,V_b)$ respectively.

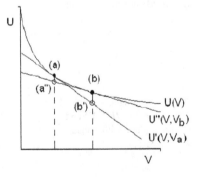

Fig. 12.2 Projection $U(V)$ of the van der Waals energy surface onto a constant entropy plane. Tangent lines $U'(V,V_a)$ at (a) and $U''(V,V_b)$ at (b) are shown. These lie below $U(V)$

The distances $a''a$ and $b'b$ must both be positive. In terms of Λ_U, this is the requirement that

$$\Lambda_U(S_a, V_a; S_b, V_b) = U_a - U_b - T_b(S_a - S_b) + P_b(V_a - V_b) > 0 \qquad (12.23)$$

and

$$\Lambda_U(S_b, V_b; S_a, V_a) = U_b - U_a - T_a(S_b - S_a) + P_a(V_b - V_a) > 0. \qquad (12.24)$$

The inequalities (12.23) and (12.24) are basic algebraic statements that must hold for any surface which lies above all of its tangent planes. This is the general geometrical stability requirement for the fundamental surface. The inequalities (12.23) and (12.24) are then the basic stability requirements with which we shall work.

Adding (12.23) and (12.24) we obtain the inequality

$$\Delta T \Delta S - \Delta P \Delta V > 0 \qquad (12.25)$$

where $\Delta T = (T_b - T_a)$, $\Delta S = (S_b - S_a)$, $\Delta P = (P_b - P_a)$, and $\Delta V = (V_b - V_a)$. Because it is obtainable solely from (12.23) and (12.24), the inequality 12.25 is also a general stability requirement. Since $T = (\partial U / \partial S)_V$ and $P = -(\partial U / \partial V)_S$, (12.25) is a relationship among the changes in the first derivatives (slopes) of the fundamental surface resulting from changes in the coordinates ΔS and ΔV.

If the states (a) and (b) are infinitesimally close we can use a linear Taylor series to write $\Delta T = (\partial T / \partial S)_V]_a \Delta S$, and $\Delta P = (\partial P / \partial V)_S]_a \Delta V$. Then (12.25) becomes

$$\left(\frac{\partial T}{\partial S} \right)_V \bigg]_a \Delta S^2 - \left(\frac{\partial P}{\partial V} \right)_S \bigg]_a \Delta V^2 > 0, \qquad (12.26)$$

where a is any arbitrary state point on $U(S, V)$. The inequality (12.25) is satisfied if the inequalities (12.17) and (12.18), which are the curvature conditions, are satisfied. Therefore the inequality (12.25) is actually our curvature requirement in a general form.

For a mixture the development is the same as that presented here. Corresponding to (12.25) we obtain

$$\Delta T \Delta S - \Delta P \Delta V + \sum_\lambda \Delta \mu_\lambda \Delta n_\lambda > 0 \qquad (12.27)$$

for a mixture, where $\Delta n_\lambda = n_\lambda^{(2)} - n_\lambda^{(1)}$ and $\Delta \mu_\lambda = \left(\mu_\lambda^{(2)} - \mu_\lambda^{(1)} \right)$. For states infinitesimally close to one another we may write $\Delta \mu_\lambda = (\partial \mu_\lambda / \partial n_\lambda)_{S,V,n_\beta \neq n_\lambda}]_1 \Delta n_\lambda$. Corresponding to (12.26) we then have

$$\begin{aligned} (\partial T / \partial S)_{V,\{n_\lambda\}} \big]_1 \Delta S^2 - (\partial P / \partial V)_{S,\{n_\lambda\}} \big]_1 \Delta V^2 \\ + \Sigma_\lambda \, (\partial \mu_\lambda / \partial n_\lambda)_{S,V,n_\beta \neq n_\lambda} \big]_1 \Delta n_\lambda^2 > 0. \end{aligned} \qquad (12.28)$$

The inequality (12.28) is valid if

$$\left(\frac{\partial \mu_\lambda}{\partial n_\lambda}\right)_{S,V,n_\beta \neq n_\lambda}\Bigg]_1 = \left(\frac{\partial^2 U}{\partial n_\lambda^2}\right)_{S,V,n_\beta \neq n_\lambda}\Bigg]_1 > 0 \,\forall \lambda. \qquad (12.29)$$

We recall that in Sect. 5.3.3 we were able to deduce the curvatures of the potentials based on our claim that the specific heats and the compressibilities are positive. The claim was valid, but we presented no scientific evidence for the claim. Here we do present the necessary scientific evidence. We have found that the requirements of stability based on the second law result in the curvature requirements.

We now seek a requirement involving second derivatives based on the inequalities (12.23) and (12.24). To introduce second derivatives we carry out a Taylor expansion of the energies U_a and U_b to second order in ΔS and ΔV. The energy U_b is found from a second order Taylor expansion around a as

$$U_b = U_a + T_a(S_b - S_a) - P_a(V_b - V_a)$$
$$+ \frac{1}{2}\left(\frac{\partial^2 U}{\partial S^2}\right)_V\Bigg]_a \Delta S^2 + \frac{1}{2}\left(\frac{\partial^2 U}{\partial V^2}\right)_S\Bigg]_a \Delta V^2 + \left(\frac{\partial^2 U}{\partial S \partial V}\right)\Bigg]_a \Delta S \Delta V \quad (12.30)$$

and U_a is found from a second order Taylor expansion around b as

$$U_a = U_b + T_b(S_a - S_b) - P_b(V_a - V_b)$$
$$+ \frac{1}{2}\left(\frac{\partial^2 U}{\partial S^2}\right)_V\Bigg]_b \Delta S^2 + \frac{1}{2}\left(\frac{\partial^2 U}{\partial V^2}\right)_S\Bigg]_b \Delta V^2 + \left(\frac{\partial^2 U}{\partial S \partial V}\right)\Bigg]_b \Delta S \Delta V \quad (12.31)$$

Combining the basic stability requirement (12.24) with (12.30) we have

$$+ \frac{1}{2}\left(\frac{\partial^2 U}{\partial S^2}\right)_V\Bigg]_a \Delta S^2 + \frac{1}{2}\left(\frac{\partial^2 U}{\partial V^2}\right)_S\Bigg]_a \Delta V^2 + \left(\frac{\partial^2 U}{\partial S \partial V}\right)\Bigg]_a \Delta S \Delta V > 0. \quad (12.32)$$

And Combining (12.23) with (12.31) we have

$$\frac{1}{2}\left(\frac{\partial^2 U}{\partial S^2}\right)_V\Bigg]_b \Delta S^2 + \frac{1}{2}\left(\frac{\partial^2 U}{\partial V^2}\right)_S\Bigg]_b \Delta V^2 + \left(\frac{\partial^2 U}{\partial S \partial V}\right)\Bigg]_b \Delta S \Delta V > 0. \quad (12.33)$$

Because the two points a and b are arbitrary (12.32) and (12.33) are forms of the general inequality

$$\left(\frac{\partial^2 U}{\partial S^2}\right)_V \Delta S^2 + \left(\frac{\partial^2 U}{\partial V^2}\right)_S \Delta V^2 + 2\left(\frac{\partial^2 U}{\partial S \partial V}\right) \Delta S \Delta V > 0 \qquad (12.34)$$

that must be satisfied for any point on the fundamental surface [101, 152, 162].

Since $(\partial^2 U/\partial S^2)_V > 0$ and $(\partial^2 U/\partial V^2)_S > 0$ based requirement that the surface $U(S,V)$ lies above its tangent planes, the inequality (12.34) is preserved after multiplication by either $(\partial^2 U/\partial S^2)_V$ or $(\partial^2 U/\partial V^2)_S$. The results of multiplying the inequality (12.34) by $(\partial^2 U/\partial S^2)_V$ and completing the square results in

$$\left(\left(\frac{\partial^2 U}{\partial S^2}\right)_V \Delta S + \left(\frac{\partial^2 U}{\partial V \partial S}\right) \Delta V\right)^2 + \left(\left(\frac{\partial^2 U}{\partial S^2}\right)_V \left(\frac{\partial^2 U}{\partial V^2}\right)_S - \left(\frac{\partial^2 U}{\partial V \partial S}\right)^2\right)(\Delta V)^2 > 0$$

$$(12.35)$$

and multiplying the inequality (12.34) by $\left(\partial^2 U/\partial V^2\right) S$ and completing the square results in

$$\left(\left(\frac{\partial^2 U}{\partial V^2}\right)_S \Delta V + \left(\frac{\partial^2 U}{\partial V \partial S}\right) \Delta S\right)^2 + \left(\left(\frac{\partial^2 U}{\partial S^2}\right)_V \left(\frac{\partial^2 U}{\partial V^2}\right)_S - \left(\frac{\partial^2 U}{\partial V \partial S}\right)^2\right)(\Delta S)^2 > 0.$$

$$(12.36)$$

The first lines in (12.35) and (12.36) are both positive, as are the factors $(\Delta V)^2$ and $(\Delta S)^2$. Therefore in either case we must have

$$\left(\frac{\partial^2 U}{\partial S^2}\right)_V \left(\frac{\partial^2 U}{\partial V^2}\right)_S - \left(\frac{\partial^2 U}{\partial S \partial V}\right)^2 > 0 \qquad (12.37)$$

This is the general stability requirement involving the second partial derivatives, including the crossed partial derivative, that we sought.

The general requirements for stability are then the inequality (12.25), which results in inequalities (12.17), (12.18), and the general second derivative inequality (12.37). We now seek to make of these general conditions that can be transported easily to other potentials.

12.4.2 Transformed Relationships

We recognize the left hand side of (12.37) as the determinant of the matrix

$$\begin{bmatrix} \left(\frac{\partial^2 U}{\partial S^2}\right)_V & \left(\frac{\partial^2 U}{\partial S \partial V}\right) \\ \left(\frac{\partial^2 U}{\partial S \partial V}\right) & \left(\frac{\partial^2 U}{\partial V^2}\right)_S \end{bmatrix} = \begin{bmatrix} \left(\frac{\partial T}{\partial S}\right)_V & \left(\frac{\partial T}{\partial V}\right)_S \\ -\left(\frac{\partial P}{\partial S}\right)_V & -\left(\frac{\partial P}{\partial V}\right)_S \end{bmatrix}. \qquad (12.38)$$

Where we have used (12.15) and (12.16). The inequality (12.37) then becomes

$$\left(\frac{\partial^2 U}{\partial S^2}\right)_V \left(\frac{\partial^2 U}{\partial V^2}\right)_S - \left(\frac{\partial^2 U}{\partial S \partial V}\right)^2 = \det \begin{bmatrix} \left(\frac{\partial T}{\partial S}\right)_V & \left(\frac{\partial T}{\partial V}\right)_S \\ -\left(\frac{\partial P}{\partial S}\right)_V & -\left(\frac{\partial P}{\partial V}\right)_S \end{bmatrix}$$

$$= -\det \begin{bmatrix} \left(\frac{\partial T}{\partial S}\right)_V & \left(\frac{\partial T}{\partial V}\right)_S \\ \left(\frac{\partial P}{\partial S}\right)_V & \left(\frac{\partial P}{\partial V}\right)_S \end{bmatrix} > 0. \ (12.39)$$

If we write the Pfaffians of $T(S,V)$ and $P(S,V)$ in matrix form we have

$$
\begin{bmatrix} dT \\ dP \end{bmatrix} = \begin{bmatrix} \left(\dfrac{\partial T}{\partial S}\right)_V & \left(\dfrac{\partial T}{\partial V}\right)_S \\ \left(\dfrac{\partial P}{\partial S}\right)_V & \left(\dfrac{\partial P}{\partial V}\right)_S \end{bmatrix} \begin{bmatrix} dS \\ dV \end{bmatrix}. \tag{12.40}
$$

And the matrix appearing in (12.39) produces the Pfaffians of T and P expressed as functions of (S,V). This is the general form of the transformation from the differentials (dS, dV) to the differentials (dT, dP). The determinant of the square matrix appearing in (12.40) is the *Jacobian matrix* or *Jacobian* of the transformation and is defined as

$$
\frac{\partial(T,P)}{\partial(S,V)} \equiv \det \begin{bmatrix} \left(\dfrac{\partial T}{\partial S}\right)_V & \left(\dfrac{\partial T}{\partial V}\right)_S \\ \left(\dfrac{\partial P}{\partial S}\right)_V & \left(\dfrac{\partial P}{\partial V}\right)_S \end{bmatrix}. \tag{12.41}
$$

Jacobians are discussed in mathematics texts [cf. [17, 34]] and in thermodynamics text [36].

With the abbreviation in (12.41) we can write the inequality (12.39) as

$$
\frac{\partial(T,P)}{\partial(S,V)} < 0 \tag{12.42}
$$

Using (A.45) from the appendix, the Jacobian $\partial(T,P)/\partial(S,V)$ can be written as

$$
\frac{\partial(T,P)}{\partial(S,V)} = \left(\frac{\partial T}{\partial S}\right)_P \left(\frac{\partial P}{\partial V}\right)_S \tag{12.43}
$$

$$
= \left(\frac{\partial P}{\partial V}\right)_T \left(\frac{\partial T}{\partial S}\right)_V \tag{12.44}
$$

With (12.43) and (12.44) the inequality (12.39) produces the two equivalent stability requirements

$$
-\left(\frac{\partial T}{\partial S}\right)_P \left(\frac{\partial P}{\partial V}\right)_S > 0 \tag{12.45}
$$

or

$$
-\left(\frac{\partial P}{\partial V}\right)_T \left(\frac{\partial T}{\partial S}\right)_V > 0. \tag{12.46}
$$

The inequalities (12.45) and (12.46) are equivalent expressions of the general condition involving the second partial derivatives (12.37), as is the inequality (12.42). These provide forms of the stability requirement that can be applied readily to other potentials.

Example 12.4.1. The Gibbs energy. Because the Gibbs energy is a characteristic function of (T,P), we use A.46 to write the inequality (12.42) as

$$\frac{\partial (S,V)}{\partial (T,P)} = \frac{1}{\partial (T,P)/\partial (S,V)} < 0. \tag{12.47}$$

Then the stability requirement (12.42) is

$$\frac{\partial (S,V)}{\partial (T,P)} = \det \begin{bmatrix} \left(\dfrac{\partial S}{\partial T}\right)_P & \left(\dfrac{\partial S}{\partial P}\right)_T \\ \left(\dfrac{\partial V}{\partial T}\right)_P & \left(\dfrac{\partial V}{\partial P}\right)_T \end{bmatrix}$$

$$= \det \begin{bmatrix} -\left(\dfrac{\partial^2 G}{\partial T^2}\right)_P & -\left(\dfrac{\partial^2 G}{\partial P \partial T}\right) \\ \left(\dfrac{\partial^2 G}{\partial T \partial P}\right) & \left(\dfrac{\partial^2 G}{\partial P^2}\right)_T \end{bmatrix} < 0 \tag{12.48}$$

or

$$\left(\frac{\partial^2 G}{\partial T^2}\right)_P \left(\frac{\partial^2 G}{\partial P^2}\right)_T - \left(\frac{\partial^2 G}{\partial P \partial T}\right)^2 > 0. \tag{12.49}$$

Using (5.29) and (5.40) the inequality (12.49) yields

$$\frac{nC_P}{T}\kappa_T - \beta^2 V > 0. \tag{12.50}$$

With the help of (5.32) we see that (12.50) is the requirement that $C_V > 0$. That is the requirement that the specific heat at constant volume is a positive quantity (vanishing only in the limit as $T \to 0$) is a consequence of thermodynamics. It is not a requirement imposed on thermodynamics.

12.5 Summary

In this chapter we have established the requirements of stable thermodynamic equilibrium based on the consequences of the second law. These are reflected in the form of the potentials and in the positive nature of the specific heats and compressibilities.

That all of the potentials are minima at stable equilibrium is the important result. We shall base most of the remaining sections in this book on particularly the extremum properties of the Gibbs energy at stable equilibrium.

We also found that the magnitudes of caloric and mechanical thermodynamic properties are a result of the conditions of stable equilibrium. Based on the reader's understanding of the meaning of physical law this may come as no surprise. This discovery is, nevertheless, fundamental.

Exercises

12.1. We have already established (see (5.41)) that $\left(\partial^2 F/\partial V^2\right)_T$ is equal to $1/V\kappa_T$, and is, therefore > 0. Begin now with the general expression (A.45) Show that $\left(\partial^2 F/\partial V^2\right)_T$ is given by

$$\left(\frac{\partial^2 F}{\partial V^2}\right)_T = \left[\left(\frac{\partial^2 U}{\partial V^2}\right)_S \left(\frac{\partial^2 U}{\partial S^2}\right)_V - \left(\frac{\partial^2 U}{\partial V \partial S}\right)\right] \bigg/ \left(\frac{\partial^2 U}{\partial S^2}\right)_V,$$

which is > 0 by the fundamental stability requirement.

12.2. Use the general expression (A.45) to show that the relationship between the compressibilities κ_T and κ_S is

$$\kappa_T - \kappa_S = \frac{T\beta^2 V}{nC_P},$$

which we previously established.

12.3. Use the general expression (A.45) to show that the relationship between the specific heats C_P and C_V is

$$n\left(C_P - C_V\right) = \frac{T\beta^2 V}{\kappa_T}$$

12.4. Show that the general curvature relationship

$$\left(\frac{\partial^2 U}{\partial S^2}\right)_V \left(\frac{\partial^2 U}{\partial V^2}\right)_S - \left(\frac{\partial^2 U}{\partial V \partial S}\right)^2 = \frac{T}{nC_V}\left(\frac{1}{\kappa_S V} - \frac{T\beta^2}{n\kappa_T^2 C_V}\right) > 0$$

is valid for the van der Waals fluid.

12.5. Two sealed, diathermal vessels are connected by a rubber hose, which may also be considered diathermal. The vessels have volumes V_1 and V_2. One vessel (1) is on the laboratory floor and the other (2) is hanging from a high ceiling. Let the floor be an elevation z_1 and the ceiling z_2. The total internal energy does not change, and there are no leaks from the system. Maximize the entropy to find the ratio of the pressures and the mol numbers in each vessel.

Chapter 13
Equilibrium of Chemical Reactions

13.1 Introduction

In this chapter we shall consider the thermodynamic consequences of chemical change occurring in a closed system. We begin our study with thermodynamic equilibrium in spatially homogeneous systems. We will add to these stability conditions the requirement of the conservation of mass and atomic species in reactions.

Because of the importance of gas reactions in industrial applications as well as for the sake of simplicity we shall consider explicitly only reactions in gaseous systems. We will begin with the ideal gas and then systematically increase the complexity of our subject to the treatment of nonideal gases.

The primary results of our study of chemical equilibrium will be the introduction of the equilibrium constant of a reaction. Adhering to thermodynamical considerations will allow us to develop equations relating the equilibrium constant of a reaction to measurable thermodynamic properties of substances. We will then be able to use our understanding of statistical mechanics from Chaps. 8 and 9 to study the dependence of the equilibrium constant on the details of the molecular interactions.

In the final section of the chapter we consider the entropy production by reactions out of equilibrium, which is related to the chemical affinity.

13.2 Stability of Reactions

For most laboratory applications, which include applications to biological systems, and industrial applications the thermodynamic variables of importance are temperature and pressure. In the preceding chapter we found that the thermodynamic requirement for stability in a spatially homogeneous, multicomponent system at constant temperature and pressure is that the Gibbs energy is a minimum. So our problem is to minimize the Gibbs energy for the system subject to constraints imposed by the requirements of any possible chemical changes.

We consider a general chemical reaction specified by (11.42) , which we repeat here for continuity

C.S. Helrich, *Modern Thermodynamics with Statistical Mechanics*,
DOI 10.1007/978-3-540-85418-0_13, © Springer-Verlag Berlin Heidelberg 2009

$$\alpha_A A + \alpha_B B + \cdots \rightleftharpoons \alpha_D D + \alpha_E E + \cdots . \tag{13.1}$$

The double arrows in (13.1) indicate that the reaction may proceed in either direction. Chemical equilibrium results when the transformation of reactants into products is balanced by the transformation of products into reactants.

If we again introduce the coefficients v_λ that have the magnitude of the corresponding stoichiometric coefficients, but are positive for products and negative for reactants we have the stoichiometric condition in (11.43) for this reaction

$$\sum_\lambda^r v_\lambda M_\lambda = 0. \tag{13.2}$$

in which M_λ is the molar mass of the λth substance and r is the number of separate components taking place in the reaction, which include reactants and products. The requirement of mass conservation for the reaction results in

$$\sum_\lambda^r n_\lambda M_\lambda = M, \tag{13.3}$$

If there is a number N of reactions taking place simultaneously in our system we have a stoichiometric relationship of the form (13.1) and a stoichiometric condition

$$\sum_\lambda^{r^{(\alpha)}} v_\lambda^{(\alpha)} M_\lambda = 0 \quad \alpha = 1, 2, \cdots, N. \tag{13.4}$$

for each reaction. In (13.4) $r^{(\alpha)}$ is the number of components and $v_\lambda^{(\alpha)}$ is the coefficient v_λ for the αth reaction. We may sum (13.4) over the set of simultaneous reactions to obtain

$$\sum_\alpha^N \sum_\lambda^{r^{(\alpha)}} v_\lambda^{(\alpha)} M_\lambda = 0. \tag{13.5}$$

We can, however, only write a single expression for the conservation of mass for the coupled reactions, which will be identical to (13.3), because we cannot separate the number of mols of a particular component into fractions involved in each reaction. Whether or not we can separate the stoichiometric Eq. (13.1) into a set of possible separate reactions that may be occurring simultaneously is, therefore, of no consequence in determining the condition of stability.

Regardless of the number of reactions that may be simultaneously involved, our problem is then to minimize the Gibbs energy $G(T, P, \{n_\lambda\})$ for the multicomponent system subject to the requirement that the mass of the system is conserved, expressed in (13.3), and that the temperature and pressure are constant. To do this we use the method of Lagrange undetermined multipliers. In principle this method will produce the requirements for a weak extremum without determining whether the Gibbs energy is a maximum or a minimum. However, the studies of the preceding chapter on thermodynamic stability have shown that the stability requirement, i.e.

the minimum of the Gibbs Function at an extremum, is guaranteed. Our extremum will then be a minimum.

The subsidiary function, $\eta\,(T,P,\{n_\lambda\})$ (see Sect. A.4, (A.22)), which has a minimum for the same set of variables that minimize $G(T,P,\{n_\lambda\})$, is

$$\eta\,(T,P,\{n_\lambda\}) = G(T,P,\{n_\lambda\}) + \gamma\left(\sum_\lambda^r n_\lambda M_\lambda - M\right),$$

where γ is a Lagrange undetermined multiplier. The extremum condition requires that

$$0 = dG(T,P,\{n_\lambda\}) - \gamma\sum_\lambda^r M_\lambda dn_\lambda, \tag{13.6}$$

and the constraint in (13.3), are simultaneously satisfied. Under conditions of constant temperature and pressure the differential of the Gibbs energy is

$$dG(T,P,\{n_\lambda\}) = \sum_\lambda^r \mu_\lambda dn_\lambda.$$

Then (13.6) becomes

$$\sum_\lambda^r (\mu_\lambda - \gamma M_\lambda)\,dn_\lambda = 0. \tag{13.7}$$

The variations dn_λ in (13.7) are independent so (13.7) requires that

$$\mu_\lambda = \gamma M_\lambda \tag{13.8}$$

for each constituent of the reaction. We may obtain the undetermined multiplier γ algebraically by combining (13.8) with the original mass conservation constraint (13.3). But it is simpler to use the stoichiometric condition (13.2) as the equivalent of mass conservation. Multiplying (13.8) by the stoichiometric coefficient, ν_λ, and using (13.2), we have

$$\sum_\lambda^r \nu_\lambda \mu_\lambda = 0, \tag{13.9}$$

and there is no need to solve for γ.

We recognize (13.9) as the requirement that the chemical affinity, defined in (11.76), must be zero. That is the entropy production rate must vanish. This condition is contained in the stability requirement we have used here.

To progress any farther we must have an expression for the chemical potential. In the next section we shall develop equations for the chemical potential that can be applied to the conditions that are encountered in chemical reactions in gases.

13.3 Chemical Potential

This section will be divided into three subsections. In the first subsection we shall treat the ideal gas. In the remaining subsections we shall introduce temperature dependent contributions to the specific heat modifications to the thermal equation of state to account for collisions among molecules. The final result will be a chemical potential which can be used under conditions present in gas reactions.

13.3.1 Ideal Gases

We obtained the Gibbs energy for an ideal gas mixture in Sect. 4.7.2. The result is (4.77), which we repeat here for continuity.

$$G(T,P,\{n_\lambda\}) = \sum_\lambda n_\lambda \left[C_{P,\lambda} \left(T - T \ln \frac{T}{T_{\text{ref}}} \right) + RT \ln P_\lambda \right.$$
$$\left. - T s_{\text{ref},\lambda} - RT \ln P_{\text{ref}} \right]. \qquad (13.10)$$

We also obtained the chemical potential in (4.80), which we repeat here as well

$$\mu_\lambda (T,P,\{n_\lambda\}) = C_{P,\lambda} \left[T - T \ln \frac{T}{T_{\text{ref}}} \right] + RT \ln P_\lambda$$
$$- T s_{\text{ref},\lambda} - RT \ln P_{\text{ref}}. \qquad (13.11)$$

It will be convenient in our discussion to define the function $\mu_\lambda^* (T)]_{\text{ideal}}$ as

$$\mu_\lambda^* (T)]_{\text{ideal}} = C_{P,\lambda} \left(T - T \ln \frac{T}{T_{\text{ref}}} \right) - T s_{\text{ref},\lambda} - RT \ln P_{\text{ref}}. \qquad (13.12)$$

This function $\mu_\lambda^* (T)]_{\text{ideal}}$ depends only on the temperature and is independent of the pressure in the system. It is also a function only of the physical properties of the gas λ and the reference state $(T_{\text{ref}}, P_{\text{ref}})$ provided the entropy in the reference state is known. So we can obtain values for $\mu_\lambda^* (T)]_{\text{ideal}}$ if the physical properties of the gas are known. It is convenient to separate this term from the pressure dependence $RT \ln P_\lambda$ appearing in the chemical potential. The reason will become clear with the form of the law of mass action below.

With (13.12) the chemical potential for the ideal gas (13.11) becomes

$$\mu_\lambda (T,P,\{n_\lambda\}) = \mu_\lambda^* (T)]_{\text{ideal}} + RT \ln P_\lambda. \qquad (13.13)$$

We may obtain the chemical potential of a pure ideal gas of species λ by setting $n = n_\lambda$ and $P_\lambda = P$ in (13.13). This has no affect on $\mu_\lambda^* (T)]_{\text{ideal}}$. The result is

$$\mu_{0,\lambda}(T,P) = \mu_\lambda^*(T)\big]_{ideal} + RT \ln P. \tag{13.14}$$

The Gibbs energy for a mixture of ideal gases can then be written in either of the forms

$$G(T,P,\{n_\lambda\}) = \sum_\lambda n_\lambda \mu_\lambda(T,P,\{n_\lambda\}) \tag{13.15}$$

$$= \sum_\lambda n_\lambda \mu_{0,\lambda}(T,P_\lambda). \tag{13.16}$$

The simplicity of (13.13), (13.14), (13.15), and (13.16) is a result of the fact that the mixing of ideal gases only produces a mixing entropy (4.73).

Using (13.13) in (13.9) we see that the requirement for chemical equilibrium in a mixture of ideal gases is

$$-\frac{1}{RT}\sum_\lambda^r \nu_\lambda \, \mu_\lambda^*(T)\big]_{ideal} = \ln \frac{\displaystyle\prod_{products} P_\lambda^{\nu_\lambda}}{\displaystyle\prod_{reactants} P_\lambda^{\nu_\lambda}}. \tag{13.17}$$

We now define the equilibrium constant $K_P(T)$ for a mixture of ideal gases as

$$\ln K_P(T) = -\frac{1}{RT}\sum_\lambda^r \nu_\lambda \, \mu_\lambda^*(T)\big]_{ideal}. \tag{13.18}$$

Using (13.18), (13.17) becomes

$$K_P(T) = \frac{\displaystyle\prod_{products} P_\lambda^{\nu_\lambda}}{\displaystyle\prod_{reactants} P_\lambda^{\nu_\lambda}}. \tag{13.19}$$

Equation (13.19) is known as the *law of mass action*. Because chemical reactions result from intermolecular collisions, a mixture of ideal gases is not, in principle, capable of producing chemical reactions. Nevertheless (13.19) is attractive because of its simple form and because it has an intuitive ring to it.

For example if we add a small amount of one of the reactants, keeping the temperature and pressure constant, (13.19) requires that the partial pressures of one or more of the products must increase. If there were collisions among the molecules, some of which resulted in chemical transformations, an increase in the number of molecules of one reactant should increase the number of product molecules. This would increase the partial pressures of those product molecules.

The simple form of (13.19) is a result of the fact that the chemical potential in (13.13) contains the term $RT \ln P_\lambda$. Isolation of the term $RT \ln P_\lambda$ is the motivation for the separation of the function $\mu_\lambda^*(T)\big]_{ideal}$ in (13.12). Because the function

$\mu_\lambda^*(T)]_{\text{ideal}}$ is dependent only on the properties of the gas λ we can consider the equilibrium constant $K_P(T)$ to be a known quantity.

In the next sections we shall attempt to preserve the simple form of (13.19) as we modify the ideal gas to incorporate phenomena that we expect to be of importance in real applications.

13.3.2 Nonideal Caloric Effects

Even if we neglect collisions a significant increase in temperature means that the specific heats of the gases become temperature dependent (see Fig. 9.11) because transitions among vibrational energy states become possible. In this section we shall extend our definition of the ideal gas to include monomolecular variations in the specific heat resulting from vibrations. As we found in Sect. 9.13.5, (9.123) this extension alone will have no affect on the thermal equation of state of the gas. In that sense the gas may still be considered ideal.

The graph in Fig. 9.11 is of $C_V(T)$. But the relationship between the specific heats

$$C_P - C_V = R$$

is a result of the ideal gas thermal equation of state alone. Therefore $C_P(T)$ has the same form as $C_V(T)$ for a collisionless gas which obeys the thermal equation of state of the ideal gas, i.e. a gas in which collisions are neglected. We then write

$$C_{P,\lambda}(T) = C_{P,\lambda}^0 + C_{P,\lambda}^1(T),$$

where C_P^0 is the constant (low temperature) form of the specific heat of the gas and $C_P^1(T)$ is the temperature dependent part. In general $C_{P,\lambda}^1(T)$ vanishes below a certain temperature, which is gas specific. We call this temperature $T_{\text{ref v},\lambda}$ and use this for the reference temperature for the entropy of the gas, T_{ref}. Then

$$\lim_{T < T_{\text{ref v},\lambda}} \frac{C_{P,\lambda}^1(T)}{T} = 0$$

and we may write

$$\int_{T_{\text{ref v},\lambda}}^T C_{P,\lambda}^1(T') \frac{dT'}{T'} = \int_0^T C_{P,\lambda}^1(T') \frac{dT'}{T'}.$$

If we include monomolecular variations in specific heats the chemical potential for the component λ in a mixture of ideal gases, (4.77) contains two additional terms which are integrals involving C_P^1.

$$G(T,P,\{n_\lambda\}) = \sum n_\lambda \left[C_{P,\lambda}^0 \left(T - T \ln \frac{T}{T_{\text{ref},\lambda}} \right) + RT \ln P_\lambda \right.$$
$$+ \int_0^T C_{P,\lambda}^1 (T')\, dT' - T \int_0^T C_{P,\lambda}^1 (T')\, \frac{dT'}{T'}$$
$$\left. - T s_{\text{ref},\lambda} - RT \ln P_{\text{ref}} \right]. \tag{13.20}$$

And the chemical potential (13.11) is

$$\mu_\lambda (T,P,\{n_\lambda\}) = C_{P,\lambda}^0 \left(T - T \ln \frac{T}{T_{\text{ref},\lambda}} \right) + RT \ln P_\lambda$$
$$+ \int_0^T C_{P,\lambda}^1 (T')\, dT' - T \int_0^T C_{P,\lambda}^1 (T')\, \frac{dT'}{T'}$$
$$- T s_{\text{ref},\lambda} - RT \ln P_{\text{ref}}. \tag{13.21}$$

We shall now define the function $\mu_\lambda^* (T)$ to take the place of the function $\mu_\lambda^* (T) \big]_{\text{ideal}}$ as the temperature dependent contribution to (13.21) from a nonideal gas for which the specific heat includes monomolecular terms.

$$\mu_\lambda^* (T) = \left[C_{P,\lambda}^0 \left(T - T \ln \frac{T}{T_{\text{ref},\lambda}} \right) \right.$$
$$+ \int_0^T C_{P,\lambda}^1 (T')\, dT' - T \int_0^T C_{P,\lambda}^1 (T')\, \frac{dT'}{T'}$$
$$\left. - T s_{\text{ref},\lambda} \right] - RT \ln P_{\text{ref}}. \tag{13.22}$$

With (13.22) the chemical potential of a pure ideal gas λ (13.14) becomes

$$\mu_{0,\lambda} (T,P) = \mu_\lambda^* (T) + RT \ln P. \tag{13.23}$$

That is the relationship between the chemical potential for the ideal gas in the mixture and in the pure state is unaffected by the addition of monomolecular caloric effects. Equations (13.13), (13.14), (13.15), and (13.16) are also still valid because the mixing entropy has not changed by the addition of monomolecular caloric effects.

13.3.3 Nonideal Collisional Effects

If we introduce collisional effects the thermal equation of state will be altered. The gas becomes nonideal.

In principle we can obtain the Gibbs energy $G(T,P,\{n_\lambda\})$ for a system of non-ideal gases of composition $\{n_\lambda\}$ by integrating the Pfaffian $dG(T,P,\{n_\lambda\})$ along a contour from an arbitrary reference state $(T_{\text{ref}}, P_{\text{ref}}, \{n_\lambda\})$ to the state $(T,P,\{n_\lambda\})$. An easier and more direct approach is to integrate only $(\partial G / \partial P)_{T,\{n_\lambda\}} = V(T, P, \{n_\lambda\})$. The result includes an arbitrary additive function of the temperature and the composition, which we designate here as $G_0 (T, \{n_\lambda\})$.

$$G(T,P,\{n_\lambda\}) = G_0(T,\{n_\lambda\}) + \int_{P_{\text{ref}}}^{P} V(T,P',\{n_\lambda\})\, dP'. \tag{13.24}$$

We shall then seek the form of the arbitrary function of temperature $G_0(T,\{n_\lambda\})$ such that $G(T,P,\{n_\lambda\})$ is identical to (13.10) in the limit for which the gas under consideration may be approximated as ideal.

A gas becomes ideal and $V(T,P,\{n_\lambda\}) = \sum_\lambda n_\lambda RT/P$ in the limit as the pressure approaches zero. But in this limit the integral in (13.24) does not exist because the integrand becomes infinite. We can avoid this difficulty if we add and subtract the integral of the ideal gas volume in (13.24) to obtain

$$G(T,P,\{n_\lambda\}) = G_0(T,\{n_\lambda\}) + nRT \ln \frac{P}{P_{\text{ref}}}$$
$$+ \int_{P_{\text{ref}}}^{P} \left[V(T,P',\{n_\lambda\}) - \frac{nRT}{P'} \right] dP'. \tag{13.25}$$

We now choose a reference pressure, P_{ref}, which is greater than zero, but sufficiently low for satisfactory use of the ideal gas approximation. This is a practical choice and depends on the gases involved and on the application. Specifically we shall assume that the value of P_{ref} to be sufficiently low that for the application considered we have

$$\int_{0}^{P_{\text{ref}}} \left[V(T,P',\{n_\lambda\}) - \frac{nRT}{P'} \right] dP' << \int_{P_{\text{ref}}}^{P} \left[V(T,P',\{n_\lambda\}) - \frac{nRT}{P'} \right] dP'.$$

We may then extend the lower limit of the integral in (13.25) to 0 so that we can write (13.25) as

$$G(T,P,\{n_\lambda\}) = G_0(T,\{n_\lambda\}) + nRT \ln \frac{P}{P_{\text{ref}}}$$
$$+ \int_{0}^{P} \left[V(T,P',\{n_\lambda\}) - \frac{nRT}{P'} \right] dP'. \tag{13.26}$$

A comparison of (13.26) with (13.25) in the ideal gas limit, which is now $P \to P_{\text{ref}}$, reveals that

$$G_0(T,\{n_\lambda\}) = \sum n_\lambda \left[C_{P,\lambda} \left(T - T \ln \frac{T}{T_{\text{ref}}} \right) \right.$$
$$+ \int_{0}^{T} C_{P,\lambda}^1(T')\, dT' - T \int_{0}^{T} C_{P,\lambda}^1(T') \frac{dT'}{T'}$$
$$\left. + RT \ln \chi_\lambda - T s_{\text{ref},\lambda} \right]. \tag{13.27}$$

satisfies our requirements for the arbitrary function of temperature we are seeking. With (13.22) and (13.27) the Gibbs energy for nonideal gases (13.26) is then

$$G(T,P,\{n_\lambda\}) = \sum n_\lambda \mu_\lambda^* (T) + RT \sum n_\lambda \ln P_\lambda$$
$$+ \int_0^P \left[V(T,P',\{n_\lambda\}) - \frac{nRT}{P'} \right] dP' \tag{13.28}$$

From (13.28) we can obtain the chemical potential for the component λ in a mixture of nonideal gases as

$$\mu_\lambda (T,P,\{n_\lambda\}) = \mu_\lambda^* (T) + RT \ln P_\lambda$$
$$+ \int_0^P \left[v_\lambda (T,P',\{n_\lambda\}) - \frac{RT}{P'} \right] dP'. \tag{13.29}$$

The chemical potential for the pure nonideal gas is then

$$\mu_\lambda (T,P) = \mu_\lambda^* (T) + RT \ln P + \int_0^P \left[v_\lambda (T,P') - \frac{RT}{P'} \right] dP'. \tag{13.30}$$

We note that, as in the case of ideal gases, the difference between the chemical potential of the pure nonideal gas λ and the chemical potential of the nonideal gas λ in a mixture is that $RT \ln P$ appears in the chemical potential of the pure gas and $RT \ln P_\lambda$ appears in that for the mixture. This difference again results from the mixing entropy.

In the case of the nonideal gas the volume $v_\lambda (T,P,\{n_\lambda\})$ appearing in the integrand in (13.28) is a result of the interactions among the molecules of all components. In general this is not identical to the volume $v_\lambda (T,P,n_\lambda)$ appearing in (13.30), which results from the interactions among molecules of the same gas, λ. The difference in the chemical potentials for pure nonideal gases and mixtures of nonideal gases is, therefore, not, simply an ideal gas mixing term. In this we see something of the complexity introduced through the intermolecular interactions.

13.4 Fugacity

The presence of the integral in (13.30) prevents the simple formulation of the law of mass action that we obtained for the ideal gas. The obvious mathematical solution to the difficulty is to define a term f_λ as

$$f_\lambda (T,P,\{n_\lambda\}) = P_\lambda \varphi_\lambda (T,P,\{n_\lambda\}). \tag{13.31}$$

with

$$\varphi_\lambda (T,P,\{n_\lambda\}) = \exp\left\{ \frac{1}{RT} \int_0^P \left[v_\lambda (T,P',\{n_\lambda\}) - \frac{RT}{P'} \right] dP' \right\}. \tag{13.32}$$

Then

$$\ln f_\lambda\left(T,P,\{n_\lambda\}\right) = \ln P_\lambda + \frac{1}{RT}\int_0^P\left[v_\lambda\left(T,P',\{n_\lambda\}\right) - \frac{RT}{P'}\right]dP' \qquad (13.33)$$

and the chemical potential for a mixture of nonideal gases (13.29) is

$$\mu_\lambda\left(T,P,\{n_\lambda\}\right) = \mu_\lambda^*\left(T\right) + RT\ln f_\lambda\left(T,P,\{n_\lambda\}\right). \qquad (13.34)$$

The form of the chemical potential is then the same as that we obtained for the ideal gas with the function f_λ taking the place of the ideal gas partial pressure. The thermodynamic function $f_\lambda\left(T,P,\{n_\lambda\}\right)$ is the fugacity for the component λ in the mixture and the function $\varphi_\lambda\left(T,P,\{n_\lambda\}\right)$ is called the fugacity coefficient [123]. In the ideal gas limit the fugacity is equal to the partial pressure, i.e.

$$\lim_{P\to 0} f_\lambda\left(T,P,\{n_\lambda\}\right) = P_\lambda. \qquad (13.35)$$

The fugacity was not originally introduced only to simplify chemical equilibrium studies. G.N. Lewis, who introduced the fugacity in 1901, was interested in general changes in systems, which include changes among solid, liquid and gaseous forms of a substance, as well as chemical and electrochemical reactions [104, 105]. Lewis introduced the concept of escaping tendency to describe these transformations taking place within the system. For example the transformation of ice into water is a result of the escaping tendency of the ice at a specific temperature and pressure. Lewis identified the negative change of the Gibbs energy as a measure of the escaping tendency. The melting of ice at a specific temperature and pressure occurs if the Gibbs energy of the water is less than that of the ice. Escaping tendency is still dicussed in some introductory texts [cf. [4], p. 235].

If we consider an isothermal change in the pressure from $P_A \to P_B$ in a mixture of ideal gases the change in the Gibbs energy is found by using (13.13) in (13.15). The result is

$$G\left(T,P_B,\{n_\lambda\}\right) - G\left(T,P_A,\{n_\lambda\}\right) = RT\ln\frac{\prod_\lambda P_{B,\lambda}^{n_\lambda}}{\prod_\lambda P_{A,\lambda}^{n_\lambda}}. \qquad (13.36)$$

The change in system Gibbs energy is negative if $\prod_\lambda P_{A,\lambda}^{n_\lambda} > \prod_\lambda P_{B,\lambda}^{n_\lambda}$ and the gas can be thought of as escaping to the state of lower pressure. For this state transition in a mixture of ideal gases the escaping tendency is determined by the same ratio of products of partial pressures as the requirement for chemical equilibrium in ideal gases in (13.19). The same difficulty then occurs in obtaining an expression for the escaping tendency for nonideal gas mixtures as that encountered in obtaining an expression for the chemical equilibrium condition in nonideal gases.

Lewis' solution to the difficulty was to introduce the quantity f_λ defined in (13.31). We find the change in the Gibbs energy resulting from an isothermal change from state A to state B in a mixture of nonideal gases by using (13.34) in (13.15).

The result is

$$G\left(T,P_B,\{n_\lambda\}\right) - G\left(T,P_A,\{n_\lambda\}\right) = RT\ln\frac{\prod_\lambda f_{B,\lambda}^{n_\lambda}}{\prod_\lambda f_{A,\lambda}^{n_\lambda}}. \tag{13.37}$$

The escaping tendency in mixtures of nonideal gases is then determined by the ratio of products of fugacities. There is an escaping tendency from state A to state B if $\prod_\lambda f_{A,\lambda}^{n_\lambda} > \prod_\lambda f_{B,\lambda}^{n_\lambda}$. The word fugacity comes from the Latin verb *fugere* meaning to flee ([123], p. 141).

Our form of the chemical potential in (13.34) now includes both monomolecular caloric and collisional effects. With this form of the chemical potential we can conduct a more realistic development of the law of mass action.

13.5   Activities and Mass Action

13.5.1   In Nonideal Gases

We can obtain a form of the law of mass action directly from the equilibrium condition (13.9) and the chemical potential (13.34). The resulting equilibrium constant will then be a function of $\sum_\lambda v_\lambda \mu_\lambda^*(T)$ as we found in the case of the ideal gas. The function $\mu_\lambda^*(T)$ given in (13.22) is still calculable from the properties of the gas if we know the temperature dependence of the specific heat and the entropy at the reference condition. We can, however, reformulate the problem in a way that will result in an equilibrium constant which is a function of the difference in the molar Gibbs energies of the pure products and reactants. We may then obtain the equilibrium constant for a reaction in terms of tabulated thermodynamic properties of pure substances.

We begin by obtaining the chemical potential for a pure gas λ from (13.34). To do this we drop the mixing terms in the fugacity (13.33). That is we set the partial pressure appearing in (13.33) equal to the system pressure. And we replace the specific volume of the gas λ in the mixture $v_\lambda\left(T,P',\{n_\lambda\}\right)$, appearing in the integral in (13.33), with the specific volume of the pure gas λ. The resulting value for the chemical potential of the pure gas λ in the state $\left(T,P^0\right)$ is

$$\mu_\lambda^0\left(T,P^0\right) = \mu_\lambda^*(T) + RT\ln f_\lambda^0\left(T,P^0\right), \tag{13.38}$$

in which $f_\lambda^0\left(T,P^0\right)$ is the fugacity of the pure gas λ in the state $\left(T,P^0\right)$. Since $\mu_\lambda^*(T)$ depends only on properties of the component λ and is unaffected by mixing, the form of this function is unchanged in (13.34) and (13.38). Therefore, subtracting (13.38) from (13.34) we have

$$\mu_\lambda\left(T,P,\{n_\lambda\}\right) - \mu_\lambda^0\left(T,P^0\right) = RT\ln\frac{f_\lambda\left(T,P,\{n_\lambda\}\right)}{f_\lambda^0\left(T,P^0\right)}. \tag{13.39}$$

Using (13.39) the condition for chemical equilibrium in (13.9) is

$$\exp\left[-\frac{1}{RT}\sum_\lambda \nu_\lambda\mu_\lambda^0\left(T,P^0\right)\right] = \frac{\displaystyle\prod_{\text{products}}\left(\frac{f_\lambda}{f_\lambda^0}\right)^{\nu_\lambda}}{\displaystyle\prod_{\text{reactants}}\left(\frac{f_\lambda}{f_\lambda^0}\right)^{\nu_\lambda}}. \tag{13.40}$$

The quantities $f_\lambda\left(T,P,\{n_\lambda\}\right)/f_\lambda^0\left(T,P^0\right)$ appearing in (13.40) are the activities defined by

$$a_\lambda\left(T,P,\{n_\lambda\};f_\lambda^0\right) = \frac{f_\lambda\left(T,P,\{n_\lambda\}\right)}{f_\lambda^0\left(T,P^0\right)}. \tag{13.41}$$

Written in terms of the activities the condition of chemical equilibrium in (13.40) becomes

$$K_{\text{a}}\left(T\right) = \frac{\displaystyle\prod_{\text{products}}\left(a_\lambda\right)^{\nu_\lambda}}{\displaystyle\prod_{\text{reactants}}\left(a_\lambda\right)^{\nu_\lambda}} \tag{13.42}$$

where

$$K_{\text{a}}\left(T\right) = \exp\left[-\frac{1}{RT}\sum_\lambda \nu_\lambda\mu_\lambda^0\left(T,P^0\right)\right] \tag{13.43}$$

is the equilibrium constant for the reaction. The summation appearing in this form of the equilibrium constant is the reaction (molar) Gibbs energy

$$\Delta g_{\text{rxn}}^0\left(T,P^0\right) = \sum_\lambda \nu_\lambda\mu_\lambda^0\left(T,P^0\right)$$

$$= \sum_{\text{products}}\nu_\lambda\mu_\lambda^0\left(T,P^0\right) - \sum_{\text{reactants}}\nu_\lambda\mu_\lambda^0\left(T,P^0\right). \tag{13.44}$$

This is the change in the chemical potential or molar Gibbs energy in the reaction at temperature T and pressure P^0. With Eqs. (13.44), (13.43) becomes

$$K_{\text{a}}\left(T\right) = \exp\left[-\frac{\Delta g_{\text{rxn}}^0\left(T,P^0\right)}{RT}\right]. \tag{13.45}$$

The function $\Delta g_{\text{rxn}}^0\left(T,P^0\right)$ is dependent on the properties of the pure substances involved in the reaction and independent of the mixing properties of the reaction.

We have then accomplished what we set out to do. We have a formulation of the equilibrium condition in terms of the properties of the pure components. This requires a reformulation of the equilibrium condition in terms of the activities rather than basing calculations on fugacities alone. We shall consider that (13.42) is the general form of the law of mass action for chemical applications.

We may obtain approximations to the general form of the law of mass action in the ideal gas limit. The resulting expressions are useful for approximations.

13.5.2 In Ideal Gases

In the ideal gas limit the fugacity $f_\lambda^0\left(T,P^0\right)$ is equal to the reference pressure P^0 and the activity in this limit is

$$\lim_{P,P^0 \to 0} a_\lambda \left(T,P,\{n_\lambda\};f_\lambda^0\left(T,P^0\right)\right) = \frac{P_\lambda}{P^0}.$$

Taking the ideal gas limit of (13.42) we have then

$$K_a\left(T\right)\left(P^0\right)^{\Delta v} = \frac{\displaystyle\prod_{\text{products}} (P_\lambda)^{v_\lambda}}{\displaystyle\prod_{\text{reactants}} (P_\lambda)^{v_\lambda}} \tag{13.46}$$

The general law of mass action in (13.42) then becomes (13.19) with K_P given by

$$K_P\left(T\right) = \left(P^0\right)^{\Delta v} K_a\left(T\right), \tag{13.47}$$

in which

$$\Delta v = \sum_{\text{products}} v_\lambda - \sum_{\text{reactants}} v_\lambda. \tag{13.48}$$

We can then retrieve the ideal form of the law of mass action from the general form. This exercise has also given us an expression for the constant K_P in terms of the reaction Gibbs energy through (13.47) and (13.43).

Using the identity

$$f_\lambda = \frac{\chi_\lambda P}{P_\lambda} f_\lambda, \tag{13.49}$$

we can further show that other common expressions for the law of mass action are obtained. For example, if we choose to work with mol fractions, the result is

$$K_\chi\left(P,T\right) = \frac{\displaystyle\prod_{\text{products}} \chi_\lambda^{v_\lambda}}{\displaystyle\prod_{\text{reactants}} \chi_\lambda^{v_\lambda}}, \tag{13.50}$$

with

$$K_\chi\left(T\right) = \left(\frac{P}{P^0}\right)^{-\Delta v} K_a\left(T\right). \tag{13.51}$$

And using the definition of concentration

$$C_\lambda = \frac{n_\lambda}{V} = \frac{\chi_\lambda}{v_{sys}},$$

(13.52)

the law of mass action takes the form

$$K_C(P,T) = \frac{\displaystyle\prod_{products} C_\lambda^{v_\lambda}}{\displaystyle\prod_{reactants} C_\lambda^{v_\lambda}},$$

(13.53)

with

$$K_C(P,T) = v_{sys}^{-\Delta v} \left(\frac{P}{P^0}\right)^{-\Delta v} K_a(T).$$

(13.54)

Equations (13.19), (13.50), and (13.53) are applicable in the limit in which the gases may be approximated by ideal gases. As we have pointed out, this approximation logically excludes the collisions necessary for chemical reactions to occur. Nevertheless, this approximation may prove expedient in practical applications when approximate solutions are adequate or as an aid to our intuition. In any application, however, the thermodynamic formulation of chemical equilibrium, in which the equilibrium constant is given in terms of the reaction Gibbs energy, is based on the chemical activities of the components. The chemical activities do include collisions.

13.6 Temperature Dependence

From (13.43) we see that the temperature dependence of K_a is determined by the temperature dependence of the chemical potentials $\mu_\lambda^0(T,P^0)$. We can write chemical potential of the pure substance as

$$\mu_\lambda^0\left(T,P^0\right) = \mu_\lambda^0\left(T^0,P^0\right)$$
$$+ \int_{T^0}^{T} C_{P,\lambda}(T')\, dT' - T \int_{T^0}^{T} C_{P,\lambda}(T') \frac{dT'}{T'} - \left(T-T^0\right) s_\lambda^0\left(T^0,P^0\right)$$
$$+ \int_0^{P^0} \left[v_\lambda(T,P') - v_\lambda\left(T^0,P'\right) - \frac{R(T-T^0)}{P'} \right] dP'$$

(13.55)

in which $C_{P,\lambda}(T)$ is the complete specific heat and $\mu_\lambda^0(T^0,P^0)$ is the chemical potential for the pure substance λ at the reference state (T^0,P^0). If we assume that the ideal gas approximation may be used for the volumes $v_\lambda(T,P)$ and $v_\lambda(T^0,P)$ the integrand in the integral appearing in the last line of (13.55) vanishes. We have included only monomolecular effects in the specific heat $C_P(T)$. The form of the chemical potential we use in this calculation is then independent of molecular collisions and the form of $K_a(T)$ will be as well.

From (13.55) the dependence of the equilibrium constant K_a on temperature is given by

$$K_a(T) = \exp\left[-\frac{1}{RT}\Delta g^0_{rxn}\left(T^0, P^0\right)\right.$$
$$-\frac{1}{RT}\int_{T^0}^T \Delta C_{P,rxn}\left(T'\right)dT' + \frac{1}{R}\int_{T^0}^T \Delta_{rxn} C_P\left(T'\right)\frac{dT'}{T'}$$
$$\left.+\frac{\left(T-T^0\right)}{RT}\Delta s^0_{rxn}\left(T^0, P^0\right)\right]. \tag{13.56}$$

In (13.56) we have written

$$\Delta C_{P,rxn}(T) = \sum_\lambda \nu_\lambda C_{P,\lambda}(T), \tag{13.57}$$

and

$$\Delta s^0_{rxn}\left(T^0, P^0\right) = \sum_\lambda \nu_\lambda s^0_\lambda\left(T^0, P^0\right) \tag{13.58}$$

and

$$\Delta g^0_{rxn}\left(T^0, P^0\right) = \sum_\lambda \nu_\lambda \mu^0_\lambda\left(T^0, P^0\right) \tag{13.59}$$

are the standard reaction entropy the standard reaction Gibbs energy. The standard reaction Gibbs energy is the change in Gibbs energy per mol of the pure substances involved in the reaction evaluated at the standard state conditions $\left(T^0, P^0\right)$. The standard state is defined by the temperature $T^0 = 298.15\,\mathrm{K}$ $(25\,^\circ\mathrm{C})$ and the pressure $P^0 = 100\,\mathrm{kPa}$ $(1\,\mathrm{bar})$. The pressure P^0 is very close to one atmosphere, which was once the designation of standard pressure. Except for very precise work the small difference in thermodynamic properties at 1 bar and 1 atm may be neglected.

The standard molar Gibbs energies $g^0_\lambda\left(T^0, P^0\right) = \mu^0_\lambda\left(T^0, P^0\right)$ for elements in their standard state are defined to be zero. Specifically $g^0_{element}$ $(T^0 = 298.15$ K, $P^0 = 100\,\mathrm{kPa}) \equiv 0$ for the element in the naturally occurring phase characteristic of the element. The naturally occurring phase may be solid (commonly occurring crystalline form), liquid, or gas. The standard Gibbs energy of a compound at $\left(T^0, P^0\right)$ is given by the value of $\Delta g^0_{rxn}\left(T^0, P^0\right)$ for the reaction in which the compound is formed from the constituent elements in their standard states at $\left(T^0, P^0\right)$.

Values for $K_a(T)$ can then be found from tabulated thermodynamic data (or a NIST database). We can then obtain $K_a(T)$ from a calculation based on (13.56). But an alternative approach based on the derivative of $\ln K_a(T)$ is often more practical.

From (13.56) we have

$$RT^2\frac{d}{dT}\ln K_a(T) = \left[\Delta g^0_{rxn}\left(T^0, P^0\right) + T^0 \Delta s^0_{rxn}\left(T^0, P^0\right)\right] + \int_{T^0}^T \Delta C_{P,rxn}\left(T'\right)dT' \tag{13.60}$$

With the identification of the standard reaction enthalpy as

$$\Delta h_{rxn}^0\left(T^0,P^0\right) = \left[\Delta g_{rxn}^0\left(T^0,P^0\right) + T^0\Delta s_{rxn}^0\left(T^0,P^0\right)\right],$$

(13.60) becomes

$$RT^2\frac{d}{dT}\ln K_a\left(T\right) = \Delta h_{rxn}^0\left(T^0,P^0\right) + \int_{T^0}^{T}\Delta C_{P,rxn}\left(T'\right)dT'$$
$$= \Delta h_{rxn}^0\left(T,P^0\right). \tag{13.61}$$

Then

$$\frac{d}{dT}\ln K_a\left(T\right) = \frac{\Delta h_{rxn}^0\left(T,P^0\right)}{RT^2} \tag{13.62}$$

This is the van't Hoff[1] equation. Because measurements of reaction enthalpy are common, the van't Hoff equation is a very practical expression for the dependence of the equilibrium constant $K_a\left(T\right)$ on temperature.

13.7 Model Systems

The fugacity coefficient for the species λ in (13.32) is a function of the specific volume $v_\lambda\left(T,P,\{n_\lambda\}\right)$ of the component λ in the mixture. The specific volume is then the term we must model.

In a real system $v_\lambda\left(T,P,\{n_\lambda\}\right)$ will depend upon the environment of the molecules. For example in our treatment of statistical mechanics we saw that the dependence of the partition function on the system volume enters through the integration of $\exp\left(\sum_{\text{all molecules}}\phi_{int}/k_BT\right)$ over spatial coordinates.

Here ϕ_{int} is the intermolecular potential and is a function of the distance between molecules. We may assume that ϕ_{int} is generally dependent on the identity of the interacting molecules. There may be any number of molecules interacting with a specific molecule at any instant. The integration must then be carried out over different ranges depending on collision partners. This results in a volume dependence which is sensitive to the microscopic environment of the molecules.

In ideal mixtures of gases the force of interaction between any two molecules is assumed to be independent of the identities of the molecules. The molecules of different components are still considered distinguishable in that the masses and structures may be different. These differences affect the specific heats but they do not affect the dependence of the specific volumes on types of collisions.

The specific volume of a component λ is then independent of the identities of the other components. The properties of ideal mixtures of gases approximate those

---

[1] Jacobus Henricus van 't Hoff (1852–1911) was a Dutch physical chemist who was professor at Amsterdam and Honorary Professor at Berlin. He was the first winner of the Nobel Prize in chemistry (1901). Van 't Hoff was convinced of the power of imagination in Science.

of mixtures of nonideal (real) gases when the gases in the mixture are similar to one another, such as, for example, mixtures of ethanol and methanol.

For the calculation of the fugacity our choice of a model determines the form of the specific volume. In general there are two approaches that can be considered. We can specify a general thermal equation of state as fundamental or we can develop an thermal equation of state based on considerations of the statistical mechanics of interacting molecules. We shall consider use of the van der Waals thermal equation of state as an example of the first approach and the use of a virial expansion for the approach based on statistical mechanics.

13.7.1 Van der Waals Fluid

The van der Waals thermal equation of state (see Sect. 8.3.1, (8.20)) is the result of the addition of long range attractive forces and impenetrability to the noninteracting point molecules of the ideal gas. If we evaluate the integral in (13.33) using the specific volume from (8.20) we obtain for the fugacity of a single molecular species

$$\ln f = \ln \frac{RT}{(v-b)} + \frac{b}{(v-b)} - \frac{2a}{RTv},$$

(13.63)

which is a function of (T,v). But the fugacity is defined as a function of (T,P). We may separately calculate $v(T,P)$ numerically from (8.20) and use these values in (13.63) to obtain the fugacity as a function of (T,P).

Alternatively we may choose to obtain an approximate form of v from (8.20) for use in (13.63). If we neglect $a/v^2 << P$ we have the approximate result

$$f(T,P) \approx P \exp \left[\frac{P}{RT} \left(b - \frac{2a}{RT + b(P)} \right) \right].$$

(13.64)

13.7.2 Virial Approximation

We encountered the virial equation of state in Sect. 8.4.2, (8.36). Written in terms of specific volume (8.36) is

$$Z = \frac{Pv}{RT} = 1 + B(T)\frac{1}{v} + C(T)\frac{1}{v^2} + \cdots.$$

(13.65)

Alternatively we may choose to write the virial equation in terms of pressure as

$$Z = \frac{Pv}{RT} = 1 + B'(T)P + C'(T)P^2 + \cdots,$$

(13.66)

with

$$A' = A$$
$$B' = \frac{1}{A}B$$
$$C' = \frac{1}{A^2}C - \frac{1}{A^3}B^2$$
$$\vdots$$

$$(13.67)$$

The coefficients $B(T)$, $C(T)$, $\cdots$ (or $B'(T)$, $C'(T)$, $\cdots$) are called the second, third, ... virial coefficients respectively. As the pressure vanishes, or as the specific volume becomes very large, the virial thermal equation of state, in either form, becomes that of the ideal gas.

A systematic introduction of forces between molecules into a statistical mechanical study of gases results in a series expansion in the inverse specific volume ([139], Chap. 9). The methods of statistical mechanics then provide approximations for the calculation of the virial coefficients which are based on specific interaction potentials between the molecules [cf. [97]]. One advantage to this approach is that we may then obtain insight into the affect a specific choice of the interaction potential may have on the fugacity and, therefore, on conditions of chemical equilibrium or affinity.

If we use the virial thermal equation of state in the form (13.66), the fugacity for a nonideal gas is

$$f(T,P) = P \exp\left(\frac{1}{RT}\left[B'(T)P + \frac{1}{2}C'(T)P^2 + \cdots\right]\right). \qquad (13.68)$$

In many practical applications the third and higher virial coefficients are of less importance and we shall ignore them. We consider only the second virial coefficient here.

The second virial coefficient involves only binary interactions, i.e. collisions between two molecules. We shall, however, consider two forms of the interaction potential between molecules. The first of these is the hard sphere potential and the second is the so-called Lennard-Jones 6–12[2] potential in which

$$\mathscr{V}(r) = 4\varepsilon\left[\left(\frac{\sigma}{r}\right)^{12} - \left(\frac{\sigma}{r}\right)^6\right]. \qquad (13.69)$$

The graphical form of (13.69) is given in Fig. 8.4. It has a minimum equal to $-\varepsilon$ at $r = 2^{\frac{1}{6}}\sigma$. For $r > 2^{\frac{1}{6}}\sigma$ the Lennard-Jones potential produces a weak attractive intermolecular force and for $r < 2^{\frac{1}{6}}\sigma$, a strong repulsive force. The potential (13.69) then represents the basic characteristics of the potential energy of interaction between two molecules.

---

[2] Sir John Edward Lennard-Jones (1894–1954), English theoretical physicist, who produced the seminal work on molecular orbital theory in 1929. He suggested the interaction potential that bears his name in 1924.

Equation (13.69) is a specific form of the generalized Lennard-Jones potential given by

$$u(r) = \frac{\lambda}{r^n} - \frac{\nu}{r^{-6}} \qquad n > 6. \tag{13.70}$$

In (13.70) the dependence of the attractive part of the potential is preserved as proportional to r^{-6}. The form of the repulsive part is less critical. The attractive part of the potential in (13.70) is based on what are called London[3] dispersion forces. The London dispersion forces arise from a distortion of the charge distributions of the colliding molecules due to electrostatic interactions between the electrons on the two molecules. This distortion results in an energy of interaction proportional to r^{-6}. Neither theoretical nor experimental investigations have provided a definitive value for n in (13.70). Values of $10 \leq n \leq 14$ produce results that are as good as those obtained when $n = 12$ ([38], pp. 321–324).

If the intermolecular potential is modeled by (13.69) the second virial coefficient is ([139], p. 502)

$$B(T^*) = \left(\frac{2\pi\sigma^3}{3} \right) \sum_{n=0}^{\infty} \alpha_n \left(\frac{1}{T^*} \right)^{\frac{2n+1}{4}}, \tag{13.71}$$

where T^* is a dimensionless temperature defined as

$$T^* = \frac{k_B T}{\varepsilon}.$$

In Fig. 13.1 we have plotted the dimensionless form of the second virial coefficient

$$B^*(T^*) = \frac{B\left(\frac{k_B T}{\varepsilon} \right)}{\left(\frac{2\pi\sigma^3}{3} \right)}$$

for both the Lennard-Jones 6–12 potential and the square well potential.

We have drawn a horizontal line in Fig. 13.1 at $B^*(T^*) = 0$. The temperature at which the second virial coefficient vanishes is called the Boyle temperature. The Boyle point is defined as that point for which Boyle's law (Pv =constant when T is constant) is valid. From the virial thermal equation of state in (13.65) we see that Boyle's law is valid when $B' = 0$ (or $B = 0$) provided C' (or C) is neglected. The magnitude of C' at the temperature for which $B' = 0$ will determine the range of pressures over which Boyle's law is valid. This range may be large. This is another indication of the complexities resulting from molecular interactions.

Experimental measurements of the second virial coefficient for non-polar molecules (e.g. A, N_2, Ne, CH_4) agree very well with theoretical predictions obtained using the Lennard-Jones 6–12 potential (cf. [38], p. 325, Fig. 15.13). But there is not good agreement with predictions obtained using the square well

---

[3] Fritz Wolfgang London, 1900–1954. In 1939 London immigrated to the United States to become professor of theoretical chemistry at Duke University. London's theory of the chemical binding of homopolar molecules marked the beginning of modern quantum mechanical treatment of the hydrogen molecule and is considered one of the most important advances in modern chemistry.

Fig. 13.1 Second Virial
Coefficient B^* (T^*) for
Lennard-Jones 6-12 potential
and square well potential

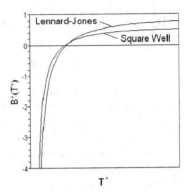

potential, except at lower temperatures. This is because the square well potential
has a hard, impenetrable core. This impenetrable core is not an appropriate model
when collisional energies are sufficient to result in an overlapping of the electron orbits of the colliding molecules. The Lennard-Jones potential, which permits a closer
approach of the molecules at high collisional energies, produces better agreement at
higher energies (temperatures).

In Fig. 13.2 we show the results of sample calculations of the fugacity coefficient
using the virial and van der Waals equations of state. The calculations with the virial
thermal equation of state include both the Lennard-Jones 6–12 potential and the
square well potential. These curves are presented only for comparison. Data used
were for oxygen. The curves in Fig. 13.2 are provided only for comparison of the
fugacity coefficient obtained from model calculations.

Fig. 13.2 Fugacities for
virial equation of state with
Lennard-Jones 6–12 and
square well potentials, and
for van der Waals equation of
state

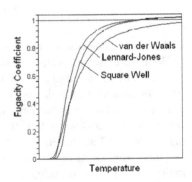

13.7.3 Mixtures of Nonideal Gases

To obtain expressions for the fugacity that will be useful in studies of chemical equilibrium we should in principle consider equations of state for mixtures of nonideal
gases. The general virial thermal equation of state for mixtures of nonideal gases is
of the form

$$PV = \sum_{\lambda} n_{\lambda} RT + \sum_{\lambda,\rho} \frac{n_{\lambda} n_{\rho}}{n} B_{\lambda,\rho}(T)\left(\frac{1}{V}\right) + \cdots,$$

where the terms $B_{\lambda,\rho}(T)$ are the second virial coefficients. The study of mixtures of fluids continues as an active area of interest [cf. [97]]. The computations have become practicable, particularly since the last decade of the twentieth century, because of the capabilities of small computers. Most studies are based on hard spheres and molecules with square well potentials. We shall not pursue the details of this work here.

13.8 Coupled Reactions

We can construct the theory of coupled reactions from the equations already obtained in Sect. 13.2. Particularly the mass conservation (13.3) and the Gibbs energy for the system are not changed by the identification of coupled reactions in the general scheme in (13.1). The problem of finding an extremum of the Gibbs energy is, therefore, unchanged and the stability requirement is still (13.8). As in case of the single reaction we may again use the stoichiometric condition with the stability requirement to eliminate the Lagrange multiplier.

Combining (13.8) with (13.5) we have

$$\sum_{\alpha}^{N} \sum_{\lambda}^{r^{(\alpha)}} v_{\lambda}^{(\alpha)} \mu_{\lambda} = 0. \tag{13.72}$$

for the coupled set of N reactions α. With (13.39) and (13.41), (13.72) becomes

$$-\frac{1}{RT} \sum_{\alpha}^{N} \sum_{\lambda}^{r^{(\alpha)}} v_{\lambda}^{(\alpha)} \mu_{\lambda}^{0}\left(T, P^{0}\right) = \ln \prod_{\alpha}^{N} \prod_{\lambda}^{r^{(\alpha)}} a_{\lambda}^{v_{\lambda}^{(\alpha)}}\left(T, P, \{n_{\lambda}\}; f_{\lambda}^{0}\left(T, P^{0}\right)\right) \tag{13.73}$$

We define the equilibrium constant for the αth reaction by the equation

$$RT \ln K_{a}^{(\alpha)}(T) \equiv -\sum_{\lambda}^{r^{(\alpha)}} v_{\lambda}^{(\alpha)} \mu_{\lambda}^{0}\left(T, P^{0}\right), \tag{13.74}$$

and the equilibrium constant for the combined reaction as

$$\tilde{K}_{a}(T) = \prod_{\alpha}^{N} K_{a}^{(\alpha)}(T). \tag{13.75}$$

Then the equilibrium condition (13.73) is

$$\tilde{K}_{a}(T) = \prod_{\alpha}^{N} \prod_{\lambda}^{r^{(\alpha)}} a_{\lambda}^{v_{\lambda}^{(\alpha)}}\left(T, P, \{n_{\lambda}\}; f_{\lambda}^{0}\left(T, P^{0}\right)\right) \tag{13.76}$$

for the coupled set of reactions.

Comparing (13.73) and (13.76) we see that

$$\tilde{K}_a\left(T\right) \equiv \exp\left[-\frac{\Sigma_\alpha^N \Sigma_\lambda^{r^{(\alpha)}} v_\lambda^{(\alpha)} \mu_\lambda^0\left(T,P^0\right)}{RT}\right]. \tag{13.77}$$

The reaction Gibbs energy for the αth reaction is

$$\Delta g_{\text{rxn}}^{0(\alpha)}\left(T,P^0\right) = \sum_\lambda^{r^{(\alpha)}} v_\lambda^{(\alpha)} \mu_\lambda^0\left(T,P^0\right). \tag{13.78}$$

And for the entire reaction the reaction Gibbs energy is

$$\begin{aligned}
\Delta g^0\left(T,P^0\right)_{\text{rxn}} &= \sum_\alpha^N \Delta g_{\text{rxn}}^{0(\alpha)}\left(T,P^0\right) \\
&= \sum_\alpha^N \sum_\lambda^{r^{(\alpha)}} v_\lambda^{(\alpha)} \mu_\lambda^0\left(T,P^0\right).
\end{aligned} \tag{13.79}$$

That is, for the coupled reaction,

$$\ln \tilde{K}_a\left(T\right) = -\frac{\Delta g_{\text{rxn}}^0\left(T,P^0\right)}{RT},$$

which is the same as (13.45). The temperature dependence of $\tilde{K}_a\left(T\right)$ may, then, be obtained from a relation of the form (13.56) and a van't Hoff equation also holds for $\tilde{K}_a\left(T\right)$.

13.9   Chemical Affinity

We encountered the chemical affinity as the driving force for chemical reactions in Sect. 11.3.3. For a single reaction the affinity is defined in (11.76), which we repeat here for continuity

$$A^{(\alpha)} = -\sum_\lambda^{r^{(\alpha)}} v_\lambda^{(\alpha)} \mu_\lambda, \tag{13.80}$$

and the entropy production due to a single chemical reaction α in an extent of reaction $d\xi^{(\alpha)}$

$$d\xi^{(\alpha)} = \frac{dn_\lambda^{(\alpha)}}{v_\lambda^{(\alpha)}}. \tag{13.81}$$

(see(11.45)) is

$$d_i S^{(\alpha)} = \frac{1}{T} A^{(\alpha)} d\xi^{(\alpha)}. \tag{13.82}$$

Summing (13.82) over the N coupled reactions we have the entropy production

$$d_i S = \frac{1}{T} \sum_\alpha^N A^{(\alpha)} d\xi^{(\alpha)}, \qquad (13.83)$$

(see (11.78)).

Using (13.39) the affinity of the αth reaction (13.80) becomes

$$A^{(\alpha)} = RT \ln K_a^{(\alpha)}(T) - RT \ln \frac{\prod\limits_{\text{products}} (a_\lambda)^{v_\lambda^{(\alpha)}}}{\prod\limits_{\text{reactants}} (a_\lambda)^{v_\lambda^{(\alpha)}}}. \qquad (13.84)$$

If $A^{(\alpha)} \geq 0$ the reaction proceeds. The equilibrium constant $K_a^{(\alpha)}$ is a function only of the Gibbs energies of the pure components evaluated at the reaction temperature and the standard pressure and is independent of the properties of the system that vary as the reaction proceeds, such as partial pressures and specific volumes. The second term in (13.84) is functionally dependent on these varying properties. We can only define activities if we speak of thermodynamic temperatures and pressures in the system. Assuming that we can, (13.84) implies that the αth reaction will proceed as long as

$$K_a^{(\alpha)}(T) > \frac{\prod\limits_{\text{products}} (a_\lambda)^{v_\lambda^{(\alpha)}}}{\prod\limits_{\text{reactants}} (a_\lambda)^{v_\lambda^{(\alpha)}}} \qquad (13.85)$$

Similarly as long as

$$\tilde{K}_a(T) > \prod_\alpha^N \frac{\prod\limits_{\text{products}} a_\lambda^{v_\lambda^{(\alpha)}} \left(T, P, \{n_\lambda\} ; f_\lambda^0 \left(T, P^0\right)\right)}{\prod\limits_{\text{reactants}} a_\lambda^{v_\lambda^{(\alpha)}} \left(T, P, \{n_\lambda\} ; f_\lambda^0 \left(T, P^0\right)\right)} \qquad (13.86)$$

the coupled set of reactions will proceed.

The reactions reach equilibrium and entropy production ceases if all affinities $A^{(\alpha)}$ vanish. If $A^{(\alpha)} = 0$ for all α, then, summing (13.80) on α we have

$$\sum_\alpha^N A^{(\alpha)} = -\sum_\alpha^N \sum_\lambda^{r^{(\alpha)}} v_\lambda^{(\alpha)} \mu_\lambda = 0, \qquad (13.87)$$

which is (13.72). Therefore, the requirement of thermodynamic stability for coupled reactions is equivalent to the condition that entropy production vanishes.

We then have at least a consistent picture of chemical reactions based rather solidly on thermodynamics. There are only two difficulties with our treatment.

The first of these is that we must be able to speak meaningfully of thermodynamic functions of state such as temperature and pressure. The second difficulty is that we have only discussed a linear theory. These are related. The claim that we can speak meaningfully about thermodynamic properties implies that we have thermodynamic equilibrium locally, i.e. in very small volumes. The dimensions of these volumes must be small compared to the length scale on which large variations occur. Under these conditions we expect a linear theory to be applicable.

13.10 Summary

The most important single result of this chapter is the law of mass action in terms of activities contained in (13.42). This is a consequence of the thermodynamic requirements of stability in a system in which molecular interactions occur at energies, i.e. temperatures, sufficient for chemical reactions to take place.

To obtain a form of the equilibrium constant abased on activities we first required the concept of fugacity, which is based on the collisional dependence of the thermal equation of state. Using our understanding of the microscopic picture of matter from Chaps. 8 and 9 we were able then to connect the details of our picture of molecular interactions to the calculation of activities.

In the development we have considered increasingly complex systems adding the various phenomena in steps. And then in our consideration of models we have indicated how we would go about actually including some aspects of collisions. Much of this we could have done without first introducing chemical reactions. But chemical reactions have provided a motivation for these studies. Study of high speed gas flows or plasmas provide similar motivation for investigating areas that bring us into contact with fundamental questions involving thermodynamics.

Finally our studies of affinities have revealed the importance of entropy production in any considerations whatsoever. The affinity is the driving force in so far as it determines the entropy production, which is the case in the linear approximation. Our representation of the affinity has included caloric, collisional, and ideal gas mixing effects. We have separated these in our stepwise development of the Gibbs energy. But we should not expect that such a separation characterizes the actual physical situation at high temperatures and pressures. Entropy production is not a simple concept.

Exercises

13.1. Consider the general dissociation reaction $AB \rightleftarrows A+B$. You have 0.250 mol of AB in a 0.01 m^3 at a pressure of 0.0899 MPa and a temperature of 298 K.

(a) How many mols have dissociated?

(b) What is the dissociation constant K_P for the reaction?

The gas NO_2 dimerizes to produce N_2O_4. In a closed vessel we have introduced a sample of NO_2 to a density of $1.64\,gl^{-1} * 10^{-3}\,1m^{-3} = 1.64 \times 10^{-3}\,gm^{-3}$. The molecular mass of NO_2 is $(= 14.007 + 2(15.9994))$ $46.006\,g\,mol^{-1}$. The total pressure of the gas is $411\,torr$ at $25\,°C$. Find the constant, K_P, for the dimerization.

13.2. Show that for the general virial expansions (13.65) and (13.66) that the transformation between the coefficients is, indeed, that given in (13.67). To carry this out we recognize that each expansion is a series in which the terms become smaller as orders increase. At each order in the expansion, then, the next order is neglected.

13.3. Use (13.33) to show that the fugacity for the virial equation of state is given by (13.68).

13.4. Show that the fugacity for a van der Waals gas is given by (13.63). Hint: This is most easily accomplished using (13.33).

13.5. Phosgene is a gas first synthesized by the British chemist, John Davy, in 1812. It has a nauseating odor resembling green corn or hay and attacks the lungs. It was used as a poison gas in WWI. Phosgene dissociates according to the reaction

$$COCl_2(g) \rightleftarrows CO(g) + Cl_2(g).$$

A vessel is filled with Phosgene, $COCl_2$ to a pressure of $0.014172\,MPa$ The temperature is $298\,K$. The gas does not dissociate at this temperature. The vessel and gas is then raised to a temperature of $686.65\,K$ at which point the pressure of the gas in the vessel is $0.045275\,MPa$.

(a) Has any of the gas dissociated at $686.65\,K$?
(b) What is the equilibrium constant, K_P at $686.65\,K$?

13.6. Consider a less noxious gas than that above, $NOCl$, but in the same experiment. The gas is prepared in a vessel with a pressure of $0.014172\,MPa$ and the temperature is $298\,K$. Assume that the gas does not decompose at this temperature. The vessel and gas is then raised to a temperature of $686.65\,K$. The pressure of the gas in the vessel is $4.5275 \times 10^{-2}\,MPa$. What is the equilibrium constant, K_P at $686.65\,K$? The dissociation is according to

$$2NOCl(g) \rightleftarrows 2NO(g) + Cl_2(g).$$

In this case the dissociation is slightly different from that of Phosgene.

13.7. Assume that you wish to have 35% of the gas in the preceding problem to be NO at 413.5 °C. What is the pressure of the gas in the vessel?

13.8. In Exercise 13.3 the equilibrium constant, K_P, was found to be $14.916\,\text{torr} = 1.9626 \times 10^{-2}\,\text{atm} = 1.9886 \times 10^{-3}\,\text{MPa}$ for the dissociation of NOCl at $686.65\,\text{K}$. Obtain from this the equilibrium constants for the mol fraction, K_χ, and for the molar concentration, K_C, at $413.5\,°\text{C} = 686.65\,\text{K}$ and a pressure of $156.99\,\text{torr} = 0.20656\,\text{atm} = 20.930\,\text{kPa}$. Carefully note the units on the equilibrium constants.

13.9. A 21 vessel which contains NO (nitrous oxide) gas at a pressure of 155 torr and a temperature of 413.5 °C. Into this you introduce 0.0125 mol of NOCl. What is the equilibrium composition?

13.10. Consider a vessel maintained at 298 K. Into this vessel equal mols of nitrogen (N_2) and hydrogen (H_2) are introduced each from pressure lines at $3.0398 \times 10^{-2}\,\text{MPa}$. The mixture will form ammonia according to $N_2 + 3H_2 \rightleftharpoons 2NH_3$. The equilibrium constant for the reaction at 298 K is $K_P = 6.89 \times 10^5\,\text{atm}^{-2} = 6.7110 \times 10^7\,\text{MPa}^{-2}$.
What is the partial pressure of ammonia in the vessel after equilibrium is established at 298 K?
[Note that if the nitrogen and hydrogen do not interact the pressure would be $3.0398 \times 10^{-2}\,\text{MPa}$.]

13.11. Find the reaction Gibbs energy for Exercise 13.9 on the formation of ammonia.

13.12. The reaction of Fe_2O_3 with CO is important industrially. It is the principle reaction in a blast furnace producing metallic iron from ore. The reaction is

$$Fe_2O_3(s) + 3CO(g) \rightleftharpoons 2Fe(s) + 3CO_2(g).$$

The molar Gibbs energies of formation for the compounds are

$$\Delta G_f^0\,[Fe_2O_3] = -177.1\,\text{kcal} = -42.368\,\text{kJ}$$
$$\Delta G_f^0\,[CO] = -32.81\,\text{kcal} = -7.8493\,\text{kJ}$$
$$\Delta G_f^0\,[CO_2] = -94.26\,\text{kcal} = -22.55\,\text{kJ}$$

Find the equilibrium constant, $K_a\,(T,P)$ at standard conditions.

13.13. The reaction forming the complex ion $SnBr^+$ was studied by C.E. Vanderzee [157]. Data are tabulated in Table 13.1 These data are plotted in Fig. 13.3

Table 13.1 Equilbrium constant SnBr

T/K	K_χ
273.15	4.3
298.15	5.38
308.15	5.75
318.15	6.16

Fig. 13.3 Plot of $\ln K_\chi$ vs $1/T$

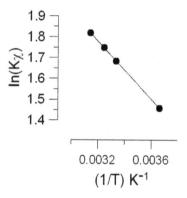

Find ΔG_{rxn}^0, ΔH_{rxn}^0, and ΔS_{rxn}^0 for the reaction.

Chapter 14
Chemical Kinetics

14.1 Introduction

We first encountered chemical reactions when we considered irreversibility. That is chemical reactions are then integral to thermodynamics, as long as we do not insist on compartmentalizing science.

In the preceding chapter we considered stable equilibrium of chemically reacting systems. We shall now move to what may be considered to be the more interesting study of the dynamics of chemical reactions. In this we shall make no claim of completeness. Specifically we do not consider the practicalities of finding reaction order. Our interest will be almost exclusively in the study of the physics of chemical reaction rates.

We will base our treatment on the statistical mechanics and kinetic theory we have developed in Chaps. 8 and 9. We will consider specifically collisional and transition state theories (TST) of rate constants for reactions. Both of these yield some understanding of chemical reactions. TST is particularly important as an active area of research at the beginning of the 21st century [8, 15, 163].

14.2 Kinetic Equations

We choose a simple reaction of the general form

$$A + B \rightarrow C + D. \tag{14.1}$$

some of the basic ideas of kinetics. If we denote the concentrations of the components as [A], [B], ... the equations for the rates of change of the concentrations of the components take the general form

$$\frac{d}{dt}[A] = f_A\{[A],[B],[C],[D]\}$$

$$\vdots$$

$$\frac{d}{dt}[D] = f_D\{[A],[B],[C],[D]\} \tag{14.2}$$

where $f_A \cdots f_D$ are functions of the concentrations.

C.S. Helrich, *Modern Thermodynamics with Statistical Mechanics*,
DOI 10.1007/978-3-540-85418-0_14, © Springer-Verlag Berlin Heidelberg 2009

Equations (14.2) are statements that the rate of change of the concentration of a particular component is a function of the concentrations of all components present. These equations are first order differential equations in the time t. They may be linear or nonlinear in the concentrations of the components. Our goal is to discover the mathematical form of the functions $f_A \cdots f_D$, including the values of any numerical terms appearing in these functions, and to obtain an acceptable molecular picture of the steps which lead from the reactants to the products.

Two fundamental concepts emerge in determining the form of the functions $f_A \cdots f_D$. These are reaction order and molecularity.

Reaction order is a concept that comes from the fact that the functions $f_A \cdots f_D$ are generally sums of terms of the form

$$k_j [A]^{\alpha_j} [B]^{\beta_j} [C]^{\gamma_j} [D]^{\delta_j}. \tag{14.3}$$

Such a term is of order α_j in $[A]$, of order β_j in $[B]$ and so on. The total order of the term is $\alpha_j + \beta_j + \gamma_j + \delta_j$. The numbers α_j, β_j, ... are, however, not necessarily integers. So the order of a reaction may involve decimals.

Molecularity is the description of manner in which the molecules interact in the reaction. This includes the number and identity of the molecules involved. The molecularity is not usually known for a specific reaction. Determining the molecularity may be the object of the chemical kinetic studies of the reaction.

The molecularity will affect the reaction order. For example it is possible that molecules of the components A and B collide and that the result is a shifting in the molecular bonding and the emergence of the product molecules of components C and D. If this is a simple binary collision we expect the equations for $d[A]/dt$ and $d[B]/dt$ to have terms proportional to the negative product of $[A]$ and $[B]$. That is we expect the rate of collisions between molecules of components A and B to the product of the concentrations of A and B. This is basically Boltzmann's postulate (or Ansatz) about the rate of collisions in a gas (see Sects. 8.2.2 and 14.3). It will result in terms which are first order in $[A]$ and $[B]$ or second order overall. The contribution is negative in the equations for $d[A]/dt$ and $d[B]/dt$ since the collision results in a loss of molecules of both $[A]$ and $[B]$.

The complexity of the reaction steps and the different rates in these steps may, however, make a direct relationship between molecularity and reaction order difficult to discern. Even if we are convinced that a binary collision must be the source of the reaction between A and B, the reaction will include requirements beyond those of the simple binary collision. Boltzmann's assumption also ignores correlations between molecules, which may be important in reactions.

Experimental measurements are of fundamental importance in chemical kinetics. When possible we design experiments so that our experimental measurements deal directly with concentrations as functions of time. Spectroscopic emission and absorption measurements are particularly suited for determining concentrations. Spectroscopic measurements also have no direct affect on most reactions, particularly if low intensity light sources are used. Tunable lasers are often used for excitation of spectral lines. Occasionally indirect measurements are made such as of electrical conductivities when some components are ions. Nuclear magnetic resonance, electron spin resonance, electron paramagnetic resonance, and mass spectrometry are also used to detect the emergence of certain components in a reaction. To observe certain aspects of a reaction some components may be added in large amounts so that their concentrations are essentially constant.

None of these considerations is inherently limited to gases. However, if we want to study collisions between molecules with any confidence we must turn to gases. In the next sections we shall develop a representation of binary collisions among gas molecules and present a fundamental theory for collisionally induced chemical reactions.

14.3 Collision Rates

In 1872 Ludwig Boltzmann presented a kinetic theory of gases which contained a formulation for the rate at which collisions occurred between molecules [12]. The proposal that led him to his formulation of molecular collisions known as the collision rate Ansatz (*Stosszahlansatz* in German). This Ansatz is that the rate of collisions between two molecules is proportional to the product of the probabilities that each molecule is separately located within the infinitesimal volume in which the collision occurs. In terms of molecular concentrations this means that the collision rate is proportional to products of concentrations.

The Stosszahlansatz is a statistical statement, as Boltzmann realized. Collisions are more accurately modeled by the Stosszahlansatz as the importance of interactions decreases and the average distance a molecule travels between collisions (mean free path) increases. This is the situation encountered at high temperatures and low to moderate pressures and is basically the situation that produced the nonideal gas considered in the preceding chapter.

Chemical bonds form between molecules if the combined molecule results in an electronic configuration that has a lower energy than that of the separate molecules. The details of this are not easily demonstrated in any but the simplest situations. The difficulty arises from the fact that the electronic configurations must be treated quantum mechanically. The quantum theory provides no understanding of chemical bonding in terms of a classical pictures of electrons.

For complex molecules the formation of chemical bonds depends on relative molecular orientations. Therefore, bonds form as a result of collisions between molecules with appropriate relative orientations. To form bonds the electron orbits on the colliding molecules must interpenetrate. This requires energy that must be supplied by the relative kinetic energy of the colliding molecules. The collisional kinetic energy must then exceed a certain threshold for the collision to result in a chemical bond. We may then expect that only a fraction of the collisions between any two types of molecules will result in a chemical bond between those molecules.

The simplest collisionally dependent reaction, with the stoichiometry of (14.1), we can consider is that in which an appropriate collision between molecules of components A and B results in a rearrangement of the bonds to form molecules of the products C and D. Because single molecules of components C and D are produced from each reactive collision, the rates of increase of the numbers of molecules of C and of D are the same and we need only consider the kinetic equation for one of these. From considerations of the Stosszahlansatz and those of the preceding paragraph, we may expect that the rate at which collisions between molecules of A and molecules of B collide to form a molecules of C and D is proportional, in a spatially homogeneous system, to the product of the concentrations of A and of B. That is

$$\frac{d}{dt}[C] = k[A][B], \tag{14.4}$$

in which the number k is the so-called rate constant of the reaction. The rate constant depends on the dynamics of the collision and the way in which energy is transferred from the reactants to the products after the collision.

The formation of a chemical bond generally involves first the breaking of some of the bonds of the reactants, the rearranging of atoms, and then the formation of the bonds of the products. After a collision which fulfills the requirements for the formation of a chemical bond, the two molecules form a dynamic complex which passes through a series of configuration states. The state of the dynamic complex can be thought of as a representative point in a phase space (see Sect. 9.4). As the state of the complex changes this point moves along a reaction coordinate in the phase space. The complex is a conservative dynamical system. So the representative point moves on a constant energy surface. Since the reaction requires the breaking of previous bonds and the rearranging of atoms and electrons to form new bonds, this constant energy surface can have a rich structure. The energy released by the newly bound system will appear as kinetic energy (heat) of the product molecule(s). Considerable effort has gone into studies of energy surfaces of dynamic complexes and motion along the reaction coordinate, particularly in the last two decades of the 20th and beginning of the the 21st century (cf. the reviews [8, 15, 163]). The result has been a great increase in our understanding of the details of the formation of molecular bonds.

Because chemical bonding can only be understood in quantum mechanical terms our calculations must be quantum mechanical. This presents no fundamental difficulty. But we must acknowledge the limits of a molecular quantum mechanical description. Molecular orbits are modeled by a linear combination of atomic orbitals (LCAO). And our understanding of the basic structure of molecules is based on x-ray diffraction measurements, which requires crystallization of the molecules. Particularly in the case of biological molecules this may present a resulting picture of the molecular form which is not a completely accurate representation of physiological reality.

The reaction path followed by the complex can be modeled rather simply by introducing some concepts that capture the physics of the breaking and reformation of bonds. The result is the concept of the activated complex and transition state theory. We shall outline these here and obtain approximations to the rate constant for the bimolecular reaction.

14.4   Activated Complex

Here we shall neglect any effects of orientation and consider only the energy requirements. We represent the energy that must be supplied by the collision as a potential energy barrier that must be crossed by the colliding molecules. If the collision energy is equal to the barrier energy an activated complex (a model of the dynamic complex) is formed. The activated complex undergoes a transition to the final product molecules. The potential energy of the molecules then drops to a value which is less than the initial potential energy of the two colliding molecules. The situation is shown pictorially in Fig. 14.1. The barrier energy ε_a is called the activation energy and ΔU is the energy difference between the bond energies of the reactants and the products. In this model the complicated reaction path is replaced by a motion up one side of the barrier and down the other. We then write the stoichiometric relationship (14.1) as

$$A + B \rightarrow (AB)^{\ddagger} \rightarrow C + D, \tag{14.5}$$

where $(AB)^{\ddagger}$ is the activated complex. The rate of production of products is equal to the rate of production of the activated complex, which is determined by the rate of collisions with energy greater than or equal to the barrier energy. We, therefore, to a study of the collisions.

The Stosszahlansatz states that the number of collisions between molecules of components A and B occurring in a very small volume ΔV in a time interval Δt is proportional to the product of the average numbers of these molecules that are in ΔV during the time interval Δt. The molecules are in motion and pass through the volume ΔV. The number of collisions that will occur in ΔV during the time interval Δt increases with the number of molecules that pass through ΔV in time

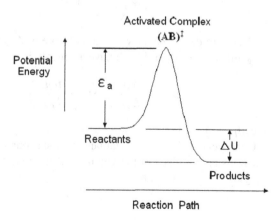

Fig. 14.1 Activation energy
of the activated complex

Δt. This number is proportional to the product of the speed of the molecules and the time interval Δt. The speed is proportional to the square root of the thermodynamic temperature. So the collisional rate should be proportional to $T^{\frac{1}{2}}\Delta t$. The probability that a molecule of component A is in ΔV at any time is $[A]\Delta V$ and that a molecule of component B is in ΔV is $[B]\Delta V$. The Stosszahlansatz then states that the number of collisions occurring in ΔV in the time Δt between molecules of types A and B is proportional to $[A][B]T^{\frac{1}{2}}\Delta V^2 \Delta t$.

In our model a collision will result in a chemical reaction if the relative kinetic energy is at least equal to the barrier energy ε_a. So we must include a factor proportional to the probability that the collisional kinetic energy is equal to or exceeds ε_a. If we consider a collision between two molecules 1 and 2 with approximately equal masses and a total kinetic energy $\mathscr{E}$, the relative kinetic energy available in the collision is (see exercises)

$$\frac{1}{2}(1 - 2\sqrt{\eta_1\eta_2}\cos\vartheta)\mathscr{E} \tag{14.6}$$

where $\vartheta \geq \pi/2$ is the angle between the velocities for a collision and the energies of the two molecules are $\eta_1\mathscr{E}$ and $\eta_2\mathscr{E}$ with $\eta_1 + \eta_2 = 1$. We must then require at least that the sum of the molecular kinetic energies exceeds ε_a.

We can find the relative number of molecules that have energies greater than or equal to E' from our statistical mechanical results. From (9.52) the relative number of molecules with an energy E_ε is

$$\frac{N_\varepsilon}{N} = \left[\sum_v \exp\left(-\frac{E_v}{k_B T}\right)\right]^{-1}\exp\left(-\frac{E_\varepsilon}{k_B T}\right). \tag{14.7}$$

The relative number of molecules with energy greater than or equal to E' is then

$$n(E \geq E_\varepsilon) = \left[\int_0^\infty \exp\left(-\frac{E'}{k_B T}\right)dE'\right]^{-1}\int_{E_\varepsilon}^\infty \exp\left(-\frac{E'}{k_B T}\right)dE'$$

$$= \exp\left(-\frac{E_\varepsilon}{k_B T}\right). \tag{14.8}$$

Therefore the probability that a molecule of A will have an energy in excess of $\eta_A\varepsilon_a$ and a molecule of B will have an energy in excess of $\eta_B\varepsilon_a$ is proportional to

$$\exp\left(-\frac{\eta_A\varepsilon_a}{k_B T}\right)\exp\left(-\frac{\eta_B\varepsilon_a}{k_B T}\right) = \exp\left(-\frac{\varepsilon_a}{k_B T}\right), \tag{14.9}$$

since $\eta_A + \eta_B = 1$.

Consistent with the Stosszahlansatz, then, our result for the of the number of collisions taking place in the volume ΔV in the time interval Δt with the required energies is proportional to

$$[A]\,[B]\,T^{\frac{1}{2}}\exp\left(-\frac{\varepsilon_a}{k_B T}\right)\Delta V^2\Delta t,$$

and the rate at which collisions resulting in chemical reactions take place in the volume ΔV is then proportional to

$$[A]\,[B]\,T^{\frac{1}{2}}\exp\left(-\frac{\varepsilon_a}{k_B T}\right)\Delta V^2. \tag{14.10}$$

Because our system is spatially homogeneous all volumes are equivalent and the factor ΔV may be dropped in Eq. (14.10). The rate constant k appearing in (14.4) is then

$$k = k_0 T^{\frac{1}{2}}\exp\left(-\frac{\varepsilon_a}{k_B T}\right). \tag{14.11}$$

The combination $k_0 T^{\frac{1}{2}}$ is known as as the *preexponential factor* and is often designated as A.

The expression for the rate constant in (14.11) includes only the limitations on the Boltzmann collision term arising from the potential energy barrier that must be crossed to form the activated complex and begin the complex process resulting in the products. None of the details of the motion of the activated complex along the reaction coordinate were considered in the derivation of (14.11). At this stage in our development we have then a collisional theory.

The form of the temperature dependence of the rate constant in (14.11) was first proposed by Arrhenius[1] in 1889. The original relationship proposed by Arrhenius lacked the factor[2] $T^{\frac{1}{2}}$, which we obtained from the rate of passage of the molecules through a volume.

From (14.11) we have

$$\ln k = \ln\left[k_0 T^{\frac{1}{2}}\right] - \frac{E_a}{RT}, \tag{14.12}$$

in which $E_a = N_A\varepsilon_a$, where N_A is Avogadro's number. We should then be able to determine the activation energy from experimental data for the dependence of the rate constant on temperature.

A graph of $\ln k$ as a function of T^{-1} is called an Arrhenius plot. The slope of this graph is

$$R\frac{d}{d(1/T)}\ln k = -E_a - \frac{1}{2}RT. \tag{14.13}$$

In Fig. 14.2 we present Arrhenius plots for two simple reactions. In each plot the solid line includes the factor $T^{\frac{1}{2}}$ and the line with the circles does not. For the reaction $SO+O_2 \rightarrow SO_2+O$ the activation energy is $27\,kJ\,mol^{-1}$ and for the reaction $NO+Cl_2 \rightarrow NOCl+Cl$ the activation energy is $85\,kJ\,mol^{-1}$. In each plot the

---

[1] Svante August Arrhenius (1859–1927) was a Swedish physicist, who is often considered to be a chemist. He is one of the founders of physical chemistry.

[2] A more complete calculation for hard spheres produces the factor $T^{\frac{1}{2}}$ in the rate of collisions.

Fig. 14.2 Arrhenius
plots for the reactions
$SO+O_2 \rightarrow SO_2+O$ **(a)** and
$NO+Cl_2 \rightarrow NOCl+Cl$ **(b)**

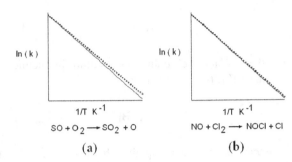

temperature range is 373-400 K. The factor $T^{\frac{1}{2}}$ has very little effect if the range of temperatures considered is not too great and if the activation energy is not too small. The curve in the Arrhenius plot for the reaction $SO+O_2 \rightarrow SO_2+O$ is more evident because of the lower value of the activation energy.

Deviations from linearity in the Arrhenius plot also result from quantum mechanical tunneling. Classically colliding molecules cannot pass the barrier of the activation energy unless the energy of the collision is equal to or greater than the barrier energy. This is not the case if the molecules are treated quantum mechanically. In a quantum treatment a particle has a nonzero probability of passing through a potential energy barrier that exceeds the energy of the particle. Quantum mechanical tunneling effectively lowers the potential barrier.

If we choose to consider details of the interaction we must, however, move beyond collision models alone. For example a description of tunneling requires a representation of the colliding particle and the barrier structure. Modern research on reactions includes Detailed investigations of this nature. But these are beyond the scope of our discussion. Our interests are in thermodynamics. Transition state theory is a thermodynamic theory which also considers some of the dynamics of chemical interaction.

14.5 Transition State Theory

Transition state theory, known to chemists as TST, is the most important and fruitful theory on which we can base understanding of the dynamic details of chemical reactions. The original theory was published simultaneously by Henry Eyring[3] at Princeton University and by M.G. Evans and Michael Polanyi[4] at the University

---

[3] Henry Eyring (1901–1981) was a Mexican-born American theoretical chemist. His major scientific contribution was to the development of the theory of chemical reaction rates and intermediates.

[4] Michael Polanyi (1891–1976) was a Hungarian–British polymath. He was hospitalized after service as a physician in the Austro-Hungarian Army in WWI and during convalescence completed a doctorate in physical chemistry. Polanyi held a chair in physical chemistry at the University of Manchester and contributed fundamental ideas to the philosophy of science, in which he opposed the mechanical approach of the positivist movement.

Fig. 14.3 Saddle point. The directions along the reaction coordinate leading to products and the direction of oscillation are indicated by *arrows*

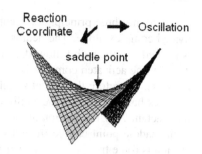

of Manchester. TST retains two of the concepts from the activated complex theory. There is an activation energy which must be overcome by the colliding molecules and there is an activated complex formed by the colliding molecules. But the theory is not a collisional theory. It is a statistical thermodynamic theory, which rests firmly on the ideas of Boltzmann and of Gibbs and requires no additional hypotheses.

There is no need to consider any detailed picture of the potential energy surface characteristic of the interaction. We do, however, know some things about the general appearance of the surface. The activated complex exists at the peak of the activation energy along the reaction coordinate. But if this is a point of equilibrium it must be a minimum point for other coordinate variations. Therefore, the location of the activated complex is a saddle point on the energy surface. For illustration purposes a saddle point is shown in Fig. 14.3.

In Fig. 14.4 we illustrate a section of a potential energy surface on which we have drawn constant energy contours. In this figure we have indicated the location of the saddle point on the energy surface. For our illustration we have chosen the general coordinates on the energy surface to be the unspecific pair q and p. These coordinates may be, for example, location and momentum of the center of the complex. We have drawn a dotted line on the energy surface to illustrate a possible path of the representative point of the dynamic complex. The activated complex is the state of the dynamic complex at the saddle point. This illustration should suffice for our understanding of basic TST. Similar plots may be found in reference [114].

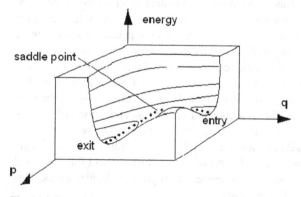

Fig. 14.4 Potential energy surface for a reaction. Varaibles q and p are general coordinates for the dynamic complex. *The dotted line* is a possible trajectory of the representative point on the surface

There are three principal elements of TST. (1) Attention is focused on the activated complex alone. The details of the process by which the activated complex is formed, i.e. the path followed by the dynamic complex, are irrelevant in the theory. (2) The activated complex is in equilibrium with the reactants, but not with the products. Therefore the rate at which the products are removed from the activated complex has no effect on the equilibrium relation among the activated complex and the reactants. (3) The motion of the activated complex along the reaction coordinate at the saddle point can be treated statistically as a free translational motion. The motion is force free because the derivatives of the potential, which yield the forces, vanish at the saddle point.

TST is fundamentally a microscopic theory. The three elements of the theory will be explicitly used to develop TST in the first of the next sections. The theory can also be cast in macroscopic, thermodynamic terms. This will be carried out in the second section.

14.5.1   Microscopic TST

The first element of the theory is the point of departure from the collision theory. If we can apply the methods of equilibrium statistical mechanics to the activated complex then we can neglect the history of the system. That is we will no longer need to consider the actual dynamics that have led to the formation of the dynamic complex, which was the primary focus of the collision theory. We are not claiming that collisions do not exist. But we are developing a theory in which it is no longer necessary to consider the mechanics of collisions.

This is an important step and is a major change in the approach to the problem of understanding chemical reactions. The collisional approach is an element of what, in microscopic physics, is called kinetic theory. In kinetic theory we are interested in the time development of the system based on models of molecular interactions. Equilibrium statistical mechanics does not treat the time development of the system and assumes that all we know about the system is a collection of macroscopic quantities. The use of statistical mechanics is justified if the number of molecules being considered is large and the time scale on which the system parameters is changing is long compared to the time required to establish a local statistical equilibrium.

If there is a large number of reacting combinations of molecules we can assert that there is then a large number of dynamic complexes. We can then claim that within this collection of dynamic complexes there is also a large subset in the state we call the activated complex. This is sufficient to justify an application of equilibrium statistical mechanics to the activated complex, even though there is a constant motion of individual activated complexes into and out of the transition state. To get a meaningful and useful mathematical expression we must connect our statistical mechanical description of the activated complex to tangible molecular states. This is done in the second element of the theory.

To develop the second element of the theory we begin with (14.5) which we rewrite here with rate constants indicated

$$A + B \xrightarrow{k^{\ddagger}} (AB)^{\ddagger} \xrightarrow{k_{\text{products}}} C + D. \tag{14.14}$$

The kinetic equations of interest are those for the reactants, the activated complex, and the products. We consider the activated complex to be a separate component consistent with the first element of the theory. For the reactants we need only consider the rate equation for $[A]$ and for the products we need only consider that for $[C]$. These are the same as those for $[B]$ and $[D]$ respectively. The equations we need are then

$$\frac{d}{dt}[A] = -k^{\ddagger}[A][B], \tag{14.15}$$

$$\frac{d}{dt}\left[(AB)^{\ddagger}\right] = k^{\ddagger}[A][B], \tag{14.16}$$

and

$$\frac{d}{dt}[C] = k_{\text{products}}\left[(AB)^{\ddagger}\right]. \tag{14.17}$$

The rate constant $k^{\ddagger}$ appearing in (14.16) is the same rate constant that we calculated previously from the collision theory provided all activated complexes transition to products. But we shall now use a statistical approach to obtain this rate constant.

The second element of the theory requires that the activated complex is in equilibrium with the reactants. From our studies of chemical equilibrium (see Sect. 13.5, (13.53)) this means that

$$K_C^{\ddagger} = \frac{\left[(AB)^{\ddagger}\right]}{[A][B]} \tag{14.18}$$

where $K_C^{\ddagger}$ is the equilibrium constant for the reaction in which the activated complex is formed. Mass conservation requires equality of $d\left[(AB)^{\ddagger}\right]/dt$ and $d[C]/dt$. From (14.16) and (14.17) we have then

$$k^{\ddagger}[A][B] = k_{\text{products}}\left[(AB)^{\ddagger}\right]. \tag{14.19}$$

With the equilibrium requirement (14.18), (14.19) results in

$$k^{\ddagger} = k_{\text{products}}K_C^{\ddagger} \tag{14.20}$$

We then have an expression for the rate constant in (14.15) in terms of the equilibrium constant $K_C^{\ddagger}$ and the rate constant for the loss of the activated complex.

We turn to the methods of statistical mechanics to compute $K_C^{\ddagger}$. In Sect. 9.7.2, (9.52), we obtained the relative numbers of molecules in an energy state. We have the three components A, B, and AB each of which has the distribution i(9.52). The ratio of the concentrations is, therefore, given by

$$\frac{\left[(AB)^{\ddagger}\right]}{[A][B]} = \frac{q_{(AB)^{\ddagger}}}{q_A q_B} \exp\left(-\frac{\varepsilon_a}{k_B T}\right). \tag{14.21}$$

The quantities $q_{A,B,(AB)^{\ddagger}}$ are the partition functions for a microcanonical ensemble. Then, using (14.21), (14.18) becomes

$$K_C^{\ddagger} = \frac{q_{(AB)^{\ddagger}}}{q_A q_B} \exp\left(-\frac{\varepsilon_a}{k_B T}\right). \tag{14.22}$$

The partition functions $q_{A,B}$ can be separated into a products of partition functions for translation, rotation, and vibration of the molecule (see Sect. 9.13). But the partition function for the activated complex requires some creative thought. The construction of a partition function for the activated complex and a theoretical expression for the rate constant $k_{products}$ will be the results of the third element of the theory.

The third element of the theory requires that we consider the motion of the representative point of the complex as a force free motion along the reaction coordinate. We have represented the activated complex as a point moving along the reaction coordinate on the potential energy surface. This is not the same as the motion of a particle along a spatial coordinate. It can be shown, however, that with some modification we can interpret motion along the reaction coordinate as a one dimensional translation of a particle with mass equal to the reduced mass μ of the complex ([130], p. 71).

The motion of the complex along the reaction coordinate, e.g. the trajectory indicated by the dotted line in Fig. 14.4, is then a representation of the time development of the complex in terms of a mass moving on a potential surface. At the saddle point the complex begins to dissociate into the products. We can still define a reduced mass and plot a trajectory even if the complex has begun to separate.

We shall assume that the activated complex exists in an undissociated state over a distance δ_R along the reaction coordinate at the saddle point. The time required for the products to be formed decreases as the speed with which the representative point traverses the distance δ_R increases. That is the rate of emergence of products from the activated complex is proportional to the average speed of the representative point at the saddle point.

In our model the average speed can be calculated from equilibrium kinetic theory using a Maxwellian distribution for the speed of the reduced mass μ. The result is

$$\text{average speed} = \left(\frac{k_B T}{2\pi\mu}\right)^{\frac{1}{2}}. \tag{14.23}$$

The rate constant $k_{products}$ is then

$$k_{products} = \frac{1}{\delta_R}\left(\frac{k_B T}{2\pi\mu}\right)^{\frac{1}{2}} \tag{14.24}$$

The partition function of the activated complex can also be separated into a translational part and a part corresponding to rotation and vibration. That is we may write

$$q_{(AB)^\ddagger} = q_{T,(AB)^\ddagger} q_{VR,(AB)^\ddagger} \tag{14.25}$$

in which $q_{T,(AB)^\ddagger}$ is the translational partition function and $q_{VR,(AB)^\ddagger}$ is the vibrational and rotational partition function. For the translational contribution we consider the reduced mass confined in a length δ_R. The partition function for this confined mass is well known quantum mechanically and is given by[5]

$$q_{T,(AB)^\ddagger} = (2\pi\mu k_B T)^{\frac{1}{2}} \frac{\delta_R}{h} \tag{14.26}$$

(see Sect. 9.13, (9.97)). Then

$$k_{products} q_{T,(AB)^\ddagger} = \left(\frac{k_B T}{2\pi\mu}\right)^{\frac{1}{2}} \frac{1}{\delta_R} (2\pi\mu k_B T)^{\frac{1}{2}} \frac{\delta_R}{h}$$
$$= \frac{k_B T}{h}. \tag{14.27}$$

Combining (14.20), (14.22), (14.25), and (14.27) we have

$$k^\ddagger = \left(\frac{k_B T}{h}\right) \left[\frac{q_{VR(AB)^\ddagger}}{q_A q_B}\right] \exp\left(-\frac{\varepsilon_a}{k_B T}\right). \tag{14.28}$$

This is the first of what are referred to as the *general Eyring equations*. The theoretical value for the preexponential factor from TST is then

$$A(T) = \left(\frac{k_B T}{h}\right) \left[\frac{q_{(VR)^\ddagger}}{q_A q_B}\right].$$

The partition functions are functions of the temperature. That is the temperature dependence of the theoretical preexponential factor is not simply $k_B T/h$.

We can obtain an experimental value for the barrier energy $E_a = N_A \varepsilon_a$ from the slope of an Arrhenius plot. From (14.28) this slope is

$$R\frac{d}{d(1/T)} \ln k^\ddagger = -E_a - RT + R\frac{d}{d(1/T)} \ln\left[\frac{q_{VR(AB)^\ddagger}}{q_A q_B}\right]. \tag{14.29}$$

This may be compared to the results of the collision theory presented in Eq. (14.13). As in Eq. (14.13) the slope in Eq. (14.29) is not directly proportional to the value of the molar activation energy E_a alone.

We can identify the sources of the individual terms appearing in (14.29) based on the development. The addition of RT to the theoretical activation energy appearing in (14.29) is a result of the breakup of the complex. The last term on the right

---

[5] As we point out in Sect. 9.13, the quantum mechanical translational partition is identical to the classical.

hand side of (14.29) comes from the requirements that a chemical equilibrium is established between the activated complex and the reactants.

In TST we do not deny that there was an encounter of the reactants with the energy barrier as they collide and form the dynamic complex. This last term is an indication that the energy involved in the formation of the activated complex must be considered in any discussion of modifications to the collision theory.

We may then speak in realistic terms about an activation energy for the TST, which is a combination of the barrier energy and the energy involved in producing the activated complex. If we define the activation energy for TST as

$$\Delta E^{\ddagger} = E_a - R \frac{d}{d(1/T)} \ln \left[\frac{q_{VR(AB)^{\ddagger}}}{q_A q_B} \right] \qquad (14.30)$$

our theoretical equation for the slope of the Arrhenius plot becomes

$$R \frac{d}{d(1/T)} \ln k^{\ddagger} = -\Delta E^{\ddagger} - RT. \qquad (14.31)$$

The slope of our Arrhenius plot then provides an energy

$$E_{exp} = \Delta E^{\ddagger} + RT. \qquad (14.32)$$

Using (14.30) and (14.32) in (14.28) we have

$$k^{\ddagger} = \left(\frac{k_B T}{h} \right) \left\{ \left[\frac{q_{VR(AB)^{\ddagger}}}{q_A q_B} \right] \cdots \right.$$
$$\left. \cdots \exp \left(1 + T \frac{d}{dT} \ln \left[\frac{q_{VR(AB)^{\ddagger}}}{q_A q_B} \right] \right) \right\} \exp \left(-\frac{E_{exp}}{RT} \right) \qquad (14.33)$$

The TST then allows a separation of the activation energy $\Delta E^{\ddagger}$ from the contribution RT arising from the breakup into products.

The preexponential factor remains a function of the temperature the form of which depends on our model of the reactant molecules and of the activated complex. We may simplify the result if we assume that the variation in temperature is not excessive. An expansion of the product of partition functions, which we write as

$$F_q(T) = \frac{q_{VR(AB)^{\ddagger}}}{q_A q_B}, \qquad (14.34)$$

produces a final form of the rate constant

$$k^{\ddagger} = \left(\frac{k_B T}{h} \right) e F_q'(0) \exp \left(-\frac{E_{exp}}{RT} \right), \qquad (14.35)$$

in which e is the base of natural logarithms and $F_q'(0)$ is $dF_q(T)/dT$ evaluated at $T = T^0$. Using (14.32) this becomes

$$k^{\ddagger} = \left(\frac{k_B T}{h}\right) K \exp\left(-\frac{\Delta E^{\ddagger}}{RT}\right) \tag{14.36}$$

in which $K = F_q'(0)$. Equation (14.36) is the standard form in which the Eyring equation is usually presented.

The simplicity of the first Eyring equation is striking considering the theoretical effort which brought it into being. It is not only a tribute to Eyring himself as a scientist. It is also an example of a bold idea which has been very fruitful. The basic approach which led to the Eyring equation can also be modified readily to encompass other pictures of the breakup of the activated complex into products.

The modern approaches consider that the breakup of the activated complex is a result of vibration. This does not conflict in any direct fashion with the original idea in which the activated complex was considered to breakup as it passed through a distance δ_R along the reaction coordinate over the saddle point. Attributing the breakup to a vibration, however, makes the physical basis for the breakup more explicit and provides a basis for spectroscopic investigations of the activated complex. Expressing the rate constant $k_{products}$ in terms of vibrations will not modify the Eyring result in any substantial way.

If we consider that the breakup of the activated complex into products is determined by a vibration we may assume that the rate at which the complex breaks up into products is proportional to the vibration frequency of the mode in question, which we shall call v. Then $k_{products} = \kappa v$ in which κ is a proportionality constant. To complete the formulation we must isolate the vibrational partition function $q_{V,(AB)^{\ddagger}}$ rather than the translational partition function from the quotient $q_{(AB)^{\ddagger}}/q_A q_B$. The vibrational partition function for a single mode is (see Sect. 9.13.1, (9.102))

$$q_{V,(AB)^{\ddagger}} = \left[1 - \exp\left(-\frac{hv}{k_B T}\right)\right]^{-1}.$$

This is approximately equal to $k_B T/hv$. Therefore in place of (14.28) we have

$$k^{\ddagger} = v \left(\frac{k_B T}{hv}\right) \left[\frac{q_{(AB)^{\ddagger}}'}{q_A q_B}\right] \exp\left(-\frac{\varepsilon_a}{k_B T}\right). \tag{14.37}$$

in which $q_{(AB)^{\ddagger}}'$ is the partition function for the activated complex after removing the single mode that we believe is responsible for the breakup. Because of the appearance of the vibration frequency in numerator and denominator, this is identical to (14.28). Therefore (14.36) follows as well.

The transition state cannot be directly observed experimentally so that the relative importance of the types of motion cannot be determined. This is due at least in part to the rules of the quantum theory. But it is also because of the practical difficulty in measuring the state of a very short lived system. Modern spectroscopic techniques are helpful in investigations of the transition state. Particularly femtosecond IR (Infrared) spectroscopy was developed precisely to gain access to structures very close

to the transition point. And some of the most modern studies of the transition state
are of vibrational states.

14.5.2 Macroscopic TST

In the basic development of TST we assumed that the system of activated complexes
was in a state of thermodynamic equilibrium in order to apply equilibrium statistical mechanics. Consistent with this assumption we may then claim that the system
of activated complexes is a thermodynamic system and apply the methods of thermodynamics to this system. We may then use Eq. (13.54) written in terms of the
activated complex, which will be

$$K_C^{\ddagger}(T) = \left(\frac{RT}{P^0}\right) K_a^{\ddagger}(T) \tag{14.38}$$

for the ideal gas, since for the formation of the activated complex $\Delta v = -1$. We
may then use (13.45) in (14.38) to obtain

$$K_C^{\ddagger}(T) = \frac{RT}{P^0} \exp\left[-\frac{\Delta g^{\ddagger}(T,P^0)}{RT}\right], \tag{14.39}$$

in which $\Delta^{\ddagger}g(T,P^0)$ is the change in the molar Gibbs energy associated with the
formation of the activated complex from the reactants. Equation (14.39) is the thermodynamic (macroscopic) equivalent of (14.22).

With

$$\Delta g^{\ddagger}(T,P^0) = \Delta h^{\ddagger}(T,P^0) - T\Delta s^{\ddagger}(T,P^0), \tag{14.40}$$

(14.39) may be written as

$$K_C^{\ddagger}(T) = \frac{RT}{P^0} \exp\left[\frac{\Delta s^{\ddagger}(T,P^0)}{R}\right] \exp\left[-\frac{\Delta h^{\ddagger}(T,P^0)}{RT}\right]. \tag{14.41}$$

To obtain the rate constant for the production of the activated complex we must
still turn to (14.20). We shall consider that the activated complex breaks up as a
result of a vibration along a particular coordinate. In thermodynamic terms we then
write $k_{products}$ as proportional to an average frequency which we define by writing
the average energy of a quantum oscillator as $h v_{ave}$. Then $k_{products} = \kappa(k_B T/h)$.
From (14.20) the rate constant for the activated complex is then

$$k^{\ddagger} = \kappa\left(\frac{k_B T}{h}\right)\frac{RT}{P^0}\exp\left[\frac{\Delta s^{\ddagger}(T,P^0)}{R}\right]\exp\left[-\frac{\Delta h^{\ddagger}(T,P^0)}{RT}\right]. \tag{14.42}$$

We may find the experimental activation energy of this expression in terms of the thermodynamic activation enthalpy $\Delta h^{\ddagger}$ and activation energy $\Delta u^{\ddagger}$ as

$$
\begin{aligned}
E_{\exp} &= -R \frac{d}{d(1/T)} \ln k^{\ddagger} \\
&= \Delta h^{\ddagger} + 2RT \\
&= \Delta u^{\ddagger} + RT
\end{aligned} \tag{14.43}
$$

Then the rate constant for the formation of the activated complex in (14.42) is

$$
k^{\ddagger} = \left(\frac{k_B T}{h} \right) \left\{ \kappa e^2 \frac{RT}{P^0} \exp \left[\frac{\Delta s^{\ddagger}(T, P^0)}{R} \right] \right\} \exp \left[-\frac{E_{\exp}}{RT} \right]. \tag{14.44}
$$

This is the *second Eyring equation*. This is the thermodynamic (macroscopic) equivalent of (14.33).

If we write the rate constant in the form

$$
k^{\ddagger} = A(T) \exp \left(-\frac{E_{\exp}}{RT} \right) \tag{14.45}
$$

we have

$$
A(T) = \kappa \left(\frac{k_B T}{h} \right) \frac{RT}{P^0} e^2 \exp \left[\frac{\Delta s^{\ddagger}(T, P^0)}{R} \right] \tag{14.46}
$$

as the preexponential factor from the macroscopic formulation of TST.

14.6   Specific Reactions

In this section we will consider two reactions in some detail. The reactions have been chosen because they reveal some of the complexity of investigations in chemical kinetics while remaining transparent in the physics.

14.6.1   Lindemann Mechanism

A first order reaction is one which satisfies the rate equation

$$
\frac{d}{dt} [A] = -k [A] \tag{14.47}
$$

A first order rate equation would result if the reaction is monomolecular. This may occur if the molecules of component A are unstable and transform naturally into products, such as in radioactive decay. A reaction may also be first order even though collisions are required to initiate the chemical transformation. This may occur if the collisions are with molecules whose concentration is large enough to be considered constant in spite of combinations.

The Lindemann (Frederick Lindemann[6]) mechanism is an example of a reaction dependent on the concentration of a background molecule. Breaking the chemical bonds of a molecule of component A to form products requires a certain energy. Lindemann proposed that the source of this energy is a collision with another molecule of the component M. That is

$$A+M \underset{k_2}{\overset{k_1}{\rightleftharpoons}} A^*+M \tag{14.48}$$

Here A^* designates a component chemically identical to A but with molecules that have sufficient energy to undergo decomposition into products. This may be thought of as an activated or energized state of the component A. The component M may be another reactant, a product or any inert substance present. Since molecules of A^* are, on the average, of higher energy than those of other components, they may also lose energy in collisions. This is referred to as collisional deactivation. The reaction in (14.48) is, therefore, reversible.

If a molecule of A^* is not collisionally deactivated it may transform into products according to

$$A^* \overset{k_3}{\rightarrow} \text{products, P.} \tag{14.49}$$

This is considered the final step in the Lindemann mechanism and proceeds at a rate proportional to the product of k_3 and $[A^*]$. The net reaction is

$$A \overset{k_\text{unimolecular}}{\rightarrow} P. \tag{14.50}$$

This net reaction has the form of a unimolecular reaction, although it is not unimolecular in detail. We have defined a unimolecular rate "constant" $k_\text{unimolecular}$ as that which would be measured experimentally from a determination of the concentrations of A and of P. As our development will show, this is not a constant, but depends on the concentration of the collision partners M.

The concentration of A^* cannot be measured. Therefore we cannot determine the rate constant k_3 experimentally. We can, however, obtain an approximation to k_3 based on physical reasoning.

In the Lindemann mechanism the final step in the formation of the products, i.e. (14.49) is assumed to be the rate determining step. The collisional activation and deactivation of the molecules A^* occurs on a time scale that is much shorter than that for this rate determining step, which is the reaction time scale. Mathematically this means that the reaction time scale is determined by the value of the rate constant k_3. We may then assume that the concentration of the molecules A^* is in a state of equilibrium determined by the conditions in the reacting mixture at any time. Mathematically this is the assumption that, on the reaction time scale,

---

[6] Frederick Alexander Lindemann, 1st Viscount Cherwell (1886–1957), was an English physicist who was an influential scientific adviser to the WWII British government.

$$\frac{d}{dt}[A^*] = 0. \tag{14.51}$$

This assumption can be used to obtain an expression for the concentration of A^* as a function of the concentrations of A and M at a particular time in the course of the reaction.

If the reaction takes place in a gas mixture we may assume that the reaction rate is proportional to a product of the concentrations of the collision partners (Stosszahlansatz). From (14.48) and (14.49) we may then write

$$\frac{d}{dt}[A^*] = k_1[A][M] - k_2[A^*][M] - k_3[A^*] \tag{14.52}$$

as the kinetic equation for $[A^*]$. With (14.51) we have

$$k_1[A][M] - k_2[A^*][M] - k_3[A^*] = 0. \tag{14.53}$$

Solving (14.53) for the (instantaneous) concentration of $[A^*]$,

$$[A^*] = \frac{k_1[A][M]}{k_2[M] + k_3}. \tag{14.54}$$

The rate of the reaction in (14.49) is the product of k_3 and the concentration of A^* in (14.54). That is

$$\frac{d[P]}{dt} = k_3 \frac{k_1[A][M]}{k_2[M] + k_3}. \tag{14.55}$$

From the formulation of Eq. (14.50) we may also write $d[P]/dt$ as

$$\frac{d[P]}{dt} = k_{\text{unimolecular}}[A], \tag{14.56}$$

Then, comparing (14.55) with (14.56) we see that

$$k_{\text{unimolecular}} = \frac{k_3 k_1[M]}{k_2[M] + k_3} \tag{14.57}$$

The formulation of the reaction as unimolecular, then, produces a rate constant that is dependent on the molecular concentration, and, therefore, on pressure. We have plotted the rate constant $k_{\text{unimolecular}}$ from (14.57) as a function of $[M]$ in Fig. 14.5. The form of $k_{\text{unimolecular}}$ may be simplified under certain limiting conditions.

If deactivation is more probable than transformation to products, then from (14.48) we have $k_2 \gg k_1$. And from (14.49) k_3 is small. Then $k_2[M] \gg k_3$ and $k_{\text{unimolecular}} \approx k_3 k_1/k_2$. Equation(14.55) then becomes

$$\frac{d[P]}{dt} = k_3 \frac{k_1}{k_2}[A]. \tag{14.58}$$

Fig. 14.5 Plot of $k_{unimolecular}$ vs [M] for Lindemann mechanism

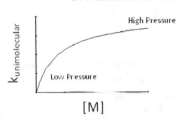

Regardless of the actual magnitudes of k_2 and k_3, this condition results if $[M] \gg k_3/k_2$. Since [M] is proportional to the system pressure, this limit is called the high pressure limit. In this limit the reaction is first order and (14.54) gives $[A^*]$ as proportional to [A].

At the other extreme transformation to products is more probable than deactivation. Then $k_3 \gg k_2 [M]$ and (14.55) becomes

$$\frac{d[P]}{dt} = k_1 [A] [M].$$ (14.59)

In this limit the reaction is second order. Regardless of the actual magnitudes of k_2 and k_3, this condition results if $[M] \ll k_3/k_2$ and will be obtained if the pressure of the system is low. This is, therefore, called the low pressure limit.

The order of the kinetics for a reaction that follows the Lindemann mechanism depends, then on the pressure of the system and may be either first or second order. These limits are shown in Fig. 14.5.

In Fig. 14.6 we have plotted $k_{unimolecular}$ from Eq. (14.57) logarithmically. Here we have used the chemical (molecular) concentration [A] as the independent variable. The concentration of the colliding partner [M] is at least proportional to and may be identical to [A]. The slope of the logarithmic plot is unity in the low pressure limit regardless of the value of k_1. The value of $k_{unimolecular}$ becomes a constant in the high pressure limit. Hermann Ramsperger showed that these predictions agreed in principle with experimental results in a study of the decomposition of azomethane [136]:

$$CH_3N_2CH_3 \rightarrow C_2H_6 + N_2.$$

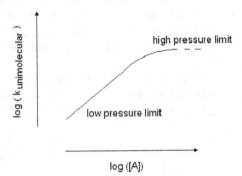

Fig. 14.6 Logarithmic plot of $k_{unimolecular}$ vs [A] for the Lindemann mechanism

We can find the failures of the Lindemann theory if we look closely at quantitative predictions. If we attempt a collision theoretical understanding of the rate constant k_1 we obtain values that are smaller than those obtained experimentally. We can trace the problem to our neglect of the internal modes of motion. There are a number of vibrational modes in a molecule which increase with the number of atoms. For example Ramsperger found it necessary to include 25 vibrational modes for azomethane. The result is an increase in rate constant k_1 from a realization that collisions raise the vibrational energy, which is understood in terms of a quantum treatment of the molecular energies (see Sect. 9.13).

Our treatment of TST has showed us that any transition to final products involves consideration of the potential energy surface of the dynamic complex and the transition from the activated complex to the products. That is the path from the molecules of the component A^* to the product must involve an activated complex $A^{\ddagger}$, which obtains from A^* in the manner discussed above. In TST the transition in (14.49) is represented as

$$A^* \xrightarrow{k_{3a}} A^{\ddagger} \xrightarrow{k^{\ddagger}} P,$$

and the rate constant k_3 is replaced by a combination of the two steps with rate constants k_{3a} and $k^{\ddagger}$. This situation was first successfully treated by Oscar K. Rice and Ramsperger and was later developed by Louis S. Kassel and Rudolph A. Marcus and is now known as the RRKM theory (cf. [130], p. 131). The rate constant $k^{\ddagger}$ is given by a vibration frequency, as we encountered above, and generally $k_{3a} \ll k^{\ddagger}$. The rate determining step is then the formation of the activated complex and $k_3 \approx k_{3a}$. The coefficient k_3 increases with energy and decreases with complexity of the molecule.

The Lindemann theory then provides the basis for the understanding of the physics of unimolecular reactions. Careful considerations of the internal structures and the requirements of the quantum theory provide a better understanding of the energy dependencies of the rate constants (coefficients) and a better understanding of the details of the physical processes involved in unimolecular reactions.

14.6.2 Enzyme Kinetics

Enzymes are proteins which function as catalysts in biological reactions. The importance of proteins in biological systems is based on their structures. The physical form of an enzyme is such that the molecule whose reaction it catalyzes, the substrate, fits into a cavity in the enzyme with bonding sites on the protein lining up with specific sites on the substrate. The role of the enzyme is fundamentally to alter the form of the saddle point on the potential surface making the transition to products more probable. Figure 14.7 is a picture of the action of an enzyme on the potential energy profile.

The solid line is the potential energy profile for the substrate S alone. The energy of the substrate initially is indicated by S and that of the substrate activated complex

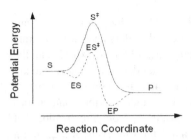

Reaction Coordinate

Fig. 14.7 Change in potential energy profile along the reaction coordinate from action of an enzyme. $S \to S^{\ddagger} \to P$ is the energy profile for the substrate alone (*solid line*). $S \to ES \to ES^{\ddagger} \to EP \to P$ is the energy profile for the enzyme-substrate complex (*dashed line*)

is indicated by $S^{\ddagger}$. The energy of the products in either case is indicated by P. A combination of the substrate with the enzyme E to produce ES is indicated by the dip in the energy profile at ES. The new activated complex of the combined enzyme-substrate ES is indicated by $ES^{\ddagger}$. In the transition to the state indicated by EP the substrate undergoes a change to the product while the enzyme is unchanged. The last step is the release of the product from the enzyme.

In thermodynamic terms the reaction that carries the substrate to products results in a decrease in the Gibbs energy. But the modern understanding of biochemistry and cellular biology provides a much richer and more detailed description of the role of enzymes. In addition to modifying the potential energy surface, enzymes orient the substrate molecule, which facilitates interactions. Our discussion here will, however, only consider the reaction rate equations in what is known as the Michaelis-Menten scheme.[7]

We can write the stoichiometric equation for the enzyme reaction referring to the picture in Fig. 14.7. We shall refer to the substrate as the component S, to the enzyme as E, and to the product as P. The combination of the enzyme and the substrate we shall refer to as the enzyme-substrate complex ES. The terminology complex for this combination is standard and we shall not deviate from this here. The enzyme-substrate complex forms an activated complex at the peak of the potential energy surface along the reaction coordinate, which we designated as $ES^{\ddagger}$ in Fig. 14.7. We then have

$$E + S \underset{k_2}{\overset{k_1}{\rightleftharpoons}} ES \overset{k_3}{\to} E + P. \tag{14.60}$$

as the stoichiometric equation.

The first step is reversible and the enzyme-substrate complex may separate to form the separate enzyme and substrate without completing the reaction. But the product P does not recombine to produce the enzyme-substrate complex. To obtain the differential equations for the reaction we first write this as the sum of two separate steps.

---

[7] Leonor Michaelis (1875–1949) and Maud Leonora Menten (1879–1960). Michaelis was born in Berlin and studied clinical medicine. Menten was born in Port Lambton, Ontario, and in 1911 became one of the first Canadian women to receive a medical doctorate. The famous joint publication appeared in German in 1913.

$$E + S \underset{k_2}{\overset{k_1}{\rightleftharpoons}} ES \tag{14.61}$$

and

$$ES \overset{k_3}{\rightarrow} E + P. \tag{14.62}$$

If we designate the initial enzyme concentration as $[E_0]$, then conservation of the enzyme molecules yields

$$[E_0] = [E] + [ES]. \tag{14.63}$$

The substrate molecules are not conserved. Some of them become product molecules. However, one product molecule results from each substrate molecule. Therefore, if $[S_0]$ is the initial concentration of substrate present, this must appear as free substrate, as enzyme-substrate complex or in the form of the product molecule. Then

$$[S_0] = [S] + [ES] + [P], \tag{14.64}$$

Normally the initial concentration of the enzyme is much lower than that of the substrate. A particular enzyme is used multiple times in a particular reaction. A single enzyme may assist the reaction of 5×10^5 substrate molecules in less than a second.

These biological reactions may take place in membranes of cells. In a membrane the molecules move as a two dimensional gas and the Stosszahlansatz may reasonably be made. For reactions occurring in the cytoplasm of the cell we may think of the enzyme and substrate molecules as moving in an inert liquid and we may again make the Stosszahlansatz for collisions between the enzyme and substrate. With the Stosszahlansatz the kinetic equations follow from (14.61) and (14.62) as

$$\frac{d[E]}{dt} = -k_1 [E] [S] + (k_2 + k_3) [ES] \tag{14.65}$$

$$\frac{d[S]}{dt} = -k_1 [E] [S] + k_2 [ES] \tag{14.66}$$

$$\frac{d[ES]}{dt} = k_1 [E] [S] - (k_2 + k_3) [ES] \tag{14.67}$$

$$\frac{d[P]}{dt} = k_3 [ES]. \tag{14.68}$$

The rate determining step in (14.65),(14.66),(14.67), and (14.68) is the release of the product from the enzyme-substrate complex. The reaction time scale is then determined by the value of k_3. On the reaction time scale we may assume that the enzyme substrate complex is in equilibrium with the free enzyme and substrate. That is,

$$\frac{d[ES]}{dt} = k_1 [E] [S] - (k_2 + k_3) [ES] \approx 0. \tag{14.69}$$

The constant

$$K_M \equiv \frac{k_2 + k_3}{k_1} \tag{14.70}$$

is called the *Michaelis constant*. From the equilibrium assumption of (14.69) we have

$$K_M \approx \frac{([E_0] - [ES]) [S]}{[ES]}, \tag{14.71}$$

using (14.63). With the assumption that $k_3 \ll k_2$

$$K_M \approx \frac{k_2}{k_1} \equiv K_S, \tag{14.72}$$

which is the *substrate constant* for the enzyme. Physically K_S is the ratio of the rate of loss of the complex without product being formed to the rate of production of the enzyme-substrate complex. Low values of K_S then imply high concentrations of the enzyme-substrate complex.

From (14.71) we have an equation for the concentration of the enzyme-substrate complex in terms of the Michaelis constant, which is consistent with the assumption of an equilibrium of the formation of the complex on the reaction time scale,

$$[ES] = \frac{[S]}{(K_M + [S])} [E_0]. \tag{14.73}$$

We may then write the rate determining step as

$$\frac{d[P]}{dt} = k_3 \frac{[S]}{(K_M + [S])} [E_0]. \tag{14.74}$$

Consistent with the assumption that [ES] is a constant we have, from (14.64), that $d[S]/dt = -d[P]/dt$. Then

$$\frac{d[S]}{dt} = k_3 \frac{[S]}{(K_M + [S])} [E_0], \tag{14.75}$$

or, as long as $d[S]/dt \neq 0$,

$$\frac{1}{d[S]/dt} = \frac{K_M}{k_3 [E_0]} \left(\frac{1}{[S]} \right) + \frac{1}{k_3 [E_0]}. \tag{14.76}$$

A plot of $(d[S]/dt)^{-1}$ versus $([S])^{-1}$ is called a *Lineweaver-Burk plot* (cf. [130], p. 213, [33], pp. 42–44). According to (14.76) this is linear. The slope and intercept of this linear plot will K_M and k_3 if $[E_0]$ is known. The intercept must be obtained from an extrapolation to an infinite value of substrate concentration, $[S]$.

14.7 Summary

Although our treatment in this chapter has not been exhaustive, we have presented relatively complete introductions to the basic physics of reactions. We have also chosen two specific types of reactions to study because of the rich details each reveals.

Our understanding of collisionally determined reactions is based on the collision theory developed by Boltzmann in 1872. In this chapter we have used Boltzmann's ideas to develop a collision theory, although we have continued to avoid the details of the full Boltzmann theory. The reader will be able to enter more complete discussions based on what is presented here.

We have devoted more space to the development of transition state theory (TST) because of its present importance in understanding the physics of reactions. In our treatment we have tried to be explicit regarding the ideas underlying the theory [98]. Our study of the microscopic description of the behavior of matter in Chaps. 8 and 9 has made this possible. Through this we have also been able to see how the results of statistical mechanics can be used in practical applications.

Exercises

14.1. C.N. Hinshelwood and W.K. Hutchinson studied the decomposition of gaseous acetaldehyde at high temperatures [73]. The reaction proceeds according to

$$CH_3CHO(g) \rightarrow CH_4(g) + CO(g).$$

As CH_3CHO is lost in the reaction two molecules are added. So the pressure increases as the reaction proceeds. Before the reaction begins the pressure of the CH_3CHO is P_0. At the end of the reaction there is no more CH_3CHO and the vessel pressure is $2P_0$. If we designate the pressure of the vessel at any time as $P(t)$ then the partial pressure of CH_3CHO is $P_{CH_3CHO} = 2P_0 - P(t)$. The pressure in the reaction vessel, held at 791 K, designated as $P(t)$, was recorded. The initial pressure in the vessel was $P_0 = 363\,torr = 4.8395 \times 10^{-2}\,MPa$. The data recorded are in Table 14.1.

Table 14.1 Pressure vs time acetaldehyde

t/S	$P/torr$
0	363
42	397
105	437
242	497
480	557
840	607
1070	627
1440	647
∞	726

You want to find first the kinetic description of the reaction and then find the rate constant. This is almost never an easy process, although there are certain patterns you may look for. In Fig. 14.8 we plotted the data for P_{CH_3CHO} in

Fig. 14.8 Plots of P_{CH_3CHO}
(a) and $1/P_{CH_3CHO}$ (b) as
functions of the time t

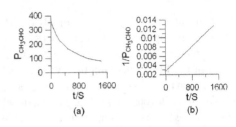

two separate ways. To analyze the data you must have a linear plot. You can
analyze nothing else.

The plot in Fig. 14.8(a) indicates that the reaction is not $dP_{CH_3CHO}/dt = $ constant. It is also not a first order reaction (see (14.47)). The plot in Fig. 14.8(b)
shows that $d(1/P_{CH_3CHO})/dt = $ constant.

(a) Reduce the data to obtain $P_{CH_3CHO}(t)$ and plot the data semilogarithmi-
cally to show that the reaction is not $dP_{CH_3CHO}/dt = kP_{CH_3CHO}$.
(b) Show that the result we have deduced graphically predicts a reaction of
second order.
(c) Plot the data as we have in Fig. 14.8(b) and determine the rate constant
from the data.

14.2. Derive (14.6) for molecules almost equal masses. Begin with the relative kinetic energy of the colliding molecules

Two masses m_1 and m_2 have a relative kinetic energy $T = (m_1 m_2) g^2/2(m_1 + m_2)$ where $\mathbf{g} = \mathbf{v}_1 - \mathbf{v}_2$ is the relative velocity. Introduce the angle between
the velocities ϑ by $\mathbf{v}_1 \cdot \mathbf{v}_2 = v_1 v_2 \cos \vartheta$ and use $\eta_1 + \eta_2 = 1$.

14.3. You are studying the Lindemann-type reaction $A \underset{k_2}{\overset{k_1}{\rightleftharpoons}} A^{\ddagger} \overset{k_3}{\rightarrow} B$.

Assume that you study the reaction using flash photolysis. That is the reaction
is initiated by absorption of a photon. And you may then measure the loss of
A by measuring the absorption of a tuned laser beam as the reaction proceeds.
You are able to gather short time data for the initial stage of the reaction. Your
data for $[A]/[A]_0$ as a function of time are in Table 14.2.

Table 14.2 $[A]/[A_0]$ versus time

Time ms	$[A]/[A_0]$
25	1.071
75	1.231
125	1.414
175	1.624
225	1.866
275	2.144
325	2.462
375	2.828

You then conduct the same experiment for temperature controlled conditions between $280 - 300\,K$. Since you know the order of the reaction you can now record data as half-times $t_{1/2}$ defined as the time for one half of A to transform into B. Your data are in Table 14.3.

Table 14.3 Temperature versus half-time

Temperature K	$t_{1/2}$ ms
280	251.95
284	251.49
288	251.05
292	250.62
296	250.20
300	249.80

(a) What is the order of the reaction? Why?

(b) What is the time constant τ for the reaction?

14.4. Conduct a parallel development to that in the text for this Lindemann reaction.

(a) What are the analogs of $k_{products}$ and $k^{\ddagger}$ in your reaction.

(b) what is your time constant in TST terms?

14.5. Turn now to the macroscopic theory of TST. Plot your data as $\ln \tau$ vs $1/T$.

(a) Obtain E_{exp} and $\Delta u^{\ddagger}$ and $\Delta h^{\ddagger}$ from your plotted data.
You now have a basic understanding of the physics of the reaction in terms of TST theory. The final step is to obtain the factor $A(T)$ in the macroscopic theory (see (14.41)). You wisely took your temperature data over a limited range. So consider T constant for the evaluation of $A(T)$.

(b) Obtain $A(T)$ for the reaction.

Consider the expression for $A(T)$ in (14.41). If you pick a midrange value for T (i.e. $290\,K$) there are only two unknowns in $A(T)$. The value of κ must be obtained from hypotheses about the breakup of the activated complex. If you know κ the function $A(T)$ becomes a measure of the activation entropy $\Delta s^{\ddagger}$. Note that $\Delta u^{\ddagger}$, $\Delta h^{\ddagger}$, and $\Delta s^{\ddagger}$ are system quantities, and not properties of single molecules.

14.6. Your experiment on the Lindemann-type reaction used flash photolysis as a measuring technique. In this technique the sample is exposed to a flash of very short duration to begin the reaction. Before and during the reaction a tuned laser, or a filtered beam of light passes through the region where the reaction is taking place. From measurement of the laser intensity before and after passing through the sample the absorption coefficient of the sample is determined. The data are then for asorption coefficient as a function of time.

If the dependence of the absorption coefficients of A and B on frequency are sufficiently different that the presence of A or B can be measured separately the experiment is then a direct determination of the loss of A or the gain of B. Unfortunately this is often not the case. The abosorption coefficients of A and B may overlap.

Consider the reaction A→2B. At the laser frequency the absorption coefficients are $k_{abs,A}$ and $k_{abs,B}$. At any time there are $n(t)$ total mols present: $n_A(t)$ mols of A and $n_B(t)$ mols of B. Your absorption measurement records an absorption coefficient $K_{abs}(t)$, which you equate to $n_A(t)k_{abs,A} + n_B(t)k_{abs,B}$. Assume that the reaction runs to completion and there is only B present as $t \to \infty$. You can then determine the total number of mols that were present in the sample.

Find $n_A(t)$ and $n_B(t)$ in terms of $K(t)$.

14.7. For the reaction $N_2O_5 \rightleftharpoons N_2O_3 + O_2$ the forward step is monomolecular and the reverse involves a collision.

(a) What is the kinetic equation resulting from the Stosszahlansatz?
(b) Drop the collisional term, assuming it to be small compared to the monomolecular term. What is the order of the resultant kinetic equation?
(c) Solve your simplified kinetic equation.
(d) You study the reaction in a vessel of fixed volume. The reaction begins at time $t = 0$ when the pressure in the reaction vessel is P_0. Find the pressure for all later times.

Chapter 15
Solutions

15.1 Introduction

Solutions are homogeneous liquid or solid mixtures. In this chapter we shall consider only liquid solutions.

Many chemical reactions of interest in the laboratory and in industry take place among components of liquid solutions. In Chap. 13 we considered chemical stability in nonideal gases, building upon the known properties of ideal gases introducing nonideality carefully in a systematic fashion. We cannot do this in a liquid solution since the molecules are never widely separated from one another, except in the vapor of the solution. Our discussion of solutions will then be of a more general thermodynamic nature from the beginning. We shall introduce empirical results for the vapor to idealize our final expressions for the chemical potentials.

15.2 Thermodynamic Equilibrium in Solutions

We shall consider a system containing a solution in equilibrium with its vapor. The solution contains a number of distinct components in the liquid state, which we designate by the subscript λ. The vapor also contains the same components. Equilibrium between the vapor and the solution requires that the evaporation of each component from the solution into the vapor is balanced, in terms of the mass transferred, by condensation of the vapor at the surface of the solution. That is the number of mols of each component in the solution and in the vapor are constants. If we designate the number of mols of the component λ in the vapor (gaseous state) as $n_{g,\lambda}$ and in the solution (liquid state) as $n_{l,\lambda}$, then

$$\sum_{\lambda} n_\lambda = \sum_{\lambda} \left(n_{g,\lambda} + n_{l,\lambda} \right) = \text{constant}. \tag{15.1}$$

where n_λ is the number of mols of the component λ in the system. With no net evaporation the volumes of the solution V_l and of the vapor V_g are also constants and their sum is the total system volume

C.S. Helrich, *Modern Thermodynamics with Statistical Mechanics*,
DOI 10.1007/978-3-540-85418-0_15, © Springer-Verlag Berlin Heidelberg 2009

$$V_{\text{total}} = V_1 + V_g. \tag{15.2}$$

If the system is in stable thermodynamic equilibrium the total internal energy of the system U_{total} is a constant and equal to the sum of the the internal energies of the solution U_1 and of the vapor U_g

$$U_{\text{total}} = U_1 + U_g. \tag{15.3}$$

And the total system entropy S_{total}, which is the sum of the entropy of the solution S_1 and of the vapor S_g, is a maximum. The specific conditions of equilibrium for the system can be found using the method of Lagrange undetermined multipliers (see Sect. A.4). We define the subsidiary function (see (A.22))

$$\begin{aligned} h_{\text{sol}}(U, V, \{n_\lambda\}) &= (S_1 + S_g) + \alpha_1 (U_1 + U_g) \\ &\quad + \alpha_2 (V_1 + V_g) + \alpha_3 \sum_\lambda (n_{1,\lambda} + n_{g,\lambda}) \end{aligned} \tag{15.4}$$

and seek the extremum of this function. The variables U, S, and V must satisfy the Gibbs equation for the solution

$$dU_1 = T_1 dS_1 - P_1 dV_1 + \sum_\lambda \mu_{1,\lambda} dn_{1,\lambda} \tag{15.5}$$

and the vapor

$$dU_g = T_g dS_g - P_g dV_g + \sum_\lambda \mu_{g,\lambda} dn_{g,\lambda}. \tag{15.6}$$

Using (15.5) and (15.6) the variation of h_{sol} is

$$\begin{aligned} \delta h_{\text{sol}}(U, V, \{n_\lambda\}) &= 0 \\ &= \delta S_g (1 + \alpha_1 T_g) + \delta S_1 (1 + \alpha_1 T_1) \\ &\quad + \delta V_g (\alpha_2 - \alpha_1 P_g) + \delta V_1 (\alpha_2 - \alpha_1 P_1) \\ &\quad + \sum_\lambda \delta n_{g,\lambda} (\alpha_3 + \alpha_1 \mu_{g,\lambda}) + \sum_\lambda \delta n_{1,\lambda} (\alpha_3 + \alpha_1 \mu_{1,\lambda}) \end{aligned} \tag{15.7}$$

The variations in (15.7) are independent of one another. The coefficients of the variations must then all vanish. That is

$$T_g = T_1, \tag{15.8}$$

$$P_g = P_1, \tag{15.9}$$

and

$$\mu_{g,\lambda} = \mu_{1,\lambda}. \tag{15.10}$$

Equilibrium then requires that the temperatures T_g and T_1, the pressures P_g and P_1, and the chemical potentials $\mu_{g,\lambda}$ and $\mu_{1,\lambda}$ are equal to one another.

If a general chemical reaction (13.1) takes place among the components of the solution, the analysis of equilibrium of the chemical reaction for liquid solutions will identical to that for gases, since only general thermodynamic and chemical principles are involved. Therefore we arrive again at the condition for chemical equilibrium expressed in (13.9). We then have a complete description of chemical equilibrium in solutions in (15.8), (15.9), (15.10) and (13.9).

We also have an understanding of the chemical potential of pure nonideal gases and mixtures of nonideal gases (see Sects. 13.3.3, (13.30) and (13.29)). Therefore we can make our measurements on the vapor above the solution and calculate the chemical potential for the gaseous state, which will be equal to the chemical potential for that component in the solution by (15.10). But this only results in values for the chemical potentials of the constituents in the solution. It does not provide the dependence of the chemical potential in solution on the properties of the solution.

15.3   Chemical Potential of a Solution

The Gibbs energy for a solution depends on temperature, pressure, and the molar composition, i.e. $G = G(T,P,\{n_\lambda\})$. The chemical potential of a solution is then generally

$$\mu_\lambda(T,P,\{n_\lambda\}) = \left(\frac{\partial G}{\partial n_\lambda}\right)_{T,P,n_\rho \neq n_\lambda}. \tag{15.11}$$

Because the order of partial differentiation is immaterial,

$$\left(\frac{\partial^2 G}{\partial n_v \partial n_\lambda}\right)_{T,P} = \left(\frac{\partial^2 G}{\partial n_\lambda \partial n_v}\right)_{T,P},$$

or

$$\left(\frac{\partial \mu_\lambda}{\partial n_v}\right)_{T,P,n_\rho \neq n_v} = \left(\frac{\partial \mu_v}{\partial n_\lambda}\right)_{T,P,n_\rho \neq n_\lambda}. \tag{15.12}$$

For stability we also have

$$(\partial \mu_\lambda / \partial n_\lambda)_{T,P,n_\rho \neq n_\lambda} > 0 \tag{15.13}$$

(see Sect. 12.4, (12.29)).

Laboratory studies of reactions in solution are normally conducted under conditions for which the temperature and pressure are constant. So we shall choose T and P constant for our discussion. For simplicity we shall also begin with a binary solution and label the components with the subscripts 1 and 2.

The Gibbs-Duhem equation (see Sect. 4.7.3, (4.95)) for a binary solution takes the form

$$n_1 d\mu_1 + n_2 d\mu_2 = 0. \tag{15.14}$$

Since T and P are constant, the chemical potentials are both functions only of (n_1, n_2). Then the Pfaffian for μ_1 is

$$d\mu_1 = \left(\frac{\partial \mu_1}{\partial n_1}\right) dn_1 + \left(\frac{\partial \mu_1}{\partial n_2}\right) dn_2,$$

and similarly for μ_2. Because there will be no confusion, we shall, for the sake of economy in writing, drop the subscripts on the partial derivatives in what follows. The Gibbs-Duhem equation for a binary solution with (T,P) constant is then

$$0 = \left[n_1 \left(\frac{\partial \mu_1}{\partial n_1}\right) + n_2 \left(\frac{\partial \mu_2}{\partial n_1}\right)\right] dn_1$$
$$+ \left[n_1 \left(\frac{\partial \mu_1}{\partial n_2}\right) + n_2 \left(\frac{\partial \mu_2}{\partial n_2}\right)\right] dn_2. \tag{15.15}$$

The mol numbers n_1 and n_2 are independent. Therefore (15.15) results in

$$n_1 \left(\frac{\partial \mu_1}{\partial n_1}\right) + n_2 \left(\frac{\partial \mu_2}{\partial n_1}\right) = 0 \tag{15.16}$$

and

$$n_1 \left(\frac{\partial \mu_1}{\partial n_2}\right) + n_2 \left(\frac{\partial \mu_2}{\partial n_2}\right) = 0. \tag{15.17}$$

Using the stability inequality (15.13) and (15.12), (15.16) and (15.17) result in

$$\left(\frac{\partial \mu_2}{\partial n_1}\right) = \left(\frac{\partial \mu_1}{\partial n_2}\right) < 0. \tag{15.18}$$

We now consider that the solution is dilute. We choose the subscript 1 to designate the solvent and 2 to designate the solute and consider the limit as $n_2 \to 0$. Equation (15.17) gives us information on the change in the potentials with n_2. In the limit as $n_2 \to 0$ Eq. (15.17), requires that either

$$\lim_{n_2 \to 0} \left(\frac{\partial \mu_1}{\partial n_2}\right) = 0 \text{ and } \lim_{n_2 \to 0} \left(\frac{\partial \mu_2}{\partial n_2}\right) \text{ is finite} \tag{15.19}$$

or

$$\lim_{n_2 \to 0} \left(\frac{\partial \mu_1}{\partial n_2}\right) \neq 0 \text{ and } \lim_{n_2 \to 0} \left(\frac{\partial \mu_2}{\partial n_2}\right) \to \infty. \tag{15.20}$$

The chemical potential is at least affected by a mixing term that does not vanish even as the density of the solute becomes infinitesimal (see Sect. 4.7.2, (4.73)). So it does not seem reasonable to assume that $\lim_{n_2 \to 0}(\partial \mu_1/\partial n_2) = 0$ (cf. [162], p. 400). Therefore the conditions (15.20) are the limiting conditions with which we shall construct our potentials in the solution.

For very small n_2 we can represent the dependence of μ_1 on n_2 as a power (Taylor) series in n_2

$$\mu_1 = \mu_1 (n_1, n_2)]_{n_2=0} + \alpha_1 n_2 + \alpha_2 n_2^2 + O\left(n_2^3\right)$$

where the αs are constants. The first term in the series cannot be proportional to any power of n_2 greater than unity, since $\partial \mu_1/\partial n_2$ would then be proportional to n_2 and would vanish as $n_2 \to 0$. Therefore μ_1 must be linear in n_2 in the limit of vanishing n_2. Using the inequality (15.18) we may then write

$$\lim_{n_2 \to 0} \left(\frac{\partial \mu_1}{\partial n_2} \right) = -C_1. \tag{15.21}$$

where C_1 is a positive constant at a fixed temperature and pressure, i.e. it may be a function of temperature and pressure. From (15.17) we then have

$$\left(\frac{\partial \mu_2}{\partial n_2} \right) = \frac{1}{n_2} C_1. \tag{15.22}$$

We now have two differential Eqs. (15.21) and (15.22) for the chemical potentials of the solvent and the solute, which are valid in the limit of very dilute concentration of the solute.

Molarity, the number of mols per liter of solution, is the most convenient unit of measure when solutions are prepared at the laboratory bench. A related unit which is more convenient for the purposes of calculation is the mol fraction. So for convenience we introduce the mol fractions $\chi_1 = n_1/(n_1 + n_2)$ and $\chi_2 = n_2/(n_1 + n_2)$ with $\chi_1 + \chi_2 = 1$. Then (15.21) and (15.22) become

$$\frac{\partial \mu_1}{\partial \chi_2} = -C_1 \text{ and } \frac{\partial \mu_2}{\partial \chi_2} = \frac{1}{\chi_2} C_1.$$

Integrating we have

$$\mu_1 = \mu_{\text{pure},1} - C_1 \chi_2 \tag{15.23}$$

and

$$\mu_2 = C_2 + C_1 \ln \chi_2. \tag{15.24}$$

Here $\mu_{\text{pure},1}$ is the chemical potential for the pure solvent in the liquid state, since $\mu_1 = \mu_{\text{pure},1}$ when $\chi_2 = 0$, and C_2 is a constant of integration, which may be a function of temperature and pressure. We note that in the limit of very small χ_2, $\ln(\chi_1) = \ln(1 - \chi_2) = -\chi_2$ neglecting terms of order χ_2^2. We may, therefore, assume that

$$\mu_1 = \mu_{\text{pure},1} + C_1 \ln \chi_1 \tag{15.25}$$

is the actual limiting form of (15.23) valid for $\chi_1 \approx 1$. Equations (15.24) and (15.25) satisfy the self consistency requirement (15.12) and they have a mathematical similarity to one another. We may, therefore, accept them as valid thermodynamic expressions for the chemical potentials of the solvent and solute in a dilute solution.

15.4 Empirical Laws

The pressure of the vapor in equilibrium with a solution is a measurable property that can be used to characterize the solution. The vapor is also a low density gas which may be approximated very well as an ideal gas. From Chap. 13 we know the chemical potentials of the components of a gas mixture. Because they are equal to the chemical potentials of the components in the solution, knowledge of the vapor properties is key to an understanding of the properties of the solution.

There are two empirical laws which relate the partial vapor pressures of the solute and of the solvent to the composition of the solution. These are Henry's[1] and Raoult's[2] laws.

For dilute solutions, i.e. for $\chi_2 \ll 1$, Henry's law relates the partial pressure of the solute in the vapor phase to the mol fraction of the solute in solution as

$$P_2 = K_2 \chi_2 . \tag{15.26}$$

The proportionality constant K_ν is called Henry's constant for the component ν. Here $\nu = 2$ for the single solute, the chemical identity of which is not, at this time, important.

In dilute solutions, i.e. for $\chi_1 \approx 1$, Raoult's law relates the partial pressure of the solvent in the vapor phase to the mol fraction of the solvent in solution as

$$P_1 = P^*_{\text{vap},1} \chi_1 , \tag{15.27}$$

where $P^*_{\text{vap},1}$ is the vapor pressure of the pure solvent. This is the pressure of the vapor in equilibrium with the pure liquid solvent. The situation is illustrated in Fig. 15.1. Two types of partial pressure curves are shown in Fig. 15.1 as examples. Neither Henry's law nor Raoult's law is applicable over wide ranges. We may only apply each with confidence separately at the extremes.

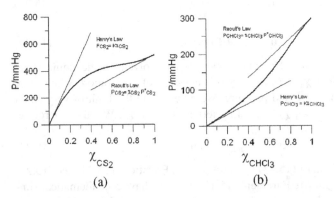

Fig. 15.1 Partial pressure in vapor as a function of mol fraction. Carbon disulfide-acetone **(a)**, chloroform-acetone **(b)**. (From R. Chang: *Physical Chemistry for the Chemical and Biological Sciences*, University Science Books, Sausalito, CA (2000). Reprinted with permission of University Science Books.)

---

[1] William Henry (1775–1836) English chemist, received a medical doctorate from Edinburgh in 1807, but ill-health interrupted his practice as a physician, and he devoted his time mainly to chemical research, especially in regard to gases.

[2] Francois Marie Raoult (1830–1901) French chemist. After various teaching posts, he held the chair in chemistry at Grenoble until his death.

If we use (15.26) in (15.24), or (15.27) in (15.25) we obtain terms $C_1 \ln P_2$ in μ_2 and $C_1 \ln P_1$ in μ_1. The chemical potential of an ideal gas λ is dependent on the partial pressure through the term $RT \ln P_\lambda$ (see (13.21)). Since the chemical potential for each component in the vapor is equal to the chemical potential for that component in the solution, we see that $C_1 = RT$.

We then have an exact form for the chemical potential of the solvent in a dilute solution

$$\mu_1 = \mu_{\text{pure},1} + RT \ln \chi_1, \tag{15.28}$$

which satisfies Raoult's law for $\chi_1 \approx 1$. Our equation for the chemical potential of the solute

$$\mu_2 = C_2 + RT \ln \chi_2 \tag{15.29}$$

is known to within a constant, which may be a function of (T, P).

To identify the constant C_2 we equate the chemical potential in the solution (15.29) to the chemical potential of the (ideal gas) vapor (13.21). The equality is completed by introducing Henry's law (15.26) to relate the partial pressure in the gas and the mol fraction in the solution. The result is

$$C_2 = C_{P,\,\text{vap},2} \left(T - T \ln \frac{T}{T_{\text{ref}}} \right)$$
$$+ h_{0,2} - T s_{\text{ref},2} - RT \ln P_{\text{ref}} + RT \ln K_2 \tag{15.30}$$

(see (13.12)).

The specific heat $C_{P,\text{vap},2}$ is that of the vapor of the solute. The constant C_2 is then well-defined function of the temperature and reference conditions. But it is not the chemical potential of a pure component. There is no absolute uniformity in the designation of this quantity. Some authors indicate this as a starred chemical potential. We shall use a designation $\mu^*_{\text{vap},\lambda}(T, P)$ for the solute λ to indicate that the vapor and not the liquid is used in the definition. Then (15.29) becomes

$$\mu_2 = \mu^*_{\text{vap},2} + RT \ln \chi_2. \tag{15.31}$$

Our development here may be repeated for any number of solutes. If we designate the solvent by the subscript 1 and all solutes by subscripts $\lambda \geq 2$, equations of the form (15.28) for the solvent and equations of the form (15.31) for the solutes are obtained from a general Gibbs energy

$$G(T, P, \{n_\lambda\}) = n_1 \mu_{\text{pure},1}(T, P) + \sum_{\lambda \geq 2} n_\lambda \mu^*_{\text{vap},\lambda}(T, P)$$
$$+ RT \left(\sum_{\lambda \geq 1} n_\lambda \ln n_\lambda - n \ln n \right), \tag{15.32}$$

where $n = \sum n_\lambda$ is the total number of mols in the solution. This form of the Gibbs energy is not exact. It is an idealization which contains approximations for the solvent (Raoult's law) and for the solute (Henry's law). What we have then is a semiempirical representation of the chemical potentials of solvent and solute in a solution.

15.5 Ideal Solutions

15.5.1 Ideal Solutions

Ideal solutions are defined as those for which Raoult's law holds for all components over the entire range of composition. Ideal solutions are then those for which the Gibbs energy for the solution is

$$G(T,P,\{n_\lambda\}) = \sum_{\lambda \geq 1} n_\lambda \mu_{\text{pure},\lambda}(T,P) + RT \left(\sum_{\lambda \geq 1} n_\lambda \ln n_\lambda - n \ln n \right). \quad (15.33)$$

The chemical potentials of the components are then

$$\mu_v(T,P,\{n_\lambda\}) = \left[\frac{\partial}{\partial n_v} G(T,P,\{n_\lambda\}) \right]_{T,P,n_{\beta \neq v}}$$

$$= \mu_{\text{pure},v}(T,P) + RT \ln \chi_v \qquad (15.34)$$

In (15.34) the chemical potentials for the individual components are similar in form to those for ideal gases. However, while all gases approach the ideal gas in the limit of vanishing pressure, solutions do not approach ideal behavior in any limiting sense. A solution is a form of condensed matter. The molecules are in constant contact and the deviation of the properties of the solution from the ideal solution here is determined by the interaction between the molecules, which is not affected by any limiting process.

It may seem more appropriate to assume that only the solvent obeys Raoult's Law and all solutes obey Henry's law and to define the ideal solution as one with different chemical potentials for solvent and solutes. The Gibbs energy we would use would then be (15.32). This choice would increase the mathematical difficulties we would encounter. And this approach still is not exact, as is evident from the graphs in Fig. 15.1.

As in any application we are always free to choose approximations that seem most appropriate. Our decision will be made based on practicality. What is suitable in the application will be decided by a balance between the mathematical difficulty involved and the accuracy we require in the calculation.

15.5.2 Lewis-Randall Ideal Solutions

An alternative approach to the definition of ideality is based on the dependence of the fugacity on composition of the solution. Ideality in this case is similar to the approximations introduced in the models of the gases discussed in Chap. 13.

We shall accept differences among molecules of the different components in terms of masses and internal structures. But we shall introduce the Ansatz that the interactions between individual molecules are independent of molecular identity. If this is the case then each fugacity deviates from its ideal gas value by an amount that is independent of composition ([123], p. 185). Specifically the fugacity coefficient (see (13.31)) for the ideal solution, designated here by the superscript 'is',

$$\varphi_\lambda^{is}(T,P) = \frac{f_\lambda^{is}(T,P,\{n_\lambda\})}{\chi_\lambda P} \tag{15.35}$$

is independent of composition. Therefore the fugacity of the ideal solution is linearly dependent on the mol fraction χ_λ. We will take (15.35) to be the defining characteristic of an ideal solution.

Since the fugacity of an ideal solution f_λ^{is} is proportional to χ_λ and is equal to the fugacity of the pure component when $\chi_\lambda = 1$,

$$f_\lambda^{is}\left(T,P_\lambda^0,\{n_\lambda\}\right) = \chi_\lambda f_\lambda^0\left(T,P_\lambda^0\right), \tag{15.36}$$

where, as before, $f_\lambda^0\left(T,P_\lambda^0\right)$ is the fugacity in a well-defined standard state. The choice of the standard state is generally arbitrary and may be conveniently chosen for the component. That is P_λ^0 may in general be different for different components. If we use what is called the *Lewis-Randall rule*, according to which the standard state is the pure substance in the phase of the solution, i.e. a liquid in our case, and at the temperature and pressure of the solution, then $f_\lambda^0\left(T,P^0\right) = f_{\text{pure},\lambda}(T,P)$ and

$$f_\lambda^{is}(T,P,\{n_\lambda\}) = \chi_\lambda f_{\text{pure},\lambda}(T,P). \tag{15.37}$$

A solution for which the fugacities of the components have this form is called a *Lewis-Randall ideal solution*.

From the general form of the chemical potential in terms of the fugacity in (13.39) and using (15.37) we may write

$$\mu_\lambda^{is}(T,P,\{n_\lambda\}) = \mu_{\text{pure},\lambda}(T,P) + RT\ln\frac{f_\lambda^{is}(T,P,\{n_\lambda\})}{f_{\text{pure},\lambda}(T,P)}$$
$$= \mu_{\text{pure},\lambda}(T,P) + RT\ln\chi_\lambda \tag{15.38}$$

That is the chemical potential for the Lewis-Randall ideal solution is the same as the chemical potential for the previous ideal solution. The two ideal solutions are then thermodynamically identical.

Our first ideal solution used the behavior of the solvent as the basis for the behavior of all components. In molecular terms this is the claim that all molecular interactions are identical. This claim is the same as the Ansatz that led us to the Lewis-Randall solution. The advantage that has been gained through consideration of both approaches is physical insight. Any thermodynamic difference in the two results then would, therefore, have come as a surprise to us.

If we introduce a standard state with pressure P_λ^0 in place of the Lewis-Randall standard state (15.38) becomes

$$\mu_\lambda\left(T,P,\{n_\lambda\}\right) - \mu_\lambda^0\left(T,P_\lambda^0\right) = RT\ln\frac{f_\lambda\left(T,P,\{n_\lambda\}\right)}{f_\lambda^0\left(T,P_\lambda^0\right)}, \tag{15.39}$$

which is the form of (13.39) for solutions.

We drop the superscript 'is' here, since the solution is no longer a Lewis-Randall ideal solution, and we drop the notation 'pure,' which is understood. The choice of a standard state with a pressure P_λ^0 specific to the component differs from the reference chosen in the case of gas mixtures (see (13.39)). The choice for solutions is dictated primarily on the basis of practicality and is based on the availability of data and on procedures that can be used to compute fugacities from those available data. The Lewis-Randall standard state may or may not fall into this category. In dealing with solutions we must be careful that we understand the reference states used by individual authors before applying results.

We may now define an activity for components of a solution as we did in the case of gases (see (13.41)).

$$a_\lambda\left(T,P,\{n_\lambda\};f_\lambda^0\right) = \frac{f_\lambda\left(T,P,\{n_\lambda\}\right)}{f_\lambda^0\left(T,P_\lambda^0\right)}. \tag{15.40}$$

There is no difference between the notation here and that for gases in (13.41). We need only be careful that we understand the reference state chosen in applications.

With (15.40) (15.39) becomes

$$\mu_\lambda\left(T,P,\{n_\lambda\}\right) - \mu_\lambda^0\left(T,P_\lambda^0\right) = RT\ln a_\lambda\left(T,P,\{n_\lambda\};f_\lambda^0\right). \tag{15.41}$$

There is then no formal mathematical difference between our treatment of chemical equilibrium in gases and now in solutions provided we base our analysis on the activities.

15.6   Activity Coefficient

In our treatment of gases we introduced a fugacity coefficient defined as the ratio of the fugacity to the partial pressure of the component (see (13.31)). In a similar vein we introduce the activity coefficient[3] γ_λ for a solution as

$$\gamma_\lambda\left(T,P,\{n_\lambda\};f_\lambda^{is}\right) = \frac{f_\lambda\left(T,P,\{n_\lambda\}\right)}{f_\lambda^{is}\left(T,P_\lambda^0,\{n_\lambda\}\right)}. \tag{15.42}$$

Here $f_\lambda^{is}\left(T,P_\lambda^0,\{n_\lambda\}\right)$ is the fugacity of the component λ in an ideal solution at the composition $\{n_\lambda\}$ and the reference pressure P_λ^0.

---

[3] The term 'activity coefficient' was proposed by Svante Arrhenius in his doctoral dissertation at Uppsala in 1884. Arrhenius' dissertation contained some 56 theses, most of which have proven correct. The modern definition of activity coefficient was given by A. A. Noyes and W.C. Bray in 1911.

Using our defining characteristic of the ideal solution as a solution for which f_λ^{is} is equal to $\chi_\lambda f_\lambda^0$, and using (15.40) in (15.42), we have

$$\gamma_\lambda \left(T,P,\{n_\lambda\};f_\lambda^0\right) = \frac{a_\lambda \left(T,P,\{n_\lambda\};f_\lambda^0\right)}{\chi_\lambda}. \tag{15.43}$$

If we choose the reference state for chemical potential to be the ideal solution and introduce (15.42), (15.39) becomes

$$\mu_\lambda \left(T,P,\{n_\lambda\}\right) - \mu_\lambda^{is} \left(T,P_\lambda^0,\{n_\lambda\}\right) = RT \ln \gamma_\lambda \left(T,P,\{n_\lambda\};f_\lambda^0\right). \tag{15.44}$$

The difference between (15.44) and (15.41) is the reference state for the chemical potential. This difference produces a logarithm of either the activity or the activity coefficient respectively. We may base our studies on either (15.44) or (15.41) and we will make the choice based on the data available and convenience of the reference point.

15.7 Excess and Mixing Properties

The difference of the two terms appearing on the left hand side of (15.44) is what is termed the excess chemical potential $\mu_\lambda^E (T,P,\{n_\lambda\})$. This is an example of a general concept that has been introduced to separate ideal contributions from the nonideal [cf. [36], pp. 274–76, [123], pp. 189–94]. Ideal contributions can be treated mathematically in a straightforward fashion. However, as we have seen in our discussion of the empirical results for vapor pressure, the behavior of solutions is not ideal. Isolating the nonideal contributions to the thermodynamic potentials has proven very fruitful in investigations of solutions. In this section we shall generalize the isolation of mixing and excess contributions to the Gibbs and the Helmholtz energies and the entropy.

There is a mixing contribution to the Gibbs and the Helmholtz energies and the entropy, but there is no mixing contribution to either the internal energy or the enthalpy. We shall designate the Gibbs and the Helmholtz energies and the entropy generically as $\Phi (X,Y,\{\chi_\lambda\})$, with the molar value of $\Phi(X,Y,\{\chi_\lambda\})$ defined by

$$\phi (X,Y,\{\chi_\lambda\}) = \frac{1}{\sum_\lambda n_\lambda} \Phi(X,Y,\{\chi_\lambda\})$$
$$= \sum_\lambda \chi_\lambda \phi_\lambda (X,Y,\{\chi_\lambda\}) \tag{15.45}$$

where $\phi_\lambda (X,Y,\{\chi_\lambda\})$ is the molar contribution of the component λ to $\phi (X,Y,\{\chi_\lambda\})$ for the solution. The functions X, and Y may be (T,P), (T,V), or (U,V) depending on the identity of Φ. We include $\{\chi_\lambda\}$ to designate that there is a mixture of components present.

We first designate the molar change of the property Φ on mixing by a superscript 'mix.' That is

$$\phi^{\text{mix}}(X,Y,\{\chi_\lambda\}) = \sum_\lambda \chi_\lambda \phi_\lambda(X,Y,\{\chi_\lambda\}) - \sum_\lambda \chi_\lambda \phi_{\text{pure},\lambda}(X,Y) \qquad (15.46)$$

in which $\phi_{\text{pure},\lambda}(X,Y)$ is the molar value of the property for the pure component and $\phi_\lambda(X,Y,\{\chi_\lambda\})$ is the molar contribution of the component λ to the solution. In the case of ideal gases the only mixing contribution came from the random distribution of the molecules. In our discussion of nonideal gases we found, however, that there are additional contributions from interactions (see (13.30)). There is no reason to expect that the sole effect of mixing in a solution will come from the randomization of the molecules, except in the case of an ideal solution.

The chemical potential for the ideal solution is (15.38). Therefore we can write the molar ideal solution value of $\Phi(X,Y,\{\chi_\lambda\})$ as

$$\phi_\lambda^{\text{is}}(X,Y,\{\chi_\lambda\}) = \phi_{\text{pure},\lambda}(X,Y) + RT\ln\chi_\lambda \qquad (15.47)$$

if Φ is G or F and

$$\phi_\lambda^{\text{is}}(X,Y,\{\chi_\lambda\}) = \phi_{\text{pure},\lambda}(X,Y) - R\ln\chi_\lambda \qquad (15.48)$$

if Φ is S. The ideal solution mixing contributions per mol are then

$$\phi^{\text{is,mix}}(X,Y,\{\chi_\lambda\}) = RT\sum_\lambda \chi_\lambda \ln\chi_\lambda \qquad (15.49)$$

if Φ is G or F and

$$\phi^{\text{is,mix}}(X,Y,\{\chi_\lambda\}) = -R\sum_\lambda \chi_\lambda \ln\chi_\lambda \qquad (15.50)$$

if Φ is S.

We now define the molar excess of the property $\Phi(X,Y,\{\chi_\lambda\})$ for the system as

$$\phi^{\text{E}}(X,Y,\{\chi_\lambda\}) = \sum_\lambda \chi_\lambda \phi_\lambda(X,Y,\{\chi_\lambda\}) - \sum_\lambda \chi_\lambda \phi_\lambda^{\text{is}}(X,Y,\{\chi_\lambda\}).$$

Using (15.47) this is

$$\phi^{\text{E}}(X,Y,\{\chi_\lambda\}) = \phi^{\text{mix}}(X,Y,\{\chi_\lambda\}) - RT\sum_\lambda \chi_\lambda \ln\chi_\lambda$$

if Φ is G or F and using (15.48) this is

$$\phi^{\text{E}}(X,Y,\{\chi_\lambda\}) = \phi^{\text{mix}}(X,Y,\{\chi_\lambda\}) + R\sum_\lambda \chi_\lambda \ln\chi_\lambda$$

if Φ is S.

These definitions of excess and mixing properties have proven very useful particularly in statistical mechanical modeling calculations. The seminal work in this area is that of John Kirkwood in 1935 and 1936 [92, 93]. More recent efforts using modern computational techniques such as Monte Carlo simulations, have made this approach very practicable [64]. Monte Carlo simulations can be readily performed for very interesting applications on microcomputers from which the excess properties may be obtained from straightforward numerical integration. The agreement with laboratory data has been good (cf. [1]).

The calculation of excess properties using the Kirkwood idea is not conceptually difficult. It is computationally intense, requiring multiple Monte Carlo simulations for each data point. The promise of the method is, however, to gain insight into the thermodynamics of complicated systems based on sophisticated models.

15.8 Summary

In this chapter we have presented a development of the binary solution that is thermodynamically sound, up to a point. In this we have used the requirements of heterogeneous and chemical equilibrium to obtain a valid thermodynamic description of a binary system in a set of differential statements involving the chemical potentials. At that point we have introduced empirical results that are applicable in limiting cases. Our integrated expressions became then the basis of a semiempirical formulation of the chemical potentials for the solution components.

From the requirements of heterogeneous equilibrium between the vapor and the liquid phases, and our understanding of the form of the chemical potential for ideal and nonideal gases, we were able to deeper insight into the details of our semiempirical formulation. We were able to standardize some of this in terms of ideal solutions.

We then introduced the concept of excess quantities and indicated some of the impact a study of excess quantities has had. This results from the fact that excess quantities can be calculated based on an established theory.

Exercises

15.1. Boiling is caused when the vapor pressure of the solution is equal to the pressure on the system. Consider that a nonvolatile substance is added to water. This may be a solution of water and sugar, for example. Raoult's law for a weak solution is $P_1 = P^*_{\text{vap},1} \chi_1$. As sugar is added to the water, χ_1 decreases and the vapor pressure decreases.

(a) Why does this lead to an increase in the boiling point? Consider you have boiling water and drop in sugar cubes. Why must you increase the temperature to return to boiling?

(b) Consider that the vapor pressure of the water is a function of the solution temperature and the mol fraction of the water, and assume that $(\partial P / \partial T)_{\text{coexist}}$ satisfies the Calusius-Clapeyron equation, what is relationship between mol fraction of the water and the boiling temperature?

Chapter 16
Heterogeneous Equilibrium

16.1 Introduction

Heterogenous systems consist of more than one phase. A phase is a state of aggregation in which a system may exist.

In a heterogeneous system the phases are separated by boundaries across which the physical properties of the system abruptly change. All substances exist in solid, liquid or gas phases. Many substances exist as well in allotropic forms, which are phases separated by differences in atomic bonding. This may result in the formation of different molecules, such as oxygen O_2 or O_3, or in the formation of different crystal structures such as graphite or diamond. Different phases in some solids are also identifiable by abrupt differences in electrical and magnetic properties, such as superconductivity or spin glasses.

In this chapter we shall present a general formulation of the thermodynamics of heterogeneous systems. We shall begin where all studies of heterogeneous equilibrium must begin: the Gibbs phase rule. Then we shall study the van der Waals model system in detail in order to develop some of the basic concepts for single component systems. We have chosen the van der Waals model system because it describes some properties of the liquid + gas transition while remaining transparent.

We shall then consider some of the basic ideas in a binary heterogeneous system. This will be treated as an example of a more complicated system.

Our primary intent here is to provide an introduction to the basic concepts of a thoroughly complex and interesting area of study.

16.2 Gibbs Phase Rule

We shall consider a system with C components. Each component is chemically distinct. We assume that we know the number of mols of each component in the system, which we designate by n_k.

In the system there are P separate phases. The mol fraction of component k in the phase j is $\chi_k^{(j)} = n_k^{(j)}/n_k$. These mol fractions are unknown (are variables), even

C.S. Helrich, *Modern Thermodynamics with Statistical Mechanics*,
DOI 10.1007/978-3-540-85418-0_16, © Springer-Verlag Berlin Heidelberg 2009

though the numbers of mols n_k of the components are known. Each component C then introduces P mol fractions, which are variables required to specify the system state. We start with these CP variables and the two thermodynamic properties, such as temperature and pressure, as variables for the heterogeneous, multicomponent system. Now we ask for the pieces of information we have that will limit this number of variables.

The first piece of information we have is mass conservation. For each component k the sum of the mol fractions in each phase j is unity

$$1 = \sum_j^P \chi_k^{(j)}. \tag{16.1}$$

So for each component we have, in fact, only $(C-1)$ variables (unknown mol fractions) in each phase. The number of variables in our system is then reduced to $P(C-1)+2$.

The next piece of information comes from the requirement that our system is in a state of stable thermodynamic equilibrium. We must now specify this piece of information in equation form so that we know how many equations we have to reduce the number of variables in our system.

If the two thermodynamic variables, which we fix for the system, are temperature and pressure, then stable thermodynamic equilibrium requires that the Gibbs energy is a minimum, i.e. $\delta G \geq 0$ (see (12.12)).[1] The Gibbs energy for the system is the sum over all phases of the Gibbs energy in each phase

$$G = \sum_j^P G^{(j)}, \tag{16.2}$$

where the Gibbs energy in each phase is

$$G^{(j)} = \sum_k^C n_k^{(j)} \mu_k^{(j)}. \tag{16.3}$$

We must then minimize the Gibbs energy (16.2) for the system under conditions of constant (T,P), and subject to the constraints (16.1) for each component. To do this we introduce a Lagrange undetermined multiplier, λ_k, for each of these constraints (see Sect. A.4). The subsidiary function (see (A.22)) used to obtain the constrained minimum (extremum) of the Gibbs energy (16.2) is

$$h = \sum_j^P G^{(j)} + \sum_k^C \lambda_k \left(n_k - \sum_j^P n_k^{(j)} \right),$$

---

[1] Had we specified (T,V) as constant for the system we would have minimized $F(T,V,\{n_\lambda\})$. The number of equilibrium equations would be the same.

At the extremum

$$dh = 0 = \sum_j^P dG^{(j)} - \sum_k^C \left(\sum_j^P \lambda_k dn_k^{(j)} \right). \tag{16.4}$$

At constant temperature and pressure $dG^{(j)}$ is, (16.3) and (4.81),

$$dG^{(j)} = \sum_k^C \mu_k^{(j)} dn_k^{(j)}.$$

The extremum condition in (16.4) then becomes

$$0 = \sum_j^P \sum_k^C \left(\mu_k^{(j)} - \lambda_k \right) dn_k^{(j)}. \tag{16.5}$$

Because the differentials $dn_k^{(j)}$ appearing in (16.5) are independent of one another, (16.5) is satisfied provided

$$\mu_k^{(j)} = \lambda_k \tag{16.6}$$

for each phase j and for each component k. For each component we then have $(P-1)$ equations

$$\mu_k^{(1)} = \mu_k^{(2)} = \cdots = \mu_k^{(P)} \quad \forall k. \tag{16.7}$$

specifying the condition for stable equilibrium.

To find the total number of variables (unknowns) in our system we then subtract the number $(P-1)$ of equilibrium conditions from the number of unknowns that we had before introducing the equilibrium conditions, which was $P(C-1)+2$. The result is the number of variables needed to specify the thermodynamic state of the system

$$P(C-1)+2-C(P-1) = C-P+2.$$

The number of variables required to specify the thermodynamic state of a heterogeneous system in thermodynamic equilibrium is called the number of degrees of freedom of the system, f. Therefore

$$f = C - P + 2. \tag{16.8}$$

Equation (16.8) was first obtained by Gibbs [57] and is called the *Gibbs phase rule*.

The simplicity of the Gibbs phase rule is remarkable because of the complexity it covers. Our counting of the numbers of variables has been to keep track of things. The constraints of mass conservation and thermodynamic stability, which we have expressed neatly in equation form, are physical laws expressing how the system behaves in a form we can comprehend. Our balance sheet finally gives us the number f in (16.8) for the final number of variables that can be altered and still maintain the basic system with C components and P phases.

For example for the condition of equilibrium between water and its vapor, with (T, P) fixed, we have $C = 1$ and $P = 2$. Then $f = 1$ and there is one remaining variable that can be changed without affecting the equilibrium of the heterogeneous system. Experiment shows that this is the system volume that varies as the water passes between the vapor and the liquid state.

16.3 Phase Transitions

In this section we shall limit our consideration of heterogeneous equilibrium to systems containing only one constituent. For the purposes of our study we shall use the van der Waals fluid as a model for the behavior of the system.

From our previous study (see Sects. 8.3.2 and 8.3.3) we realize that the van der Waals fluid provides the basis for a good qualitative description of a fluid even at the edges of the region in which both the gas and liquid phases are present. We are now in a position to consider the consequences of the requirements of thermodynamic stability on the van der Waals fluid.

16.3.1 Van der Waals Fluid

We have plotted the isotherms of the van der Waals fluid in Fig. 16.1. The insert in the upper right hand corner of Fig. 16.1 shows the corresponding isotherms for the ideal gas, which are a set of rectangular hyperbolae. The isotherm $P = P(V, T = T_C)$, labeled as T_C in Fig. 16.1, has a point of inflection for which

$$\left(\partial^2 P / \partial V^2\right)_{T=T_c} = \left(\partial P / \partial V\right)_{T=T_c} = 0.$$

The isotherms $P = P(V, T > T_C)$ and the isotherms of the ideal gas are acceptable for the representation of a real substance as far as the requirement of mechanical stability $\left(\partial P / \partial V\right)_T < 0$ is concerned. But the van der Waals isotherms $P = P(V, T < T_C)$ have minima and regions for which $\left(\partial P / \partial V\right)_T > 0$. These regions cannot represent the behavior of a real substance.

Fig. 16.1 Isotherms of van der Waals fluid. Critical temperature isotherm $T = T_C$ is labeled. Isotherms for $T \lessgtr T_C$ are indicated. Inset is ideal gas isotherms for comparison

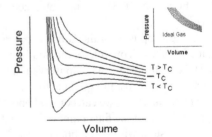

Fig. 16.2 Gibbs energy and
volume as functions of
pressure at constant
temperature for a van der
Waals fluid. Corresponding
points $a - h$ are indicated on
both plots. Multivalued region
$b - f$ includes metastable and
unstable regions

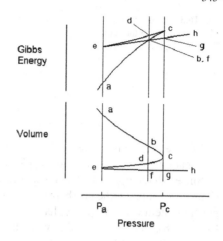

If the thermal equation of state of the substance is known, we obtain the dependence of the Gibbs energy on pressure, at constant temperature, from an integration of

$$dG = V dP.$$

In Fig. 16.2 we have plotted the Gibbs energy and the volume as functions of pressure for a van der Waals isotherm for which $T < T_C$. The pressure axes are the same for each graph and specific points $a - h$ are identified on each graph.

We see that the Gibbs energy for the van der Waals model system is multivalued between the pressures P_a and P_c. The system has the pressure P_a at the points a and e and the pressure P_c at the points c and g. The mechanical stability requirement, however, excludes the region $c - d - e$ from consideration, since there $(\partial P / \partial V)_T > 0$.

In the plot of $G(P)$ in Fig. 16.2 the minimum Gibbs energy in the region $P_a - P_c$ is that part designated by $a - b$ and then by $f - g - h$. This is the physical path followed by the system as the pressure is increased isothermally from P_a to P_c.

The system undergoes a transition from the thermodynamic state b to the thermodynamic state f without and change in temperature or pressure. The Gibbs energy remains constant during this process, as may be seen in the plot of $G(P)$ in Fig. 16.2. From the plot of $P(V)$ in Fig. 16.2, however, we see that the system volume has changed by a finite amount. This is exactly the situation considered at the end of the preceding section.

The correct form of a van der Waals isotherm, with $T < T_C$ as in Fig. 16.2, is then given in Fig. 16.3. And the correct form of the Gibbs energy is given in Fig. 16.4. There is no change in the value of the Gibbs energy between the states b and f, i.e. at (b, f), but there is a change in the slope of the Gibbs energy at (b, f). Because the chemical potential is the Gibbs energy per mol, the continuity of the Gibbs energy is equivalent to the stable equilibrium requirement in (16.7). The discontinuity in the derivative of the Gibbs energy results from the multivalued nature of the Gibbs energy for the van der Waals model in this region and the requirement of stability.

Fig. 16.3 Correct van der
Waals isotherm in the region
$T < T_C$. Coexistence region is
$b - f$

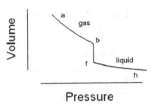

Between the points b and f a phase transition has occurred. The two states of
aggregation are the gas of region $a - b$ and the liquid of region $f - h$.

Since $(\partial G/\partial P)_T = V$, this change in the derivative of the Gibbs energy is the
change in volume indicated between the states b and f in Fig. 16.3. A single component system with two phases has a single variable or degree of freedom, as we see
from (16.8). The volume is that variable. The volume determines how much of the
system is in each phase providing a complete specification of the thermodynamic
state.

If there are n_g mols in the gas state and n_ℓ mols in the liquid state and if v_g and v_ℓ
are the specific molar volumes of the fluid in each state, then the system volume is

$$V = n_g v_g + n_\ell v_\ell$$
$$= n_g (v_g - v_\ell) + n v_\ell, \tag{16.9}$$

where $n = n_g + n_\ell$ is the total number of mols present in the system. The mol fraction
of gas present in the system when the system volume is V is then

$$\chi_g = \frac{\left(\frac{V}{n}\right) - v_\ell}{(v_g - v_\ell)}. \tag{16.10}$$

From Fig. 16.5 we see that this relationship has a simple graphical interpretation.
The numerator in (16.10) is proportional to the length of the heavy dashed line in
Fig. 16.5 and the denominator is proportional to the length of the heavy solid line.
Because of this graphical picture the mol fraction in (16.10) is found from what is
called the lever rule.

Fig. 16.4 Correct Gibbs
energy for a van der Waals
fluid in the region $T < T_C$.
The coexistence region $b - f$
of Fig. 16.3 is the single
point (b, f). The derivative of
Gibbs energy is discontinuous
at (b, f)

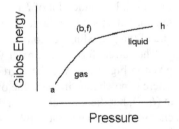

Fig. 16.5 Lever rule to find
χ_g in the coexistence region.
Heavy dashed horizontal line
is the numerator, *heavy solid
line* is the denominator in
(16.10)

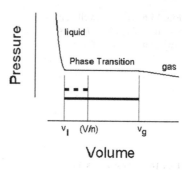

16.3.2 Maxwell Construction

We considered the so-called Maxwell construction in detail in Sect. 8.3.3. So we
have a historical and scientific perspective on what motivated Maxwell. We now
have a complete understanding of the situation in terms of the requirements of thermodynamic stability and may revisit Maxwell's brief comments based on that understanding.

Because the Gibbs energy is constant during the phase transition being considered here, the integral

$$\int_b^f \left(\frac{\partial G}{\partial P}\right)_V dP = \int_b^f V(P)\,dP \tag{16.11}$$

must be zero. Referring to Fig. 16.2, this integral may be divided into separate integrals as

$$\int_b^f V(P)\,dP = \int_b^c V(P)\,dP + \int_c^d V(P)\,dP$$
$$+ \int_d^e V(P)\,dP + \int_e^f V(P)\,dP$$
$$= 0$$

Then

$$\int_b^c V(P)\,dP + \int_c^d V(P)\,dP = \int_d^e V(P)\,dP + \int_e^f V(P)\,dP. \tag{16.12}$$

The sum of the integrals on the left hand side of (16.12) is the shaded area bcd in
Fig. 16.6 and area and the integral on the right hand side of (16.12) is the area def
in Fig. 16.6. The equality of the Gibbs energy during the phase change in a van der
Waals fluid is then equivalent to the graphical requirement that the two areas bcd
and def in Fig. 16.6 are equal. This graphical requirement is termed the Maxwell
construction.

Rotating the plot of Fig. 16.6 about two axes we have pressure plotted as a function of volume in Fig. 16.7.

Fig. 16.6 The Maxwell construction. *The shaded areas are equal*

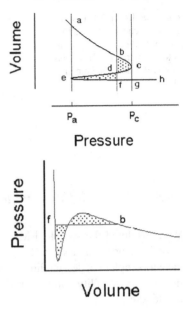

Fig. 16.7 Pressure as a function of volume with Maxwell construction. Fig. 16.6 rotated

If we plot a set of physical isotherms for the van der Waals fluid we have Fig. 16.8. In Fig. 16.8 the dots are the points on each isotherm identified by the Maxwell construction. The dots lie on a curve which separates the coexistence region, in which two phases exist in equilibrium, from the single phase regions.

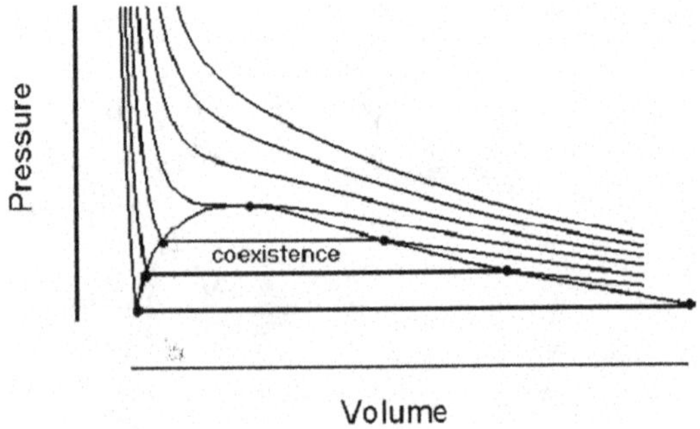

Fig. 16.8 Coexistence region for van der Waals fluid. *The horizontal lines* in the coexistence region result from the Maxwell construction

coexistence

Volume

The temperature of the isotherm with a point of inflection (see above) is the maximum temperature of the coexistence region. This is the critical temperature, T_C and the point (P_C, T_C) is the critical point.

16.3.3 Above the Critical Point

In Sect. 8.3.3 we considered what Maxwell called the masterful experiments of Thomas Andrews [2]. In these experiments Andrews considered the transition of

Fig. 16.9 Phase transition above critical point

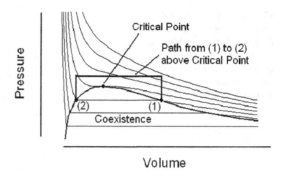

carbonic acid from a gas to a liquid state along an isotherm at a temperature above T_c. He showed that under these conditions the transition occurs with no separation into two phase regions within the system.

In Fig. 16.9 we have illustrated the situation Andrews studied experimentally. Consider that we have a system containing a van der Waals fluid in the pure gas phase designated by (1) in Fig. 16.9. We may bring the system to state (2) along a path such as, for example, the rectangular path shown in Fig. 16.9, which lies above the critical point. This path requires heating the system at constant volume to a temperature above T_C. cooling the system at constant pressure to the volume of state (2) and then cooling the system at constant volume until state (2) is reached.

At no state in this process are there two phases in equilibrium. The transition from the gas to the liquid phase cannot be observed, although the initial phase is a gas and the final a liquid. The Gibbs energy for the van der Waals fluid is single-valued at all points on the rectangular path from (1)→(2).

16.4   Coexistence Lines

The coexistence line separating phases is defined by the discontinuities in the Gibbs energy surface characterizing the phase transition. We shall continue to base our discussion of the gas to liquid $(g+l)^2$ phase transition on the properties of the van der Waals fluid.

In Fig. 16.10 we have plotted the molar Gibbs energy, $g(P, T = \text{constant})$ for a van der Waals fluid in part of the phase transition region $(0.87T_C \leq T \leq 0.93T_C)$. For each of the temperatures in Fig. 16.10 we computed the Gibbs energy, including the multivalued portion, for the van der Waals fluid as we did for Fig. 16.2. Figure 16.10 is then a set of projections of the Gibbs energy surface onto the (G, P) plane at a set of temperatures. This is a partial representation of the Gibbs energy surface.

The point of discontinuity in the derivative $((\partial g/\partial P)_T)$ of the Gibbs energy is located at the base of the multivalued portion in each projection. This discontinuity in the derivative marks the point at which a constant temperature and pressure phase

---

[2] In a phase transition the gas is often referred to as a vapor (v) and the transition as (v+l).

Fig. 16.10 Van der Waals
Gibbs energy vs pressure at
the phase transition. Multivalued regions are shown. The
coexistence line is formed by
the projection of the points of
derivative discontinuity onto
the (P,T) – plane

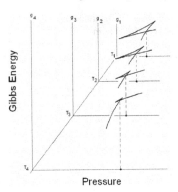

transition between the gas and liquid states of the van der Waals fluid occurs. At
the temperature T_C the discontinuity in the derivative will vanish. The projection in
Fig. 16.10 at the temperature $T_4 = 0.93T_C$ still has the discontinuity.

The pressure and temperature at which the (g+l) phase transition occurs are found
by projecting the points of discontinuity in the derivative $(\partial g/\partial P)_T$ onto the (horizontal) (P,T) – plane. These projections onto the (P,T) – plane are indicated by
vertical dashed lines in Fig. 16.10 and the values of pressure and temperature at
the phase transition by dots in the (P,T) – plane. These dots define a curve in the
(P,T) – plane known as the *coexistence line*.

The Gibbs phase rule tells us that for a single component system, a two phase
region has a single degree of freedom, i.e. is determined by a one-dimensional relationship. The coexistence line is a graphical representation of this one-dimensional
relationship.

We may also recognize the coexistence line as a projection of a portion of the
surface representing the thermal equation of state of a substance. This surface is
known simply as the $(P,V.T)$ – surface. The (P,V,T) – surface of any real substance
reveals the phase transitions as planar regions. This is shown in Fig. 16.11 where we
have plotted a portion of the (P,V,T) surface of a van der Waals fluid with a Maxwell
construction.

Fig. 16.11 (P,V,T) – surface
of a van der Waals fluid with
Maxwell construction. The
inset is a view of the planar
surface $a - b - c$ seen from
a point perpendicular to the
(P,T) – plane

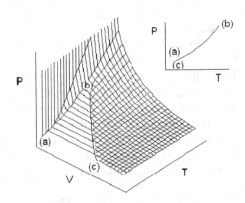

The Maxwell construction results in the planar surface bounded by the line $a - b - c$. The point (b) is the critical point. The projection of $a - b - c$ onto the $(P, T) -$ plane is a single line because the lines on the surface bounded by $a - b - c$ are isotherms. This single line, plotted in the inset in Fig. 16.11, is the coexistence line for the (g+l) phase transition.

For points on the surface bounded by $a - b - c$ the system is heterogeneous consisting of a gas phase and a liquid phase. The thermodynamic states of these two phases are specified by the end points of the isotherm. The volume of the system is the sum of the volumes of the liquid and the gas (see (16.9)).

We may find the general relation between pressure and temperature that defines the coexistence line from the requirement that the Gibbs energy is continuous across the phase change but the derivatives are not. The discontinuity in the derivative $(\partial g / \partial P)_T$, which is the specific volume v, is shown in Fig. 16.10, from which we first defined the coexistence line. The (g+l) phase transition is also marked by a discontinuity in the derivative $(\partial g / \partial T)_P$, which is the specific entropy. The (g+l) phase transition involves a considerable transfer of heat at constant temperature. There is then a marked change in the system entropy.

We shall designate the phases on the two sides of the coexistence line by the subscripts I and II. The continuity of the molar Gibbs energy at the point (T, P) is

$$g_I (T, P) = g_{II} (T, P). \tag{16.13}$$

The variation in (T, P) along the coexistence line must be such that (16.13) holds at every point. That is

$$g_I (T + \delta T, P + \delta P) = g_{II} (T + \delta T, P + \delta P) \tag{16.14}$$

at a neighboring point on the coexistence line. From a first order Taylor expansion we have

$$g_I (T, P) + \left(\frac{\partial g_I}{\partial T} \right)_P \delta T + \left(\frac{\partial g_I}{\partial P} \right)_T \delta P = g_{II} (T, P) + \left(\frac{\partial g_{II}}{\partial T} \right)_P \delta T + \left(\frac{\partial g_{II}}{\partial P} \right)_T \delta P.$$

With (16.13) this is

$$\left(\frac{\partial g_I}{\partial T} \right)_P \delta T + \left(\frac{\partial g_I}{\partial P} \right)_T \delta P = \left(\frac{\partial g_{II}}{\partial T} \right)_P \delta T + \left(\frac{\partial g_{II}}{\partial P} \right)_T \delta P. \tag{16.15}$$

Since

$$\left(\frac{\partial g_{I,II}}{\partial T} \right)_P = -s_{I,II} \text{ the specific (molar) entropy}$$

and

$$\left(\frac{\partial g_{I,II}}{\partial P} \right)_T = v_{I,II} \text{ the specific (molar) volume,}$$

Equation (16.15) becomes

$$\left. \frac{dP}{dT} \right]_{\text{coexist, I} \to \text{II}} = \frac{\Delta s_{\text{I} \to \text{II}}}{\Delta v_{\text{I} \to \text{II}}}, \qquad (16.16)$$

where

$$\Delta s_{\text{I} \to \text{II}} = (s_{\text{II}} - s_{\text{I}})$$
$$\Delta v_{\text{I} \to \text{II}} = (v_{\text{II}} - v_{\text{I}}).$$

Under conditions of constant pressure, $T ds = dh$. Therefore the molar change in entropy during the phase change at constant temperature and pressure is related to the molar change in enthalpy, $\Delta h_{\text{I} \to \text{II}}$ by

$$\Delta s_{\text{I} \to \text{II}} = \frac{\Delta h_{\text{I} \to \text{II}}}{T}, \qquad (16.17)$$

The change in molar enthalpy is the heat transferred during the phase change and is, for historical reasons, known as the *latent heat* associated with the transition. Then, combining (16.16) and (16.17) we have

$$\left. \frac{dP}{dT} \right]_{\text{coexist, I} \to \text{II}} = \frac{\Delta h_{\text{I} \to \text{II}}}{T \Delta v_{\text{I} \to \text{II}}}, \qquad (16.18)$$

Equation (16.16), which is the differential equation for the coexistence line in the (P, T)−plane, is known as the *Clapeyron equation*. This equation is general for any two phases I and II.

If the transition involves a gas as one of the phases we may simplify (16.18). The molar volume of the gas phase is much greater than the molar volume of either the liquid or the solid phase. So we may write

$$\Delta v_{\text{l} \to \text{g}} = v_{\text{g}} - v_{\text{l}} \approx \frac{RT}{P_{\text{g}}}$$

and

$$\Delta v_{\text{s} \to \text{g}} = v_{\text{g}} - v_{\text{s}} \approx \frac{RT}{P_{\text{g}}}.$$

Then for the transition (g+l) or (g+s) we have

$$\frac{dP}{dT} = \frac{\Delta h}{RT^2} P,$$

which may be integrated to give

$$\ln \frac{P_2}{P_1} = -\frac{\Delta h}{R} \left(\frac{1}{T_2} - \frac{1}{T_1} \right). \qquad (16.19)$$

for two points, (T_1, P_1) and (T_2, P_2), on the coexistence line for either boiling (g+l) or sublimation (g+s). Equation (16.19) is the *Clausius-Clapeyron equation*.

Fig. 16.12 Coexistence curves for general a substance

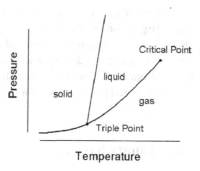

The Calusius-Clapeyron equation relates pairs of points, (P_1, T_1) and (P_2, T_2), on the coexistence line.

The general form of the coexistence curves for a substance is shown in Fig. 16.12. The triple point is the point at which all three phases are in equilibrium. The Gibbs phase rule applied at this point results in $f = 0$ and the triple point is a unique thermodynamic state obtained by the intersection of the two coexistence (one dimensional) lines (see Sect. A.1).

The critical point is the end of the liquid-gas two phase region. There is no critical point for the (l+s) transition. The liquid and solid phases are always distinct from one another.

16.5   Phase Transition Classification

Our study here has been of what is called a first order phase transition. This designation is part of a general system suggested by Ehrenfest in 1933 [77, 127] (see Sect. 10.3.4) and is based on the order of the partial derivative of the Gibbs energy that is discontinuous at the phase transition.

First order phase transitions are those for which the Gibbs energy remains constant and the first derivatives

$$\left(\frac{\partial G}{\partial T}\right)_P = -S \text{ and } \left(\frac{\partial G}{\partial P}\right)_T = V$$

are discontinuous (see (4.41) and (4.42)). Second order phase transitions are those for which the Gibbs energy and the first derivatives remain, but the second derivatives

$$\left(\frac{\partial^2 G}{\partial T^2}\right)_P = -\frac{nC_P}{T} \text{ and } \left(\frac{\partial^2 G}{\partial P^2}\right)_T = -V\kappa_T$$

are discontinuous (see (5.40)). The classifications can be continued to higher order. These are generally referred to as *multicritical transitions*, although commonly all transitions above first order are referred to as second order (cf. [77], p. 35).

First order transitions have a latent heat, or change in enthalpy, associated with them. Higher order transitions do not. L.D. Landau[3] observed (1937) that phase changes with no latent heat involved a change in the symmetry of the system. Landau associated an order parameter with the symmetry. The order parameter is a physical parameter of the system that has the value zero in the most symmetric, most disordered phase and the value unity in the most ordered phase [127].

16.6 Binary Systems

We shall consider first order phase transitions in binary, two component systems as the logical extension of the single component system. This section is not intended to be an extensive treatment of binary systems. Our intention here is to introduce the treatment of these systems using the Gibbs phase rule and the stability criterion. We shall consider only first order phase transitions so the Gibbs energy is continuous between phases. This will form the basis for our phase diagrams, just as continuity of the Gibbs energy provided a description of the coexistence line for the single component system.

The examples we use are intended for illustration of the general methods.

16.6.1 General Binary System

The Gibbs phase rule tells us that with $C = 2$, the number of degrees of freedom in a general binary system is

$$f = C - P + 2$$
$$= 4 - P.$$

For example if there are two phases present $P = 2$ and the number of degrees of freedom $f = 2$. The geometrical figure with two independent variables is a surface. There is then no coexistence line for a binary system. Rather there is a coexistence area.

We shall designate the two components by subscripts 1 and 2. The mol fractions χ_1 and χ_2 then define the composition of the system. And since their sum is unity, the Gibbs energy is a function of (T, P, χ_1). For a single phase, these are the three variables indicated by the Gibbs phase rule for a binary system.

If we consider a transition between the general phases I and II, continuity of the Gibbs energy between phases requires

---

[3] Lev Davidovich Landau (1908–1968) was a Soviet theoretical physicist. Landau made significant contributions in physics including the theory of superfluidity and the theory of second order phase transitions.

$$g_{\mathrm{I}}(T,P,\chi_1) = g_{\mathrm{II}}(T,P,\chi_1). \qquad (16.20)$$

The variation in (T,P,χ_1) must be such that (16.20) holds at the point $(T+\delta T, P+\delta P, \chi_1+\delta\chi_1)$ as well. That is

$$g_{\mathrm{I}}(T+\delta T,P+\delta P,\chi_1+\delta\chi_1) = g_{\mathrm{II}}(T+\delta T,P+\delta P,\chi_1+\delta\chi_1). \qquad (16.21)$$

From a first order Taylor expansion we have, with (16.20),

$$\left(\frac{\partial g_{\mathrm{I}}}{\partial T}\right)_{P,\chi_1}\delta T + \left(\frac{\partial g_{\mathrm{I}}}{\partial P}\right)_{T,\chi_1}\delta P + \left(\frac{\partial g_{\mathrm{I}}}{\partial \chi_1}\right)_{T,P}\delta\chi_1$$
$$= \left(\frac{\partial g_{\mathrm{II}}}{\partial T}\right)_{P,\chi_1}\delta T + \left(\frac{\partial g_{\mathrm{II}}}{\partial P}\right)_{T,\chi_1}\delta P + \left(\frac{\partial g_{\mathrm{II}}}{\partial \chi_1}\right)_{T,P}\delta\chi_1. \qquad (16.22)$$

Equation (16.22) is a general relationship among the variations δT, δP and $\delta\chi_1$ that must hold for equilibrium between two phases. Such a relationship among variations in three coordinates specifies a surface in the three dimensional space (T,P,χ_1). This is the surface on which the Gibbs energies of the two phases are identical.

The projection of this surface onto a plane specified by the constancy of one of the three coordinates is a line. For example if we pick P to be a constant the (16.22) becomes

$$\left(\frac{\partial g_{\mathrm{I}}}{\partial T}\right)_{P,\chi_1}\delta T + \left(\frac{\partial g_{\mathrm{I}}}{\partial \chi_1}\right)_{T,P}\delta\chi_1 = \left(\frac{\partial g_{\mathrm{II}}}{\partial T}\right)_{P,\chi_1}\delta T + \left(\frac{\partial g_{\mathrm{II}}}{\partial \chi_1}\right)_{T,P}\delta\chi_1, \qquad (16.23)$$

which specifies a line $T = T(\chi_1)$ in the constant $P-$ plane.

16.6.2   Liquid-Gas

Example plots of temperature as a function of mol fraction χ_1 for a binary system at two pressures are shown in Fig. 16.13 for the region in which liquid and gas phases coexist. The line separating the shaded region from the gas and liquid regions in Fig. 16.13 a and b is the projection of the plane defined by (16.22). This is the line $T = T(\chi_1)$ defined by (16.23).

Consider that we prepare a binary system of total composition $\chi_1 = \chi_a$ at a pressure $P = P_1$ and a temperature T_0, sufficiently low that only liquid is present in the system. The Gibbs phase rule for a binary system $(C=2)$ with a single phase $(P=1)$ leaves us $f = 3$. If we fix the pressure and the composition we are left with $f = 1$. That is we may vary the temperature. The situation is shown in Fig. 16.14.

We now increase the temperature of the system, at constant pressure $P = P_1$. The vertical line at constant composition is called an *isopleth*, which comes from the Greek for constant abundance. We raise the temperature along the isopleth to T_b. At this temperature gas (vapor) begins to appear above the liquid. We then have two

Fig. 16.13 Liquid-gas phase
diagram for a binary system
at two pressures

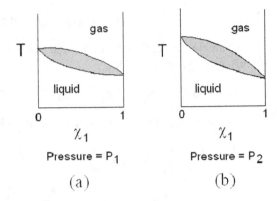

phases in the system. The Gibbs phase rule now tells us that $f = 2$. But we have
fixed the temperature at T_b and the pressure at P_1 so we have no degrees of freedom
left. The system is completely determined.

In Fig. 16.14 we see what has happened. The small amount of vapor that appeared above the liquid has the mol fraction $\chi_1 = \chi_b > \chi_a$. The gas may then be
removed and condensed to obtain a liquid with a composition $\chi_1 = \chi_b > \chi_a$ This increase in composition by boiling and subsequent condensation of the vapor is called
distillation.

If we raise the system temperature at a constant pressure $P = P_1$ to $T_c > T_b$ the
heterogeneous (two phase) system will be in a state of thermodynamic equilibrium
with the composition of the liquid and gas phases change as shown in Fig. 16.15.
The liquid and gas phases will have the compositions determined by the mol fractions χ_1 and χ_g respectively.

Thermodynamically there is no difference between the situation in Fig. 16.15 and
that in Fig. 16.14. The mols of the component of interest will simply be distributed
between the liquid at χ_1 and the gas at χ_g rather than between a liquid at χ_a and χ_b.

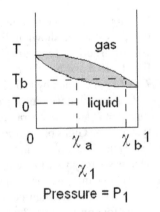

Fig. 16.14 Boiling in a binary
system. The temperature of
the liquid system at mol
fraction χ_a, pressure P_1, and
temperature T_0 is raised at
constant pressure to T_b where
boiling begins. Mol fraction
in gas is $\chi_b > \chi_a$

Fig. 16.15 Separation of a
binary system into two phases
distinct from original composition. The equilibrium binary
system at P_1 and T_c will a
liquid phase with mol fraction
$\chi_1 < \chi_a$ and a gas phase with
mol fraction $\chi_g > \chi_a$

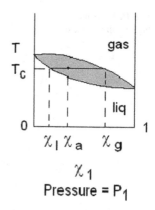

A binary system with a phase diagram of the form shown in Fig. 16.16 has an *azeotrope*, which comes from the Greek for boiling without changing. We may distill this system from a composition χ_a to $\chi_{a'}$ and from there to $\chi_{a''}$. We cannot, however, exceed the composition χ_A.

Fig. 16.16 Binary system
with an azeotrope

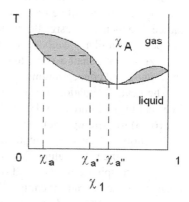

16.6.3 Solid-Liquid

The solid forms of substances are determined by the bonds formed among the atoms
of which the particular solid is composed. As a result of these bonds the structure of
a solid can be considered as being made up of certain elementary building blocks.
This results in a long range symmetry in these solids. Solids with long-range symmetry are called crystals. The building block of the crystal is the Bravais lattice.
The symmetry of the Bravais lattice defines the symmetry of the crystal. Solids for
which there is no long-range symmetry are called amorphous. Examples are glasses,
ceramics, gels, polymers, rapidly quenched melts, and films evaporated onto cold
surfaces.

Generally a binary system will solidify in at least two states distinguished by the
arrangements of the constituent atoms. We shall designate the crystalline form of the

Fig. 16.17 Binary
liquid-solid phase diagram

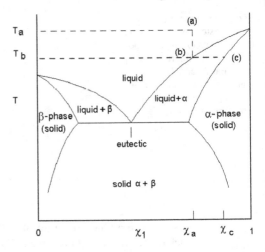

component 1 in the absence of 2 as α and the crystalline form of 2 in the absence
of 1 as β. That is, when $\chi_1 = 1$ the solid is in the form α and when $\chi_1 = 0$ it is in
the form β. A solid solution is formed if for $\chi_1 \approx 1$ the atoms of component 2 fit
within the Bravais lattice of 1 without destroying the long-range symmetry of the
lattice. We then still designate the solid phase as α, even though the composition is
not pure 1. Similarly we have a solid phase with $\chi_2 \approx 1$ which we designate as β
with $\chi_1 \neq 0$. We shall assume that the two components form a liquid solution over
the range of all concentrations.

The general situation, for a particular pressure, is illustrated in Fig. 16.17. The
lines in Fig. 16.17 are the relationships $T = T(\chi_1)$ defined by equations of the form
of (16.23) for heterogeneous equilibria.

If a liquid of general composition χ_a and temperature T_a is cooled at constant
pressure, at the temperature T_b a solid in the α-phase and of composition χ_c will
begin to precipitate. As the solid α-phase precipitates the liquid composition will
change. The temperature required for further solid precipitation is lowered as is the
composition of the solid precipitate. A liquid of the *eutectic* composition indicated
in Fig. 16.17 will freeze at a specific temperature and will produce a homogeneous
solid. If, in practice, uniform solid castings are desired, the liquid composition must
be eutectic.

16.7   Summary

In this chapter we introduced the requirements for equilibrium in systems containing
more than a single phase.

We began with a development of the Gibbs phase rule. This is an ingredient of
(almost) all thermodynamics texts. Our development was somewhat more detailed
than that provided by some authors because the quick derivations too often leave the
reader memorizing the final relationship, while still not understanding its meaning.

We elected to introduce the concept of a phase transition using the van der Waals model system This model system produces some aspects of the gas-liquid transition, as we know from our original considerations (see Sect. 8.3.3). Using the van der Waals model and the requirements of thermodynamic stability we were able to introduce the concept of coexistence of phases and of the coexistence line for the single component system. And we were able to produce a (P,V,T) surface for the van der Waals system, which bears a general resemblance to (P,V,T) surfaces fo real substances in the region of the (l+g) transition.

We extended these ideas to binary system concentrating on the liquid-gas transition. We ended the chapter with a few remarks on the solid-liquid transition in a binary system.

We intended none of this to be extensive. The ideas and the concepts are important. The final mathematical results are only important as tools for the beginning of any further work.

Exercises

16.1. We have found that a first order phase transition occurred in the van der Waals fluid in regions of mechanical instability, i.e. when $(\partial P/\partial V)_T > 0$. Based on this reasoning would you expect a fluid with the thermal equation of state

(a) $P = RT\left\{ [C_1/(V-V_0)]^2 - C_2\exp[-C_3(V-V_0)] \right\}$

(b) $P = RT\left\{ [C_1/(V-V_0)]^2 + C_2\exp\left[-C_3(V-V_0)^2\right] \right\}$

(c) $P = RT\left\{ [C_1/(V-V_0)]^2 - [C_2/(V-V_0)]\exp[-C_3(V-V_0)] \right\}$

to undergo a first order phase transition?

16.2. Does a phase change occur in the virial expansion (8.36) for a gas equation of state?

16.3. Does the Redlich-Kwong equation of state (8.46) have any unstable regions that would lead to a phase change?

16.4. In Table 16.1 are saturation data for boiling of water between $280 - 330\,\mathrm{K}$ [102]. These data are taken near room temperature and far from the critical point, which is at $647.1\,\mathrm{K}$. Use these data and the Calusius-Clapeyron equation (16.19) to obtain the enthalpy for the (g+l) transformation for water. For this problem you should plot a graph like that shown in Fig. 16.18.

16.5. In Table 16.2 are sublimation data for nitrogen between $40 - 60\,\mathrm{K}$ [102]. Use these data and the Calusius-Clapeyron equation to obtain the enthalpy for the (s+g) transformation for nitrogen. For this problem you should plot a graph like that shown in Fig. 16.19

Table 16.1 Saturation data for water (g+l)

Temperature K	Pressure MPa
280	0.00099182
285	0.0013890
290	0.0019200
295	0.0026212
300	0.0035368
305	0.0047193
310	0.0062311
315	0.0081451
320	0.010546
325	0.013531
330	0.017213

Fig. 16.18 Saturation data for water (l+g) [102]

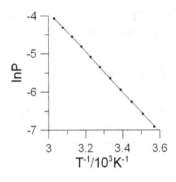

Table 16.2 Sublimation data for nitrogen (s+g)

Temperature K	Pressure MPa
40.000	0.0000064232
42.000	0.000017183
44.000	0.000042034
46.000	0.000095129
48.000	0.00020113
50.000	0.00040051
52.000	0.00075639
54.000	0.0013628
56.000	0.0023542
58.000	0.0039163
60.000	0.0062978

Fig. 16.19 Sublimation data for nitrogen (s+g) [102]

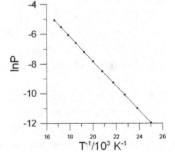

Appendix

A.1 The Ideal Gas

The ideal gas is a fictional substance that is approached by real gases as their density or pressure goes to zero. Here we shall determine the ideal gas equation of state using the ideal gas thermometer.

The ideal gas thermometer is represented in Fig. A.1. The spherical bulb containing the gas of interest is connected to a flexible mercury manometer. We can measure the gas volume by the location of the left mercury meniscus (A) and pressure by the height of the mercury column (h). We obtain the ideal gas limit by systematically bleeding off amounts of the gas (using the stopcock) between measurements and extrapolating the results to the point of vanishing density (pressure).

We may use the system in Fig. A.1 as either a constant volume or a constant pressure thermometer. For the constant volume thermometer we bring the left mercury meniscus to point A for each measurement of the pressure, which is then our thermometric property. For the constant pressure thermometer we require that h be a constant for each measurement of the volume, which will then be the thermometric property. We shall call the empirical ideal gas temperature Θ_μ, where the subscript μ is P or V depending on the thermometric property used.

We choose the empirical temperature to be linear. That is

$$\Theta_V(P) = A_V P + B_V \tag{A.1}$$

and

$$\Theta_P(V) = A_P V + B_P. \tag{A.2}$$

We shall only consider details for determining $\Theta_V(P)$.

We choose the two fixed points to be the ice and the steam points of water under a pressure of 1 atm. We choose the numerical values of the ice and steam point temperatures Θ_i and Θ_s to be 0 and 100. We have a *Celsius scale* and Eq. (A.1) is

Fig. A.1 Bulb containing
gas connected to a mercury
manometer

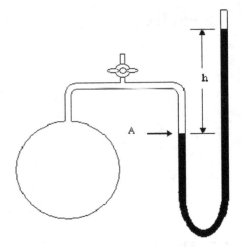

$$\Theta_V = \lim_{P_i(n) \to 0} \frac{100°C}{\left(\frac{P_s(n)}{P_i(n)}\right) - 1} \left(\left(\frac{P(n)}{P_i(n)}\right) - 1\right). \tag{A.3}$$

If we plot the pressure ratio $P_s(n)/P_i(n)$ and extrapolate the results to the condition $P_i \to 0$ we obtain graphs such as those shown in Fig. A.2. Data are from NIST [102]. The ratio $P_s(n)/P_i(n)$ approaches a constant value of 1.3661 independent of the identity of the gas in the bulb. Our experimental result for the relationship between Θ_V and the pressure for ideal gas conditions is then

$$\Theta'_V = \frac{100°C}{0.3661} \lim_{P_i(n) \to 0} \left(\left(\frac{P(n)}{P_i(n)}\right) - 1\right)$$
$$= 273.15°C \left\{ \lim_{P_i(n) \to 0} \left[\left(\frac{P(n)}{P_i(n)}\right) - 1\right] \right\}. \tag{A.4}$$

We include the prime to indicate that an extrapolation must be made. Adding $273.15°C$ to Θ'_V we can define a shifted scale, which we designate with a double prime. The units of this temperature are Kelvins, which we include.

$$\Theta''_V = 273.15\,K \left[\lim_{P_i(n) \to 0} \left(\frac{P(n)}{P_i(n)}\right) \right]. \tag{A.5}$$

The results at constant pressure are similar.

$$\Theta''_P = 273.15\,K \left[\lim_{V_i(n) \to 0} \left(\frac{V(n)}{V_i(n)}\right) \right]. \tag{A.6}$$

Measurements show that there is very little difference between Θ''_V and Θ''_P. For example midway between the ice and the steam points the temperatures obtained for helium are $\Theta''_V = 323.149\,K$ and $\Theta''_P = 323.140\,K$ and those for nitrogen are

Fig. A.2 Ratio of pressures P_s/P_i for three gases. Lines are least mean square fits to data extrapolated to $P_i = 0$

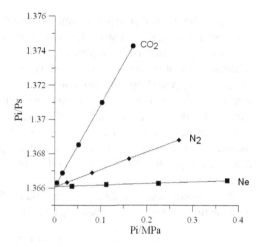

$\Theta''_V = \Theta''_P = 323.196\,\mathrm{K}$. We may then take the relationship between our empirical temperatures to be simply an equality and define a general empirical temperature for the ideal gas as $\Theta = \Theta''_V = \Theta''_P$. Designating $\Theta_i = 273.15\,\mathrm{K}$ we have

$$\Theta = \Theta_i \left[\lim_{P_i(n) \to 0} \left(\frac{P(n)}{P_i(n)} \right) \right]_{V=\mathrm{const}} \tag{A.7}$$

and

$$\Theta = \Theta_i \left[\lim_{V_i(n) \to 0} \left(\frac{V(n)}{V_i(n)} \right) \right]_{P=\mathrm{const}}. \tag{A.8}$$

Now we divide each of the Eqs. (A.7) and (A.8) by Θ, then multiply (A.7) by $1 = V(n)/V_i(n)$ and (A.8) by $1 = P(n)/P_i(n)$. We then have two expressions of the form

$$1 = \left[\lim_{n \to 0} \left(\frac{\frac{1}{\Theta}P(n)V(n)}{\frac{1}{\Theta_i}P_i(n)V_i(n)} \right) \right]. \tag{A.9}$$

The ideal gas is obtained in the limit. But n is never zero because then no gas at all is present.

Therefore, for a very small number n we have

$$\frac{1}{\Theta}P(n)V(n) = \frac{1}{\Theta_i}P_i(n)V_i(n)$$

for any temperature Θ. This can only be true if $P(n)V(n)/\Theta$ is a constant that may depend upon the number of mols present. That is

$$\frac{1}{\Theta}P(n)V(n) = \mathrm{constant}(n) \tag{A.10}$$

This is the thermal equation of state we sought for the ideal gas. We require only a determination of the constant (function of n) on the right hand side.

The ideal gas empirical temperature Θ that we have found here is (1) independent of the identity of the gas used and (2) has a single fixed point. A single fixed point scale was suggested by William Thomson (Lord Kelvin) in 1854[1] and is the reason for the designation of temperature unit as the Kelvin, K. This suggestion was adopted by the *Tenth International Conference on Weights and Measures* (1954) and the single fixed point has been chosen to be the triple point of water, which is the temperature at which solid, liquid, and gas phases of water are in equilibrium. Both the temperature (273.16 K) and the pressure ($0.61079 \pm 6.0795 \times 10^{-4}$ kPa) are fixed at this point.

We can show that the ideal gas temperature Θ in (A.10) is identical to the thermodynamic temperature T. Then Eq. (A.10) is then

$$PV = nRT, \tag{A.11}$$

where R is the universal gas constant $8.314510 \, \mathrm{J \, mol^{-1} \, K^{-1}}$.

A.2 Second Law Traditional Treatment

The traditional proof of Carnot's theorem is carried out in a number of logical steps each of which is a separate theorem. We shall present these here as theorems 1–5. Each is proven based on the requirements of the second law and using symbolic representations of Carnot engines as in Figs. 2.5 and 2.6.

Theorem A.2.1. *The efficiency of the Carnot engine is independent of working substance and depends only on the empirical temperatures of the reservoirs.*

For the Carnot cycle of Fig. 2.6 the thermodynamic efficiency is

$$\eta_C = \frac{\Theta_1 - \Theta_2}{\Theta_1}.$$

We introduce Θ_1 and Θ_2 as the empirical temperatures. These correspond to T_1 and T_2 of Sect. 2.3.3.

To prove that η_C depends only on Θ_1 and Θ_2, we assume that two Carnot engines with different efficiencies, which implies different working substances, can operate between the same reservoirs and show that this violates the second law. There are two situations that will result in a difference in efficiencies: (1) the heats transferred to the engines are the same but the works done are different or (2) the heats transferred differ and the works done are the same. In Fig. A.3 we illustrate the situation (1) in which the two Carnot engines a and b receive the same amount of heat in a cycle but produce different amounts of work. Here we choose to operate the engine b as a refrigerator.

---

[1] Kelvin's suggestion was made on the basis of Carnot's theorem, not gas thermometry.

Fig. A.3 Two Carnot engines
with different working substances

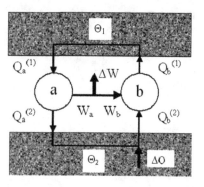

In Fig. A.3 the quantities $Q_{a,b}^{(1,2)}$ are the magnitudes of the heats transferred. The
first law results in

$$W_a = Q_a^{(1)} - Q_a^{(2)} > 0 \qquad (A.12)$$

and

$$W_b = Q_b^{(2)} - Q_b^{(1)} < 0. \qquad (A.13)$$

We require that $Q_a^{(1)} = Q_b^{(1)}$. Then $\Delta W = W_a + W_b > 0$ is the net work done by
the system. With (A.12) and (A.13), ΔW is

$$\Delta W = Q_b^{(2)} - Q_a^{(2)} = \Delta Q.$$

The combined cycle then violates the Kelvin statement of the second law.

The situation (2) in which the works done are equal and heats transferred are
different also violates the second law.

The theorem is then established.

Referring to Fig. 2.6 we see that the work done in the general Carnot cycle is
$W = Q_1 - Q_2$. The Carnot efficiency is then

$$\eta_C(\Theta_1, \Theta_2) = 1 - \frac{Q_2}{Q_1}. \qquad (A.14)$$

That is

$$\frac{Q_2}{Q_1} = f(\Theta_1, \Theta_2), \qquad (A.15)$$

and the ratio of the heats transferred to and from an engine operating on the Carnot
cycle is a function of the empirical temperatures of the reservoirs alone.

Theorem A.2.2. *The ratio of heats transferred to and from a Carnot cycle is equal
to the ratio of universal functions of the empirical temperature.*

The proof of this theorem is based on an argument that requires two Carnot engines and three reservoirs. We choose the reservoirs to be at the empirical tempera-
tures Θ_1, Θ_2, and Θ_3 with $\Theta_1 > \Theta_2 > \Theta_3$. The arrangement is shown in Fig. A.4.

Fig. A.4 Two Carnot engines
with three reservoirs

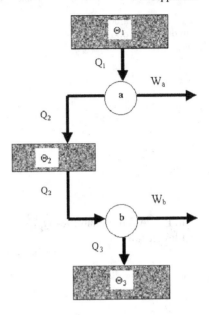

The two Carnot engines a and b combine to produce a single Carnot engine operating between Θ_1 and Θ_3. For this Carnot engine, (A.15) is

$$\frac{Q_3}{Q_1} = f(\Theta_1, \Theta_3).$$

For the individual Carnot engines in Fig. A.4,

$$\frac{Q_2}{Q_1} = f(\Theta_1, \Theta_2)$$

and

$$\frac{Q_3}{Q_2} = f(\Theta_2, \Theta_3).$$

Then

$$f(\Theta_1, \Theta_2) = \frac{f(\Theta_1, \Theta_3)}{f(\Theta_2, \Theta_3)}.$$

The Θ_3 dependence must then disappear from the ratio on the right hand side. This is the result if

$$f(\Theta_\mu, \Theta_\nu) = \frac{Q_\nu}{Q_\mu} = \frac{\tau(\Theta_\nu)}{\tau(\Theta_\mu)} \qquad (A.16)$$

for a Carnot engine. The function $\tau(\Theta)$ is a function only of the empirical temperature Θ.

The theorem is then established.

Theorem A.2.3. *There exists a positive function, T, of the empirical temperature alone such that the efficiency of a Carnot engine is*

$$\eta_C = 1 - T_{\text{low}}/T_{\text{high}}$$

This follows if we choose $\tau(\Theta)$ to be the temperature T. This was a suggestion first made by William Thomson (Lord Kelvin). Then

$$\frac{Q_\nu}{Q_\mu} = \frac{T_\nu}{T_\mu} \tag{A.17}$$

and the theorem is established.

Theorem A.2.4. *In any Carnot cycle the sum of the ratios of heats transferred to or from the cycle to the thermodynamic temperatures, T, of the corresponding reservoirs vanishes.*

To prove this theorem we rewrite (A.17) as

$$\frac{Q_\mu}{T_\mu} - \frac{Q_\nu}{T_\nu} = 0. \tag{A.18}$$

for the Carnot cycle. Using the subscript μ to designate the high temperature reservoir and the subscript ν to designate the low temperature reservoir, and recalling that the heat transferred at the low temperature is from the system and has a negative sign, (A.18) is

$$\sum_{\text{Carnot cycle}} \frac{Q}{T} = 0, \tag{A.19}$$

which establishes the theorem.

Theorem A.2.5. *In any (general) reversible cycle*

$$\oint_{\text{rev cycle}} \frac{\delta Q}{T} = 0$$

The proof of this theorem involves a thought experiment based on the Kelvin statement of the second law. First we create an arrangement that makes it possible to utilize a single reservoir at a fixed temperature, T_0, for the transfer of heat to and from the general reversible cycle. The Kelvin statement denies the possibility of transferring heat from the single reservoir and producing an equivalent amount of work, the net heat transferred must be zero. The result will be that the integral in the expression above vanishes.

A.3 Liouville Theorem

The Liouville theorem, which is a fundamental theorem of statistical mechanics, is a statement that the systems in an ensemble are neither created nor destroyed [154]. In a time dependent case the boundaries of the ensemble cloud in $\Gamma-$ space may shift. But no system representative point is lost. And the parameters for systems with representative points outside of the ensemble boundary would have to change for those systems to become part of the ensemble. This would require violation of such principles as conservation of energy.

The motion of the cloud of representative points, which is the representation of the ensemble in $\Gamma-$ space, is then analogous to the motion of a fluid. In fluid mechanics the conservation of mass is

$$\frac{\partial \rho}{\partial t} + \text{div } (\rho \mathbf{v}) = 0,$$

where ρ is the fluid density and $\mathbf{v}$ is the fluid velocity. If we carry out the analogous development for the ensemble cloud in $\Gamma-$ space, we replace the fluid density ρ with $P_\Gamma(\Omega, t)$ and the fluid velocity components $(dx/dt, dy/dt, dz/dt)$ with the components in $\Gamma-$ space $(dq_1/dt, d\,q_2/dt, \dots dp_{12N}/dt)$. We then obtain

$$\frac{\partial P_\Gamma(\Omega, t)}{\partial t} = -\sum_\mu \left[\frac{\partial}{\partial q_\mu}\left(P_\Gamma(\Omega, t)\frac{dq_\mu}{dt}\right) + \frac{\partial}{\partial p_\mu}\left(P_\Gamma(\Omega, t)\frac{dp_\mu}{dt}\right)\right]$$

$$= -\sum_\mu \left[\frac{\partial P_\Gamma(\Omega, t)}{\partial q_\mu}\frac{dq_\mu}{dt} + \frac{\partial P_\Gamma(\Omega, t)}{\partial p_\mu}\frac{dp_\mu}{dt}\right]$$

$$- \sum_\mu P_\Gamma(\Omega, t)\left[\frac{\partial}{\partial q_\mu}\left(\frac{dq_\mu}{dt}\right) + \frac{\partial}{\partial p_\mu}\left(\frac{dp_\mu}{dt}\right)\right].$$

The canonical equations are

$$\frac{dq_\mu}{dt} = \frac{\partial H}{\partial p_\mu}$$

$$\frac{dp_\mu}{dt} = -\frac{\partial H}{\partial q_\mu}. \tag{A.20}$$

Therefore,

$$\left[\frac{\partial}{\partial q_\mu}\left(\frac{dq_\mu}{dt}\right) + \frac{\partial}{\partial p_\mu}\left(\frac{dp_\mu}{dt}\right)\right] = 0$$

and

$$\frac{\partial P_\Gamma(\Omega, t)}{\partial t} = -\sum_\mu \left[\frac{\partial P_\Gamma(\Omega, t)}{\partial q_\mu}\frac{dq_\mu}{dt} + \frac{\partial P_\Gamma(\Omega, t)}{\partial p_\mu}\frac{dp_\mu}{dt}\right]. \tag{A.21}$$

This is the Liouville equation and is a mathematical expression of Liouville's theorem.

If we have an equilibrium ensemble then $\partial P_\Gamma(\Omega)/\partial t = 0$. The resultant form of Eq. (A.21) is satisfied if $P_\Gamma(\Omega)$ is a function only of the Hamiltonian, because then

$$\sum_{\mu} \left[\frac{\partial P_\Gamma(\Omega)}{\partial q_\mu} \frac{dq_\mu}{dt} + \frac{\partial P_\Gamma(\Omega)}{\partial p_\mu} \frac{dp_\mu}{dt} \right] = \sum_{\mu} \frac{\partial P_\Gamma(H)}{\partial H} \left[\frac{\partial H}{\partial q_\mu} \frac{\partial H}{\partial p_\mu} - \frac{\partial H}{\partial p_\mu} \frac{\partial H}{\partial q_\mu} \right] = 0.$$

A.4   Lagrange Undetermined Multipliers

The method of Lagrange undetermined multipliers allows us to find the extremum of a function subject to a number of constraints without confronting the problem of combining the function and the constraints algebraically.

We consider a general function of a number of variables, $f(x_1, \cdots, x_N)$.

The extremum of the function $f(x_1, \cdots, x_N)$ is that point at which the variation of f

$$\delta f = f(x_1 + \delta x_1, \cdots, x_N + \delta x_N) - f(x_1, \cdots, x_N)$$
$$= \frac{\partial f}{\partial x_1} \delta x_1 + \cdots + \frac{\partial f}{\partial x_N} \delta x_N$$

vanishes. We neglect terms $0\left(\delta x^2\right)$. If the variables $x_1, \cdots, x_N$ are independent then each of the partial derivatives independently vanishes at the extremum.

If we supplement the problem by introducing constraints

$$g_k(x_1, .., x_N) = 0 \quad k = 1, .., m$$

the variables $x_1, ..., x_N$ are no longer independent and the extremum can no longer be found by setting the partial derivatives of $f(x_1, ..., x_N)$ to zero. However, all the g_k are zero. Then the function

$$h(x_1, ..., x_N, \lambda_1, ..., \lambda_m) \equiv f(x_1, ..., x_N) + \sum_{k=1}^{m} \lambda_k g_k(x_1, ..., x_N), \tag{A.22}$$

where $\{\lambda_k\}_{k=1}^{m}$ are arbitrary constants, is always equal to f. Therefore, the function h has the same extremum as f.

The variation of h (neglecting terms of order $O\left(\delta x^2\right)$ and $O\left(\delta \lambda^2\right)$) is

$$\delta h = \sum_{j=1}^{N} \left[\frac{\partial f}{\partial x_j} \delta x_j + \sum_{k=1}^{m} \lambda_k \frac{\partial g_k}{\partial x_j} \right] \delta x_j + \sum_{k=1}^{m} g_k \delta \lambda_k, \tag{A.23}$$

and vanishes at extrema. The last term on the right hand side of (A.23) vanishes because all the $g_k = 0$.

Only $(N - m)$ of the N variations δx_j in (A.23) are independent. But there are also $m\lambda'$ s. So determining the extremum requires solving for $N - m + m = N$ independent variables. We then require N equations.

We judiciously choose the $m\lambda$'s so that the first m brackets $[\cdots]$ in (A.23) vanish. That is we choose the $m\lambda$'s such that

$$\frac{\partial f}{\partial x_j} + \sum_{k=1}^{m} \lambda_k \frac{\partial g_k}{\partial x_j} = 0 \quad j = 1, ..., m. \tag{A.24}$$

We then have

$$\delta h = \sum_{j=m+1}^{N} \left[\frac{\partial f}{\partial x_j} + \sum_{k=1}^{m} \lambda_k \frac{\partial g_k}{\partial x_j} \right] \delta x_j = 0, \tag{A.25}$$

which is a sum of $N - m$ terms of the form $A_j \delta x_j$. And there are $N - m$ independent variables left after we solve the m Eqs. (A.24). These must then be the $M - m$ variables whose variations appear in (A.25). Therefore each of the brackets $[\cdots]$ in (A.25) must vanish independently. That is

$$\frac{\partial f}{\partial x_j} + \sum_{k=1}^{m} \lambda_k \frac{\partial g_k}{\partial x_j} = 0 \quad j = m+1, .., N. \tag{A.26}$$

Now we see that the sets of Eqs. (A.24) and (A.26) are identical. Mathematically we have, therefore, simply reduced the problem to solving the set of equations

$$\frac{\partial f}{\partial x_j} + \sum_{k=1}^{m} \lambda_k \frac{\partial g_k}{\partial x_j} = 0 \quad j = 1, ..., N$$

coupled with the m constraint equations

$$g_k(x_1, ..., x_N) = 0 \quad k = 1, ..., m$$

for the $(m + N)$ variables $(x_1, ..., x_N, \lambda_1, ..., \lambda_m)$.

A.5 Maximizing $W(Z^*)$

Distributions in the neighborhood of the extremum in Eq. (9.52) will be also acceptable as approximations to the extremum $(h_N)_{ext}$ if h_N does not vary appreciably with changes in the distribution (of the occupation numbers) in the neighborhood of the extremum in Eq. (9.52). We investigate this by considering the second variation in h_N, i.e. the curvature of h_N, at the extremum. If the curvature of h_N is small in the neighborhood of the extremum, then h_N is not very sensitive to variations in the form of the distribution and many distributions close to the extremum (9.52) are acceptable as equilibrium distributions. If the curvature is large, then h_N varies considerably as the distribution is changed slightly from Eq. (9.52) and the equilibrium distribution is overwhelmingly Maxwell-Boltzmann.

To second order in variations in the occupation numbers N_i the deviation of h_N from its value at the extremum $(h_N)_{ext}$ is

$$h_N - (h_N)_{ext} = \delta h_N + \delta^2 h_N.$$

Here $\delta^2 h_N$ is the second order variation, which is dependent on $(\delta N_i)^2$. Using Eq. (9.51) we have

$$\delta^2 h_N = \frac{1}{2} \sum_{i=1}^{m} (\delta N_i)^2 \frac{\partial}{\partial N_i} \left(-\ln \frac{N_i}{N} - 1 + \alpha_N - \beta_N \varepsilon_i \right)$$

In terms of relative variations in each cell $(\delta N_i / N_i)$ this is

$$\delta^2 h_N = -\frac{1}{2} \sum_{i=1}^{m} \left(\frac{\delta N_i}{N_i} \right)^2 N_i. \tag{A.27}$$

We are interested in arbitrary relative variations in the occupation numbers. Then $|\delta N_i / N_i|$ is always greater than some minimum variation $|\delta N_i / N_i|_{min}$, and

$$\delta^2 h_N \geq -N \left(\frac{\delta N_i}{N_i} \right)^2_{min}.$$

Our partitioning of μ−space is limited by the indeterminacy principle, but is certainly such that the number of particles in a cell is extremely small compared to the total number of particles N. Therefore $|\delta N_i / N_i|_{min}$ is not vanishingly small and $N(\delta N_i / N_i)^2_{min}$ is a very large number. The curvature of h_N at the extremum is then very large and even slight deviations in occupation numbers from the Maxwell-Boltzmann distribution of Eq. (9.52) will not be appropriate equilibrium distributions. The volume $W(Z^*) = W_{max}$ for which the state Z^* is specified by a Maxwell-Boltzmann distribution is, for large numbers of particles, essentially the entire volume $\Omega_{\Delta E}$ of the $\Gamma -$ space shell of the microcanonical ensemble.

A.6 Microcanonical Volumes

To establish the validity of Eq. (9.37) we begin by separating Eq. (9.38) into integrals over a set of energy shells of thickness ΔE as

$$\int_{H<E} d\Omega = \sum_{v}^{E/\Delta E} \int_{E_v < H < E_v + \Delta E} d\Omega, \tag{A.28}$$

The number of shells $E/\Delta E$ is arbitrarily large because the shell thickness is arbitrarily small. The volume of the vth shell is

$$\Omega_{\Delta E}(E_v, V, N) = \int_{E_v < H < E_v + \Delta E} d\Omega. \tag{A.29}$$

The volumes of the shells increase with the value of the energy E_v. The volume $\Omega(E, V, N)$ is then greater than the volume of any of the individual shells, including $\Omega_{\Delta E}(E, V, N)$, but less than the product $\Omega_{\Delta E}(E, V, N)(E/\Delta E)$, since $E/\Delta E$ is arbitrarily large. Specifically

$$\Omega_{\Delta E}\left(E,V,N\right) \leq \Omega\left(E,V,N\right) \leq \Omega_{\Delta E}\left(E,V,N\right)\left(\frac{E}{\Delta E}\right). \qquad \text{(A.30)}$$

Then

$$\ln\left[\Omega_{\Delta E}\left(E,V,N\right)\right] \leq \ln\left[\Omega\left(E,V,N\right)\right] \leq \ln\left[\Omega_{\Delta E}\left(E,V,N\right)\right] + \ln\left(\frac{E}{\Delta E}\right). \qquad \text{(A.31)}$$

Now

$$\Omega\left(E,V,N\right) = \frac{1}{N!h^{fN}}\left[\int_V dq_1 dq_2 dq_3 \int_{p^2<2mE} dp_1 dp_2 dp_3\right]^N$$

$$= \frac{1}{N!h^{fN}}\left[\omega\left(V,E\right)\right]^N. \qquad \text{(A.32)}$$

For simplicity we have considered only translational motion in the directions $(1,2,3)$. Rotations and vibrations would change the form of $\omega(V,E)$, but not the power N. Then

$$\ln\Omega\left(E,V,N\right) = -\ln N! - fN\ln h + N\ln\omega\left(V,E\right)$$

$$= N\left[1 - \ln N - f\ln h + \ln\omega\left(V,E\right)\right], \qquad \text{(A.33)}$$

Where we have used Stirling's approximation in the form $\ln N! \approx N\ln N - N$. That is $\ln\Omega\left(E,V,N\right) \propto N$. In The thermodynamic limit of large N the term $\ln\left(E/\Delta E\right) = \ln N$ is negligible compared to N. In the thermodynamic limit then

$$\ln\left[\Omega_{\Delta E}\left(E,V,N\right)\right] \leq \ln\left[\Omega\left(E,V,N\right)\right] \leq \ln\left[\Omega_{\Delta E}\left(E,V,N\right)\right], \qquad \text{(A.34)}$$

or

$$\lim_{N,V\to\infty,\frac{N}{V}=\text{const}} \ln\left[\Omega_{\Delta E}\left(E,V,N\right)\right] = \ln\left[\Omega\left(E,V,N\right)\right]. \qquad \text{(A.35)}$$

A.7 Carathéodory's Principle

Carathéodory's principle provides a rigorous differential formulation of the second law of thermodynamics. It is an expression of Carnot's theorem and the Clausius inequality in terms of requirements these place on infinitesimal processes. There is no additional law of physics here. There is, however, an elegant and beautiful recasting of the second law into fundamental statements of the accessibility of states.

Carathéodory's principle has three parts. The first two of these are claims about the attainability of states by infinitesimal adiabatic processes. They are fundamental statements about the nature of infinitesimal adiabatic processes that are required for (1) the existence of the entropy as a thermodynamic function of state, and (2) a directionality for the entropy change in an irreversible process. The third is the requirement that the thermodynamic temperature is positive.

We shall present each of these parts as properties (of adiabatic processes) and shall point to their validity based on what we already know of the second law and the entropy. We make no attempt at rigorous mathematical development. Each claim is correct and based on the (experimental) second law. There is no lack of mathematical rigor behind what we present. We are simply avoiding the details. It is our hope that the reader will appreciate Carathéodory's principle based on our geometric presentation. A complete mathematical treatment can be found in A.H. Wilson's monograph [162].

In our discussion here we will accept the validity of the first law as expressed in (2.2).

Surfaces. Carathéodory's principle deals with the properties of the fundamental surface for a closed system. Our original derivation of the fundamental surface equated definitions of δQ_{rev} from the first and the second laws. The development of Carathéodory's principle will parallel our original work in that the first law will be coupled with the second law. The difference will be the way in which the second law is formulated.

Because we will deal with the fundamental surface we shall begin with a general theorem which guarantees the existence of a family of surfaces.

A general set of non-intersecting surfaces in (x, y, z) is defined by the equation

$$\Phi(x, y, z) = K, \tag{A.36}$$

where K is some arbitrary constant. For example the function

$$\Phi(x, y, z) = z - \cos\left(x^2 + y^2\right) = K$$

results in surfaces of the form shown in Fig. A.5. Each surface corresponds to a specific value of K. In this case K simply raises or lowers the surface. Specifically, the value of z at the point $(x = 0, y = 0)$ is $K + 1$.

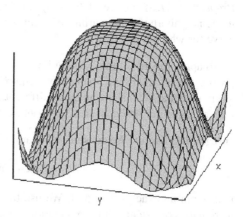

Fig. A.5 Function $z - \cos\left(x^2 + y^2\right) = K$

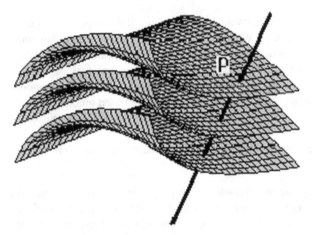

Fig. A.6 Planes of constant z and the line l

The differential of the function $\Phi(x,y,z)$ in (A.36) is the Pfaffian $d\Phi$ Each surface is then defined by $d\Phi = 0$, which is a relationship among the components (dx, dy, dz) of a differential distance along a curve on the surface.

A specific point, P, with coordinates (x_P, y_P, z_P) will lie on only one of the surfaces defined by (A.36). At the point P we can construct a line l that is not parallel to the surface containing P or to any such surface in a neighborhood of P. That is, along l $d\Phi \neq 0$ and the line l passes through, or cuts the surfaces. This is shown in Fig. A.6. The surfaces are drawn widely separated for illustration.

We may get from the surface containing the point P to any other of the surfaces in the neighborhood of P along the line l. And there is an infinity of points along any such line separating two surfaces. We then have a very simple geometric property of surfaces defined by (A.36).

Property A.7.1. If surfaces defined by Eq. (A.36) exist, there is an infinity of points in the neighborhood of an arbitrary point, P, that cannot be attained from P along a curve for which $d\Phi = 0$.

To apply this result to a general linear differential form such as (2.2), which expresses the content of the first law, we require of the concept of an integrating factor. We realize that (2.2) is not an exact differential and cannot define a set surfaces. An integrating factor will convert (2.2) to an exact differential.

Integrating Factor. We consider a linear differential form

$$\delta f = X dx + Y dy + Z dz \tag{A.37}$$

that is not an exact differential. We use the notation δf for this differential form to remain consistent with our notation for heat transfer in (2.2). In (A.37) $X, Y,$ and Z are functions of the coordinates (x,y,z), but are not partial derivatives of a function of a function $\Phi(x,y,z)$. If there exists an integrating factor, $M = M(x,y,z)$, then δf can be transformed into the differential of a function such as $\Phi(x,y,z)$ by multiplying δf by $M(x,y,z)$. That is

$$d\Phi = M\delta f$$
$$= MX dx + MY dy + MZ dz, \qquad (A.38)$$

where

$$MX = \left(\frac{\partial \Phi}{\partial x}\right)_{y,z}, \; MY = \left(\frac{\partial \Phi}{\partial y}\right)_{x,z}, \; \text{and } MZ = \left(\frac{\partial \Phi}{\partial z}\right)_{x,y}.$$

We note that (A.38) requires that the exact differential $d\Phi$ is equal to zero whenever δf is equal to zero, even though δf is not an exact differential.

Since $d\Phi = 0$ defines the set of surfaces in (A.36), we conclude that the existence of such a set of surfaces is a consequence of the existence of an integrating factor for the linear differential form δf in (A.37). We then have the theorem

Theorem A.7.1. *(integrating factor) If a linear differential form $\delta f = X dx + Y dy + Z dz$, where X, Y, and Z are continuous functions of (x,y,z), is such that in every neighborhood of an arbitrary point P there exist points that cannot be attained from P by any curve satisfying the differential equation $\delta f = 0$, then the linear differential form possesses an integrating factor.*

Mathematically the linear differential form is arbitrary. We choose to apply this theorem to the specific linear differential form expressing the first law, (2.2). The axes of the space are then thermodynamic properties and the points are thermodynamic states.

Three Parts. Carnot's theorem (2.3.1) states that such an integrating factor exists and that it is the thermodynamic temperature. Carathéodory's principle gets to this result in steps. The first part (claim) of Carathéodory's principle is that required to produce an integrating factor of (2.2).

Claim A.7.1. There are states of a system, differing infinitesimally from a given state, that are unattainable from that state by any quasistatic adiabatic process.

At this stage in the development we have only the first law in (2.2) to describe the physical behavior of our system. The claim (A.7.1.) then places additional limitations on the physical possibilities expressed in (2.2.) that are not inherent in the first law.

With claim (A.7.1) theorem (A.7.1) results in an integrating factor for the linear differential form δQ_{rev} in (2.2). The problem is a reduction to trivialities. Setting $\delta Q_{rev} = 0$ in (2.2) results in a differential form with only two variables dU and dV. Claim (A.7.1) and theorem (A.7.1) are more powerful than this! We have chosen a system that is too simple. This we simply amend by considering a more general system in which there are subsystems separate from one another, but in thermal contact so that the temperatures are equal. We may choose the subsystems to be defined by their volumes. If we choose two subsystems, we have a system with three variables: the empirical temperature ϑ and two volumes (V_1, V_2). The combination of claim (A.7.1.) and theorem (A.7.1), with $\delta Q_{rev} = 0$ then results in a set of surfaces in the space (ϑ, V_1, V_2).

If we keep the designation M for the integrating factor and Φ for the surface, then $d\Phi = M\delta Q_{rev}$. That is M is T^{-1} and Φ is the entropy, S, of Carnot's theorem. To actually establish that the integrating factor M, which from claim (A.7.1) and theorem (A.7.1) is a function of (ϑ, V_1, V_2), is actually a function only of ϑ requires additional argument. We may recall that we accomplished this in our traditional treatment (see Sect. A.2) using two Carnot engines and three reservoirs. To show that the integrating factor, M, is a function of the empirical temperature alone also requires the introduction of two systems and three reservoirs. The details of the demonstration may be found elsewhere [162].

We are now able to identify M is T^{-1} and Φ is the entropy, $S(T, V_1, V_2)$.

The second part of Carathéodory's principle is the claim

Claim A.7.2. There are states of a system, differing infinitesimally from a given state, that are unattainable from that state by any adiabatic process whatever.

Here the process causing the infinitesimal change in state may be irreversible.

To consider the consequences of claim (A.7.2) we redraw the three surfaces of Fig. A.6 in Fig. A.7. With our identification of Φ as S we now designate these as constant entropy surfaces S_a, S_b, and S_c. All states on a particular surface of constant entropy are attainable by a quasistatic adiabatic process. For example we may begin at the state (0) on the surface S_b and take the system to any other state on S_b by a reversible quasistatic adiabatic process. However, the process taking the system from the initial state (0) on surface S_b to any state (f) on surface S_c requires a general (irreversible) adiabatic process, indicated by the dashed line between the surfaces S_b and S_c in Fig. A.7.

Once we have the system in state (f) we can then take the system to any other state (f´on the surface S_c by a reversible quasistatic adiabatic process indicated by the solid line on surface S_c in Fig. A.7. The separation of the constant entropy surfaces containing the initial state (0) and the states (f) and (f´depends entirely on the process (0) – (f) and may be made arbitrarily small. The state (f´) may then be made arbitrarily close to the initial state (0). If constant entropy surfaces on both sides of the surface containing (0) are attainable by arbitrary adiabatic processes, then all

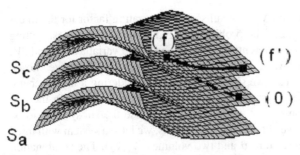

Fig. A.7 Constant entropy surfaces and an irreversible process

points in the neighborhood of the original point (0) are attainable by a combination of the original arbitrary adiabatic process and the second quasistatic adiabatic process. This violates claim (A.7.2). Therefore we conclude that arbitrary (irreversible) adiabatic processes can only carry the system to states on constant entropy surfaces on one side of the initial constant entropy surface. The initial constant entropy surface is then a bounding surface and not an intermediate surface available to adiabatic processes as a consequence of the second part of Carathéodory's principle.

The entropy, S, must then change in either a positive or negative fashion in an arbitrary adiabatic process. The algebraic sign of the change in the entropy is determined by the algebraic sign of the integrating factor, $M = T^{-1}$. There is no requirement in either parts 1 or 2 of Carathé odory's principle that this particular thermodynamic temperature must be positive. The third claim of Carathéodory's principle determines this.

Claim A.7.3. In any adiabatic process the entropy changes monotonically. If the thermodynamic temperature is defined to be positive, the entropy always increases or remains constant in such a process.

The surface (of constant entropy) onto which the representative point of the system moves in an infinitesimal irreversible process then has a higher numerical value than the surface on which the representative point was initially. If a system is in stable equilibrium no infinitesimal processes are possible within the system that will spontaneously alter the state of the system. The entropy of a system in stable equilibrium must then have a maximum value. Mathematically we may formulate this as a general principle. We state this as a theorem, which we prove in Sect. 12.2.

Theorem A.7.2. *The necessary and sufficient condition for the stable equilibrium of an isolated thermodynamic system is that the entropy is a maximum.*

Example A.7.1. We may classify irreversible adiabatic processes in terms of the work done. In a rapid expansion the substance acquires kinetic energy that is later dissipated into thermal energy. The minimum work is done in a completely unconstrained expansion, as in the Gay-Lussac experiment, and has the value zero. The maximum work is obtained when the process is quasistatic. The result is that in a general adiabatic expansion from an initial volume, V_0, to the same final volume $V_{f,exp} > V_0$, the final pressure has a value between a maximum for the unconstrained expansion and a minimum for the quasistatic expansion. The final thermodynamic state of the system is, therefore, determined by the amount of work done in the adiabatic process.

Similar considerations hold for compression to a volume $V_{f,comp} < V_0$. The least external work is required for the quasistatic adiabatic compression. A greater amount of work is required to attain the same final volume in a rapid compression because a rapid compression produces kinetic energy that will then be dissipated to thermal energy. The situation is indicated graphically in Fig. A.8. In Fig. A.8 the quasistatic (reversible) adiabatic process is the solid line. The dash and dash-dot lines represent irreversible adiabatic expansion or compression. These processes do not have intermediate equilibrium states.

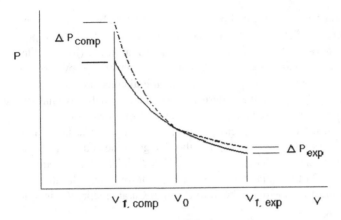

Fig. A.8 Reversible and irreversible adiabatic expansions

Ranges of final pressures, corresponding to the final states attainable, are indicated in the Fig. A.8 by ΔP_{comp} and ΔP_{exp}. Each of these final states defines an entropy $S = S(P,V)$ and to the range of states corresponds a range of final entropies attainable in arbitrary adiabatic processes beginning at the state (P_0, V_0). The final state and, therefore, the final value of the entropy depend entirely upon the details of the adiabatic process. A continuum of final entropies is then available to the system. As a consequence of the second part of Carathéodory's principle the entropy of the quasistatic adiabatic process lies at one end of this continuum of available entropies, and is not an interior value. The result, in terms of the figure above, is that the solid line representing the quasistatic adiabatic process is a limiting line for the final states available to the system in an adiabatic processes beginning at the state (P_0, V_0). All attainable states lie above this line.

A.8 Jacobians

A.8.1 Definition

We consider two sets of independent variables, (x, y) and (u, v). We consider that the variables may be transformed into one another and that the transformation is one-to-one. That is for each point (x, y) there exists a unique point (u, v) and vice-versa. The Pfaffians for $x(u, v)$ and $y(u, v)$ are

$$dx = \left(\frac{\partial x}{\partial u}\right)_v du + \left(\frac{\partial x}{\partial v}\right)_u dv \qquad (A.39)$$

and

$$dy = \left(\frac{\partial y}{\partial u}\right)_v du + \left(\frac{\partial y}{\partial v}\right)_u dv \tag{A.40}$$

These linear equations may be written in matrix form as [cf. [17], p. 239]

$$\begin{bmatrix} dx \\ dy \end{bmatrix} = \begin{bmatrix} \left(\dfrac{\partial x}{\partial u}\right)_v & \left(\dfrac{\partial x}{\partial v}\right)_u \\ \left(\dfrac{\partial y}{\partial u}\right)_v & \left(\dfrac{\partial y}{\partial v}\right)_u \end{bmatrix} \begin{bmatrix} du \\ dv \end{bmatrix} \tag{A.41}$$

The determinant of the transformation matrix in (A.41) is called the Jacobian of the transformation,

$$\frac{\partial(x,y)}{\partial(u,v)} \equiv \det \begin{bmatrix} \left(\dfrac{\partial x}{\partial u}\right)_v & \left(\dfrac{\partial x}{\partial v}\right)_u \\ \left(\dfrac{\partial y}{\partial u}\right)_v & \left(\dfrac{\partial y}{\partial v}\right)_u \end{bmatrix}.$$

We shall be primarily concerned here with transformations among sets of two variables, since our primary concern is with functions of two independent variables. The development may be generalized.

A.8.2 Differential Relations

In thermodynamics our interest is primarily in transformations among certain sets of variables, which are the characteristic variables for the thermodynamic potentials. Partial derivatives with respect to any of the variables performed while any other variable is held constant appear. These can be evaluated in terms of other partial derivatives using the concept of the Jacobian.

For example the transformation in (A.41) must also be valid along the line for which $dy = 0$. Along this line we designate the differentials dx, du and dv as $dx]_y$, $du]_y$ and $dv]_y$. Then (A.41) becomes, upon dividing by $du]_y$ and identifying partial derivatives,

$$\begin{bmatrix} \left(\dfrac{\partial x}{\partial u}\right)_y \\ 0 \end{bmatrix} = \begin{bmatrix} \left(\dfrac{\partial x}{\partial u}\right)_v & \left(\dfrac{\partial x}{\partial v}\right)_u \\ -\left(\dfrac{\partial y}{\partial u}\right)_v & -\left(\dfrac{\partial y}{\partial v}\right)_u \end{bmatrix} \begin{bmatrix} 1 \\ \left(\dfrac{\partial v}{\partial u}\right)_y \end{bmatrix}. \tag{A.42}$$

The second of these equations may be solved for the term $(\partial v/\partial u)_y$ with the result

$$\left(\frac{\partial v}{\partial u}\right)_y = -\frac{\left(\dfrac{\partial y}{\partial u}\right)_v}{\left(\dfrac{\partial y}{\partial v}\right)_u}.$$

Then (A.42) becomes

$$
\left[\begin{array}{c} \left(\dfrac{\partial x}{\partial u} \right)_y \\[2ex] 0 \end{array} \right] = \left[\begin{array}{cc} \left(\dfrac{\partial x}{\partial u} \right)_v & \left(\dfrac{\partial x}{\partial v} \right)_u \\[2ex] -\left(\dfrac{\partial y}{\partial u} \right)_v & -\left(\dfrac{\partial y}{\partial v} \right)_u \end{array} \right] \left[\begin{array}{c} 1 \\[2ex] -\dfrac{\left(\dfrac{\partial y}{\partial u} \right)_v}{\left(\dfrac{\partial y}{\partial v} \right)_u} \end{array} \right].
$$

Performing the matrix multiplication,

$$
\left(\frac{\partial x}{\partial u} \right)_y = \frac{\left[\left(\dfrac{\partial x}{\partial u} \right)_v \left(\dfrac{\partial y}{\partial v} \right)_u - \left(\dfrac{\partial x}{\partial v} \right)_u \left(\dfrac{\partial y}{\partial u} \right)_v \right]}{\left(\dfrac{\partial y}{\partial v} \right)_u} \tag{A.43}
$$

The numerator in (A.43) is the Jacobian of the transformation in (A.41). Then (A.43) can be written as

$$
\left(\frac{\partial x}{\partial u} \right)_y = \frac{\dfrac{\partial (x,y)}{\partial (u,v)}}{\left(\dfrac{\partial y}{\partial v} \right)_u}. \tag{A.44}
$$

The general form of these relationships among partial derivatives is

$$
\left(\frac{\partial x_i}{\partial y_i} \right)_{x_j} = \frac{\partial (x_i, x_j)}{\partial (y_i, y_j)} \bigg/ \left(\frac{\partial x_j}{\partial y_j} \right)_{y_i}. \tag{A.45}
$$

A.8.3 Inverse Transformation

For simplicity we write the general transformation in (A.41) in the form

$$
\left[\begin{array}{c} dx \\ dy \end{array} \right] = \left[\begin{array}{cc} A_{11} & A_{12} \\ A_{21} & A_{22} \end{array} \right] \left[\begin{array}{c} du \\ dv \end{array} \right].
$$

The A_{ij} are partial derivatives, which are continuous functions. The Jacobian is, for which we use the shorthand J, is, in terms of the A_{ij},

$$
J = A_{11}A_{22} - A_{12}A_{21}.
$$

The inverse of the transform from $\left[\begin{array}{c} du \\ dv \end{array} \right]$ to $\left[\begin{array}{c} dx \\ dy \end{array} \right]$ is

$$
\frac{1}{J} \left[\begin{array}{cc} A_{22} & -A_{12} \\ -A_{21} & A_{11} \end{array} \right].
$$

That is

$$\begin{bmatrix} du \\ dv \end{bmatrix} = \frac{1}{J} \begin{bmatrix} A_{22} & -A_{12} \\ -A_{21} & A_{11} \end{bmatrix} \begin{bmatrix} dx \\ dy \end{bmatrix}$$

The Jacobian of this transformation is designated as $\partial (u,v) / \partial (x,y)$ and is equal to

$$\frac{1}{J^2} (A_{11}A_{22} - A_{12}A_{21}) = \frac{1}{J}.$$

The Jacobian of the inverse transformation is then equal to one over the Jacobian of the first transformation. In standard notation

$$\frac{\partial (u,v)}{\partial (x,y)} = \frac{1}{\frac{\partial (x,y)}{\partial (u,v)}}. \tag{A.46}$$

A.8.4 Reversing Orders

The Jacobian

$$\frac{\partial (y,x)}{\partial (v,u)} = \det \begin{bmatrix} \left(\dfrac{\partial y}{\partial v}\right)_u & \left(\dfrac{\partial y}{\partial u}\right)_v \\ \left(\dfrac{\partial x}{\partial v}\right)_u & \left(\dfrac{\partial x}{\partial u}\right)_v \end{bmatrix}$$

is obtained from

$$\frac{\partial (x,y)}{\partial (u,v)} = \det \begin{bmatrix} \left(\dfrac{\partial x}{\partial u}\right)_v & \left(\dfrac{\partial x}{\partial v}\right)_u \\ \left(\dfrac{\partial y}{\partial u}\right)_v & \left(\dfrac{\partial y}{\partial v}\right)_u \end{bmatrix}$$

by an exchange first of rows and then of columns. Each of these operations changes the sign of the determinant with the result that

$$\frac{\partial (y,x)}{\partial (v,u)} = \frac{\partial (x,y)}{\partial (u,v)}.$$

Answers to Selected Exercises

1.4. $Q = Mg(H - H_0)$ (b) Assume that the measurement of H and the temperature coincide and that the temperature is uniform in the vessel.

1.5. $\Delta U = E_{exp} - (P_{atm} + P_b)(1/10)(4\pi R^3/3)$

1.6. As $\varepsilon \to 0$ $\Delta S_T \to 0$. You may suspect that when $\varepsilon = 0$ you have the maximum efficiency possible. That means that the production of entropy might be a loss. Then if you set $\Delta S_T = 0$ you get $Q_2/T_2 - Q_1/T_1 = 0$ for the greatest efficiency.

1.7. (a) The entropy change of the cube is $-C\Delta T/T_2$ and that of the table is $+C\Delta T/T_1$ total entropy change is $C\Delta T\left(1/T_1 - 1/T_2\right) > 0$ (b) The entropy change is, and remains, positive. (c) We lost a temperature difference. We could have used that temperature difference and heat flow to drive an engine.

1.8. When the temperature difference between engine and reservoir increases, entropy production increases. So Carnot is actually claiming we should minimize entropy production.

1.10. Motion of ions from one electrical potential to another, flow of molecules from regions of high to low concentrations, chemical reactions.

1.11. The irreversible nuclear fusion in the sun [7].

2.1. The liquid itself. The system boundary is the inside of the tube and imaginary disks at each end. The tube is excluded from the system. The system is open with mass crossing the boundary.

2.2. (a) To study combustion separate from mixing our system boundaries would be the inner walls of the combustion chamber, an imaginary boundary at the point of full mixing, and an imaginary boundary at the entrance to the nozzle. The nozzle has the additional effects of directional flow, which we initially try to avoid. The system is open. (b) The system boundary is the inner wall of the combustion chamber penetrated by the injector heads and ends where the boundary of part (a) begins, when the fuel and oxidant are mixed. The system is open.

2.3. The thermometric property will be the Pt resistance. The wire will not be under tension, which could also change resistance. The temperature scale will be chosen to be linear. This may be compared to a Celsius scale at certain points.

2.4. (b) Efficiency change is proportional to $\left(T_2/T_1\right)^2 \Delta T_{high}$ or $\left(T_1/T_2\right)^2 \Delta T_{low}$. So choose the river, if there are no PR considerations.

2.5. integrate $0 = nC_V dT/T + (P/T)dV$ using $PV = nRT$

2.6. For the isothermal heat flow: $Q_1 = nRT_1 \ln\left(V_{1,low}/V_{1,high}\right)$ and $Q_2 = nRT_2 \ln\left(V_{2,high}/V_{2,low}\right)$. Now $nRT_1 V_{1,low}^\gamma = nRT_2 V_{2,high}^\gamma$ and $nRT_1 V_{1,high}^\gamma = nRT_2 V_{2,low}$ so $V_{1,low}/V_{1,high} = V_{2,high}/V_{2,low}$ and $Q_1/Q_2 = T_1/T_2$

2.7. $P = P_0 \exp\left[-g\left(h - h_0\right)/RT\right]$

2.8. $P_0 \exp\left[gh/RT_a\right] = 3n_{He}RT_a/\left(4\pi r^3\right) - \alpha/r$ where $T_a =$ atmospheric temperature, $P_0 =$ pressure on ground, $n_{He} =$ mols He.

2.9. $\Delta T = gh/\left(\alpha T_0 + \beta\right)$

2.10. $S - S_0 = n_A\left(3/2\right)R\ln\left(\frac{T_F}{T_A}\right) + n_B\left(5/2\right)R\ln\left(\frac{T_F}{T_B}\right) + \left(n_A + n_B\right)R\ln\left(2\right)$

2.11. PV plot is a rectangle. (a) Work on: C→D, work by: A→B, no work: B→C, D→A. (b) Net work is done by the system. $W_{net} = \left(P_2 - P_1\right)\left(V_2 - V_1\right)$ (c) Heat in: D→A, A→B. Heat out: B→C, C→D (d) zero (e) We need the thermal and caloric equations of state, or one of the potentials for the working substance.

2.12. (a) isothermal $W_{12} = nRT\ln\left(V_2/V_1\right)$ (b) $W_{12} = \left[1/\left(1 - \gamma\right)\right]\left[P_2V_2 - P_1V_1\right]$

2.13. $\Delta T = \left(P\Delta t - (1/2)kx^2\right)/\left(nC_V\right)$

2.14. (a) High temperature: $\Delta S_1 = -Q_1/T_1$, low temperature: $\Delta S_2 = Q_2/T_2$. (b) Total change in entropy for gas in cycle is zero. The entropy changes on the isothermal legs are: $\Delta S_1 = Q_1/T_1$ for the high temperature, $\Delta S_2 = -Q_2/T_2$ for the low temperature. (c) Total change in entropy is zero. This is seen by adding results in (a) and (b). For the gas alone, since the total entropy change must vanish, $\Delta S_{T,gas} = Q_1/T_1 - Q_2/T_2 = 0$, which is Carnot's theorem. Then we have $\Delta S_{T,res} = Q_2/T_2 - Q_1/T_1$.

2.15. (a) Integrating the Gibbs equation around the cycle: $\oint T dS = \oint dU + \oint P dV$

with $\oint dU = 0$, this is $\oint T dS = W$. (b) Heat in A→B: $Q_{AB} = T_A (S_B - S_A)$
Heat out C→D: $Q_{CD} = T_C (S_D - S_C)$ Note: $T_A > T_C$ (c) zero (d) Work is area: $W = (T_A - T_D)(S_A - S_B)$ the efficiency is $\eta = W/|Q_{AB}| = (T_A - T_D)/T_A$
This is a Carnot cycle.

2.16. (a) $\mathcal{V} = \sqrt{C_P (T_1 - T_2)}$ (b) The nozzle has converted loss in thermal energy into a gain in kinetic energy. (c) The increase in momentum of the gas passing through the nozzle causes an increase in the momentum of the rocket in the other direction.

2.17. (a) The oil bath temperature is constant at $273.15 + 27 \approx 300\,\text{K}$. The resistor produces heat $Q = I^2 R \Delta t = (100 \times 10^{-3})^2 100 (50) = 50\,\text{J}$. There is no change in the entropy of the resistor because the state of the resistor has not changed. The bath receives 50 J of heat at 300 K. There will be a slight (negligible) change of state. The entropy will increase by $\Delta S_{bath} = 50\,\text{J}/300\,\text{K} = (1/6)\,\text{J}\,\text{K}^{-1}$ (b) The heat produced by the resistor is the same. But now there is a temperature rise of the resistor. The final temperature is $T_F = 300\,\text{K} + 50\,\text{J}/(17/4)\,\text{J}\,\text{K}^{-1} = 311.77\,\text{K}$. The entropy change of the resistor is $\Delta S = m C_P \ln (T_F/300\,\text{K}) = 0.16355\,\text{J}\,\text{K}^{-1}$. There is no entropy change of the bath.

2.18. $Q = P_0 V/(\gamma - 1)(2V_B/V)^{-\gamma}[1 - (3/2)(2V_B/V)]$

2.19. (a) $\Delta S = R \ln (10)$ (b) $W = T_1 \Delta S$

2.20. (a) $Q_{300} = -4800\,\text{J}$ exhausted, Q_{200} absorbed. (b) $S_{400} = -3\,\text{J}\,\text{K}^{-1}$, $S_{300} = 4\,\text{J}\,\text{K}^{-1}$, $S_{200} = -1\,\text{J}\,\text{K}^{-1}$ (c) $S_{total} = 0$

2.22. Assume no heat loss from the water bath. Neglect heat capacity of glass. First law analysis gives $(n_w C_V^w + n_g C_V^g) \Delta T = \int_T^{T+\Delta T} dT (\partial U_g/\partial T)_V$ where $n_{g,w}$ are mols of gas and water, $C_V^{(w)}$ is the specific heat of water.

2.25. (a) Turbine blades get damaged in wet steam, and the efficiency will drop. Also pumping a liquid-vapor mix is to be avoided. pure liquids are better. (b) Work done by the system in e→f and on system in g→h. (c) Heat in is the area under the top boundary of the cycle. Heat out is the area under the lower boundary of the cycle. (d) The area within the cycle $\oint T dS$. (e) Boiling is the heat input along the top horizontal line. Condensation is the heat exhaust along the lower horizontal line.

2.26. Boiler: steady state so $d\mathcal{E}/dt = 0$. There is no thermodynamic work done in the boiler so $\dot{W} = 0$. The steam enters and leaves the boiler slowly and there is negligible change in height. So we set $\sum_i e_i dn_i/dt = 0$. We then have

$\dot{Q} = -\sum_i h_i dn_i/dt$. The rate of heat transfer to the steam then equals the mass flow rate times the increase in thermal enthalpy. Enthalpy out > enthalpy in. Turbine: steady state so $d\mathcal{E}/dt = 0$. The boiler is assumed to be isentropic, so $\dot{Q} = 0$. The steam enters and leaves the turbine at high speed. But the thermal enthalpy changes are much larger than kinetic energy changes. So we neglect these. And there is negligible change in height. So we set $\sum_i e_i dn_i/dt = 0$. We then have $\dot{W}_s = \sum_i h_i dn_i/dt$. The rate at which work is done by the turbine then equals the mass flow rate times the drop in thermal enthalpy. Enthalpy in > enthalpy out.

2.27. This is now a refrigerator, or an air conditioner. The evaporator (the earlier condenser) is in the colder region and the heat absorbed there is exhausted to the warmer region.

3.1. a) yes, b) no, c) no.

3.2. $f = \int_{1,y=1}^{x} (2x-1)dx + \int_{1,x=x}^{y} (3x^2y^2 - 4xy^3)dy = x^2y^3 - xy^4$ this checks.

3.3. (b) $P/T = nR/V$ then $[\partial (P/T)/\partial U]_V = [\partial (nR/V)/\partial U]_V = 0$ note $V = $ constant. Using the cyclic permutation $[\partial (1/T)/\partial V]_U [\partial (U)/\partial (1/T)]_V = -[\partial (U)/\partial V]_T$ The left hand side vanishes. (c) $[\partial (1/T)/\partial V]_U = -ab/UV^2$ and $[\partial (P/T)/\partial U]_V = 0$ so no.

3.4. $g = -(Z - x^2)^2/(4x)$

3.5. (a) $\int_{V_0}^{V} dV/V = \int_{T_0, P=P_0}^{T} dT/T - \int_{P_0, T=T}^{P} dP/P$ produces $PV/T = P_0V_0/T_0$ b) $\ln(V/V_0) = \beta_0 (T - T_0) + (1/2)A_\beta (T^2 - T_0^2)$ $+ (1/3)B_\beta (T^3 - T_0^3) + L_\beta P_0 (T - T_0) - \kappa_0 (P - P_0) - (1/2)A_\kappa (P^2 - P_0^2) - (1/3)B_\kappa (P^3 - P_0^3) - L_\kappa T (P - P_0)$

3.7. $H = P^2/(2m) + V(x)$

3.8. $(\partial H/\partial P)_T = -nC_P (\partial T/\partial P)_H = -nC_P\mu_J$

4.1. (a) $F = -bVT^4/3$ (b) $U = U_0 (V_0/V)^{1/3}$ (c) $V = V_0T_0^{4/3}T^{-4/3}$ The assumption is that there is no heat flow from the universe and that the universe is in equilibrium. We know nothing about the boundary of the universe. Equilibrium is questionable.

4.2. $\Delta Q_{mix} = T\Delta S_{mix} = -nRT \sum_\lambda^n \chi_\lambda \ln(\chi_\lambda)$ This is a transfer of heat out of the system to maintain $T = $ constant.

4.3. $C_V = [1/(2C_1)] [\ln(P/P_0) + 2]^{-1} T$

4.4. (a) $P = nRT/(V - B) - nRTD/V^2$ this approaches the ideal gas as D and B become vanishingly small. (b) $U = A\alpha T^2 \exp(\alpha T)$

4.6. (a) $G = -[A/(64P)] T^4 - TS_0$ (b) $64P^2V = AT^4$ (c) $C_P = [3A/16n] T^3/P$

4.7. (a) $PV = nRT - A(T)P$ (b) $C_P = (PT/n)d^2A/dT^2$ (c) $S = PdA/dT - nR\ln(P/P_0)$ (d) $\beta = (nR - PdA/dT)/(nRT - AP)$ (e) $\kappa_T = nRT/(nRT - AP^2)$

5.2. $P(V,T) = \Gamma C_V T/V + (\varepsilon/2V_0) \left((V_0/V)^5 - (V_0/V)^3\right)$

5.3. $U - U_R = nC_V (T - T_R) + \varepsilon \left((1/8)(V_0/V)^4 - (1/4)(V_0/V)^2\right) - \varepsilon ((1/8)(V_0/V_R)^4 - (1/4)(V_0/V_R)^2)$

5.4. $S = nC_V \ln(T) + nC_V \frac{\Gamma}{n} \ln(V)$ and $F = nC_V T (1 - \ln(T)) + \varepsilon \left(\frac{1}{8}(V_0/V)^4 - \frac{1}{4}(V_0/V)^2\right) - \Gamma C_V T \ln(V)$

5.5. $U(S,V) = nC_V T_R (V/V_R)^{-\Gamma/n} \exp[S/(nC_V)] + \varepsilon\left(\left(\frac{1}{8}\right)(V_0/V)^4 - \left(\frac{1}{4}\right)(V_0/V)^2\right)$

5.6. $C_P = C_V \left(1 + \Gamma/n + 3\varepsilon(2nC_V T)(V_0/V)^2\right)$

5.7. yes

5.10. $u = \sqrt{\gamma R T}$

5.12. (b) no (c) yes

5.13. (a) For 5.12 c $S = C_1 C_2 \ln U - C_2/V$ (b) $F(T,V) = C_1 C_2 (T - \ln T) + C_2 T/V - C_1 C_2 \ln C_1 C_2$ (c) $P = C_2 T/V^2$

5.14. (a) $\beta = 1/T$ (b) $\kappa_T = v/RT$

5.15. (a) $a = b/3$ (b) $V = \exp\left(aT^3/P\right)$

7.2. yes

7.3. yes. Be careful with the logarithm in the limit.

8.5. $C_P - C_V = Tn^2 R^2 \left[nRT - 2n^2 a (V - nb)^2 /V^3\right]^{-1}$

8.6. $C_P - C_V = R + (2/P^2) R^2 (T/V^3) an^3$, and $C_P - C_V = R + 2(Rn/z) an^2 / (PV^2)$

8.7. $u = \sqrt{[RT + b(P + a\rho^2)]/(1 - \rho b) - 2a\rho + (RT/C_V)\left(R/(1 - \rho b)^2\right)}$, yes it reduces to $u = \sqrt{\gamma R T}$

8.8. $\Delta T \approx -n_g^2 a/(2n_w C_V^{(w)})$ where $n_{g,w}$ are mols of gas and water, $C_V^{(w)}$ is the specific heat of water, and a is the van der Waals constant.

8.9. $\mu_J C_P = -\left(nbP - 2n^2 a/V + 3n^3 ab/V^2\right) / (P - n^2 a/V^2 + 2n^3 ab/V^3)$ or $\mu_J C_P = -\left[RTv^3 b - 2av(v - b)^2\right] / \left[RTv^3 - 2a(v - b)^2\right]$

8.10. (A)→(B) or (C) increases T; (A)→(D) decreases T; (B)→(C) or (D) decreases T; (C)→ (D) decreases T. Stay in the region for which $\mu_J > 0$.

8.11. The final state is in the lower left hand corner of the Fig.

8.12. Begin by pumping nitrogen to high (20 MPa) pressure. Cool this with external refrigeration. But then to drop the temperature even more, pass it through a counter flow heat exchanger with the recirculated gas from the liquifaction vessel. The recirculated gas enters the pump with the fresh nitrogen.

8.14. $T_i = 2a(V - nb)^2 / (RV^2 b)$

8.15. yes

8.16. $F(T,V) = nC_V T (1 - \ln T) - nRT \ln(V - nb) - n^2 a/V$

8.17. (a) $C_P - C_V = R\left[1 - 2an(V - nb)^2 / (RTV^3)\right]$ (b) $2aP/(R^2 T^2) \approx 10^{-2}$ for CO_2.

8.18. $U(S,V) = nC_V (V - nb)^{1-\gamma} \exp(S/nC_V) + n^2 a/V$

8.19. $\mu_j C_P = -(B' - T dB'/dT) - P(C' - T dC'/dT) + \cdots$

8.20. (a) $\beta = RV^2 (V - nb) / \left[RV^3 - 2na(V - nb)^2\right]$

9.2. (a) $Q = m^{N/2} (2\pi k_B T/k_S)^N$ (b) $F = -Nk_B T \left[\ln (2\pi m^{1/2} k_B/k_S) + \ln T \right]$ (c)
$U = Nk_B T$ (d) $C_V = Nk_B$ (e) $S = Nk_B \ln \left[2\pi m^{1/2} e/(Nk_S) \right] + Nk_B \ln U$

9.4. that, for all practical purposes, the canonical ensemble is an ensemble of constant energy.

9.5. (a) $F = -Nk_B T \ln \left[\exp (\Delta \varepsilon/(2k_B T)) + \exp (-\Delta \varepsilon/(2k_B T)) \right] - N \ln N$, (b) $\langle H \rangle = N (\Delta \varepsilon/2) \{ 1 - \exp [\Delta \varepsilon/(k_B T)] \} / \{ 1 + \exp [\Delta \varepsilon/(k_B T)] \}$ (c) $C_V = (N/k_B)$ $(\Delta \varepsilon/T)^2 \exp [\Delta \varepsilon/(k_B T)] / \{ 1 + \exp [\Delta \varepsilon/(k_B T)] \}^2$ (d) yes

9.7. (a) $Q = (2\pi/h)^N (1/N!) \left[\frac{2}{u} \sinh u \right]$, (b) $\chi_m = N\mu^2 (\coth u - 1/u)$, where $u = \mu \mathscr{H}/(k_B T)$

9.10. $\langle M \rangle = N\mu_B \{ (2S+1) \coth [(2S+1) \alpha/2] - \coth (\alpha/2) \}$

9.12. (b) $v_{mp} = \sqrt{2k_B T/m}$

9.13. (a) $\bar{v} = \sqrt{8k_B T/(\pi m)}$ b) $v_{rms} = \sqrt{3k_B T/m}$

9.14. $(1/4) n_p \bar{v}$

9.15. $P = n_p k_B T = n_{mol} RT$

9.16. $(3/2) n_p k_B T$

9.17. $Nk_B \{ (3/2) \ln T + \ln (V/N) + (3/2) \ln (2\pi k_B/m) - 3/2 \}$

9.18. $P_\sigma = \exp [-PV/(k_B T)] \exp [-\lambda N_\sigma/k_B - H_\sigma/(k_B T)]$ where λ is a yet undetermined multiplier.

10.1. (a) $Q_{BE}^\mu = \prod_j 1/\{ 1 - \exp [-\varepsilon_j/(k_B T)] \}$ (b) $\langle n_j \rangle = 1/\{ \exp [\varepsilon_j/(k_B T)] - 1 \}$

(polarization would include a factor of 2) (c) $S = -(N/T) \sum_j \varepsilon_j/\{ \exp [\varepsilon_j/(k_B T)] - 1 \} - Nk_B \sum_j \ln \{ 1 - \exp [-\varepsilon_j/(k_B T)] \}$ (d) yes

10.6. The energy of each boson is $(2\hbar^2 \pi^2/mL^2) (n_x^2 + n_y^2 + n_z^2)$, which for a gas of N bosons for the ground state is $6N (\hbar^2 \pi^2/mL^2)$.

10.7. $u_{FD} \approx \frac{1}{5\pi^2} \left(\frac{2m}{\hbar^2} \right)^{\frac{3}{2}} \mu_{FD}^{\frac{5}{2}}$

11.1. $\Delta H_{rxn} [C(s) + 2H_2(g) \rightarrow CH_4(g)] = -74.4 \, \text{kJ mol}^{-1}$

11.3. $L_{ee} \Delta \varphi = L_{eq} \ln (T_{meas}/T_{ref}) \approx L_{eq} (\Delta T/T_{ref})$

11.4. This result is found by summing over all currents, noting that the Cl ion is negatively charged, and setting the numerator of the resulting expression to zero.

12.5. $P_2/P_1 = \exp [-Nmg (z_2 - z_1)/RT]$, and $n_2/n_1 = \exp [-Nmg (z_2 - z_1)/RT]$

13.1. (a) $x = $ mols AB dissociated. Then $P = (0.25 + x)(RT/V)$ or $x = 0.11284 \, \text{mol}$.
(b) $K_P = P_A P_B/P_{AB} \, \text{MPa} = 8.3456 \times 10^{-3} \, \text{MPa}$

13.5. If there is no dissociation the pressure in the vessel is $P_{nd} = P_0 (T_F/T_0) = 0.014172 \, \text{MPa} (686.65 \, \text{K}/298 \, \text{K}) = 0.032655 \, \text{MPa}$. Let $x = $ amount the $COCl_2$ pressure decreases from dissociated. For each molecule of $COCl_2$ there are two molecules of other gases added. So the total pressure is $P_T = (P_0 - x) + x + x = P_0 + x$ then $x = (0.045275 - 0.032655) \, \text{MPa} = 0.01262 \, \text{MPa}$. The partial pressures $P_{Cl_2} = P_{CO} = 0.01262 \, \text{MPa}$ and $P_{CoCl_2} = (0.032655 - 0.01262)$ MPa $= 0.020035 \, \text{MPa}$. Then $K_P = (0.01262)^2/0.020035 = 7.9493 \times 10^{-3} \, \text{MPa}$

13.6. $K_P = 1.9626 \times 10^{-2} \, \text{atm} = 1.9886 \times 10^{-3} \, \text{MPa}$

13.7. 3.9192×10^{-6} MPa

13.8. $K_\chi(T,P) = 9.5012 \times 10^{-2}$, $K_C(T,P) = 8.022 \times 10^{-4}$ mol m^{-3}

13.10. The ratio P_R = initial pressure/final pressure is $P_R = P_0/P_F = 1.9661$ then $P_{NH_3} = (1/3)(P_R - 1)(P_0/P_R) = 4.9790 \times 10^{-3}$ MPa.

13.12. $\Delta G_f^0 = 3\left(\Delta G_f^0[CO_2]\right) - \Delta G_f^0[Fe_2O_3] - 3\left(\Delta G_f^0[CO]\right) = -48.830$ kJ mol^{-1}.
Then $K_a(T) = \exp\left[48.83 \times 10^3 \text{J mol}^{-1}/\left(8.314510 \text{J mol}^{-1} \text{K}^{-1}(273.15 \text{K})\right)\right]$
$= \exp(21.501) = 2.1765 \times 10^9$

13.13. $\Delta G_{rxn}^0 = -4171.3$ J mol^{-1}, $\Delta H_{rxn}^0 = 5781.9$ J mol^{-1}, $\Delta S_{rxn}^0 = 33.383$ J mol^{-1} K^{-1}

14.1. (b) $d\left(1/P_{CH_3CHO}\right)/dt = -\left(1/P_{CH_3CHO}^2\right)d\left(P_{CH_3} CHO\right)/dt = k$ results in d $\left(P_{CH_3CHO}\right)/dt = -kP_{CH_3CHO}^2$, which is second order.

14.3. (a) First order because $\ln([A]/[A_0]) \propto t$.
(b) $\tau = 360.67$ ms

14.4. (a) k_1 is the analog of $k_{product}$, k_1 is the analog of $k^\ddagger$.
(b) In microscopic terms $\tau^{-1} = -(1/\delta_R)[k_B T/(2\pi\mu)]\left(q_A^\ddagger/q_A\right)\exp[-\varepsilon_a/(k_B T)]$.

14.5. (a) $E_{exp} = 36R = 300$ kJ mol^{-1}. $\Delta u^\ddagger = E_{exp} + RT = [300 + 8.3145(290)]$ kJ mol$^{-1} = 2711.2$ kJ mol^{-1}. $\Delta h^\ddagger = E_{exp} + 2RT = [300 + 2(8.3145)(290)]$ kJ mol$^{-1} = 5122.4$ kJ mol^{-1}.
(b) $A(T) = 5.76716$

14.6. $n_A = (-2K_{abs} + n_0 k_{abs,B})/(-2k_{abs,A} + k_{abs,B})$ and $n_B = (K_{abs} - n_0 k_{abs,A})/(-2k_{abs,A} + k_{abs,B})$

14.7. (a) $d[N_2O_5]/dt = -k_1[N_2O_5] + k_2[N_2O_3][O_2]$
(b) First order.
(c) $[N_2O_5] = n_0 \exp(-k_1 t)$ where $n_0 = [N_2O_5]_0$ = initial mols N_2O_5.
(d) $P(t) = n_0[2 - \exp(-k_1 t)]P_0$
From a linear fit to the plot $k = 6.87 \times 10^{-6}$ s^{-1}.

15.1. (a) Since $P_{H_2O}^*$ is a constant, $P_{H_2O} = \chi_{H_2O}P_{H_2O}^*$ says that a decrease in χ_{H_2O} from an increase in sugar must decrease the vapor pressure of water. To boil (again) we must increase the vapor pressure, which requires higher temperature.
(b) Writing a Pfaffian for the change in pressure $(dP = 0)$ we have $P_{H_2O}^*$ $d\chi_{H_2O} = -\left[\Delta H_{vap}/(RT^2)\right]\chi_{H_2O}P_{H_2O}^*$. Integrating with χ_{H_2O}(initial) $= 1$, we have $\ln(\chi_{H_2O}) = (\Delta H_{vap}/R)(1/T - 1/T_0)$.

16.1. (a) Mechanical instability occurs when $(C_2C_3/C_1^2)(V - V_0)^3 \exp[-C_3(V - V_0)] > 1$ (b) $(\partial P/\partial V)_T$ is always < 0 so there is no mechanical instability (c) Mechanical instability occurs when $\left[(C_2C_3/C_1^2)(V - V_0)^2 + (C_2/2C_1^2)(V - V_0)\right]\exp[-C_3(V - V_0)] > 1$

16.2. no

16.3. no

References

1. M.R. Ali, K.H. Cheng, J. Huang: PNAS **104**, 5372 (2007)
2. T. Andrews: Phil. Trans. Roy. Soc. Lon. **159**, 575 (1869)
3. J.R. Anglin, W. Ketterle: Nature **416**, 211 (2002).
4. R.P. Bauman: *Modern Thermodynamics with Statistical Mechanics* (Macmillan, New York 1992)
5. I.P. Bazarov, V.M. Faires: *Thermodynamics* (Pergamon, New York, 1964)
6. C. Beck, F. Schlögl: *Thermodynamics of Chaotic Systems* (Cambridge University Press, Cambridge 1993)
7. H.A. Bethe: Phys. Rev. **55**, 103 (1939)
8. P.G. Bolhuis, D. Chandler, C. Dellago, P.L. Geissler: Ann. Rev. Phys. Chem. **53**, 291 (2002)
9. L. Boltzmann: Wien. Ber. **76**, 373 (1877)
10. L. Boltzmann: *Vorlesungen über Gastheorie* (Barth, Leipzig 1896) translation as *Lectures on Gas Theory* (University of California Press, Berkeley 1964, reprint Dover, New York 1995)
11. L. Boltzmann: Wien. Ber. **63**, 679 (1871)
12. L. Boltzmann: Wien. Ber. **66**, 275 (1872)
13. M. Born, P. Jordan: Z. für Physik **34**, 858 (1925)
14. S.N. Bose: Zeitschrift für Physik **26**, 178 (1924)
15. J.M. Bowman, G.C. Schatz: Ann. Rev. Phys. Chem. **46**, 169 (1995)
16. S.G. Brush: *Kinetic Theory VI: The Nature of Gases and Heat* (Pergamon Press, Oxford 1965)
17. R.C. Buck: *Advanced Calculus* (McGraw-Hill, New York 1956)
18. S.H. Burbury: Phil. Mag. **30**, 301 (1890)
19. Bureau International des Poids et Mesures: *Supplementary Information for the International Temperature Scale of 1990* (Sèvres, BIPM 1990).
20. H.B. Callen: *Thermodynamics and an introduction to Thermostatistics* (Wiley, New York 1985)
21. S. Carnot: *Reflexions sur la puissance motrice du feu et sur les machines proper à développer cette puissance* (Bachelier, Paris 1824) English translation S. Carnot: In: *Reflections on the motive power of fire and other papers*

on the second law of thermodynamics, ed. by E. Mendoza (Dover, New York 1960)

22. A.H. Carter: Classical and Statistical Thermodynamics (Prentice Hall, Upper Saddle River, NJ 2001)

23. C. Cercignani: Ludwig Boltzmann: The Man Who Trusted Atoms (Oxford University Press, Oxford 1998)

24. D. Chandler: Introduction to Modern Statistical Mechanics (Oxford University Press, New York 1987)

25. S. Chapman, T.G.Cowling: The Mathematical Theory of Non-Uniform Gases (Cambridge University Press, Cambridge 1970)

26. P.L-G. Chong, M. Olsher: Soft Mat. 2, 85 (2004)

27. R. Clausius: Annalen der Physik und Chemie 79, 368, 500 (1850)

28. R. Clausius: Annalen der Physik und Chemie 125, 353 (1865)

29. R. Clausius: The Mechanical Theory of Heat – with its Applications to the Steam Engine and to Physical Properties of Bodies (van Voorst, London 1865)

30. R. Clausius: Mechanical Theory of Heat (van Voorst, London 1867)

31. E.G.D. Cohen, T.H. Berlin: Physica 26, 717 (1960)

32. E. Cornell: J. Res. Natl. Inst. Stand. Technol. 101, 419 (1996)

33. A. Cornish-Bowden: Fundamentals of Enzyme Kinetics (Portland Press, London 2004)

34. R. Courant: Differential and Integral Calculus, VI and II (Wiley Interscience, New York 1936)

35. R. Courant, K.O. Friedrichs: Supersonic Flow and Shock Waves (Interscience, New York 1963).

36. F.H. Crawford, W.D. van Vorst: Thermodynamics for Engineers (Harcourt, Brace and World, New York 1968)

37. W.H. Cropper: Great Physicists (Oxford University Press, New York 2001)

38. N. Davidson: Statistical Mechanics (McGraw-Hill, New York 1962)

39. T. De Donder: Lecons de Thermodynamique et de Chemie Physique (Gauthier-Villars, Paris.1920)

40. U.K. Deiters, K.M. De Reuck: Pure Appl. Chem. 69, 1237 (1997)

41. C. Dieterici: Wied. Ann. 65, 826 (1898)

42. C. Dieterici: Wied. Ann. 69, 685 (1899)

43. P.A.M. Dirac: Proc. Roy. Soc. 112, 661 (1926)

44. P.A.M. Dirac: The Principles of Quantum Mechanics, 4th edn. (Oxford University Press, London 1958)

45. P.A.M. Dirac: Recollections of an Exciting Era. In: History of Tentieth Century Physics, Proceedings of the International School of Physics "Enrico Fermi", Course LVII (Academic, New York 1977)

46. P. Ehrenfest, T. Ehrenfest: The Conceptual Foundations of the Statistical Approach in Mechanics, No. 6, Vol IV:2:II, Encyklopädie Der Mathematischen Wissenschaften (Teubner, Leipzig 1912, reprint Dover, New York 1990)

47. A. Einstein: Ann. Phys. 17, 132 (1905)

48. A. Einstein: Ann. Phys., 17, 549 (1905)

49. G.F.R. Ellis: 'On the Nature of Emergent Reality'. In: Re-emergence of Emergence. ed. by P. Clayton, P.C.W. Davies (Oxford University Press, New York 2006) pp. 79–110. *Before the Beginning, Cosmology Explained* (Boyars, Bowerdean, London New York 1994).

50. J.W. Stout, Jr: **4**, 662 *Encyclopedia Britannica* (Benton, Chicago 1969)

51. J.R. Ensher, D.S. Jin, M.R. Matthews, C.E. Wieman, E.A. Cornell: Phys. Rev. Lett. **77**, 4984 (1996).

52. J. J. Erpenbeck: Phys. Rev. E **48**, 223 (1993)

53. R.P. Feynman: *Statistical Mechanics* (Addison-Wesley, Reading 1998)

54. R.H. Fowler: *Statistical Mechanics, 2nd edn.* (Cambridge University Press, Cambridge 1936, reprinted 1966)

55. R. Fox: *Sadi Carnot: Reflexions on the Motive Power of Fire, a Critical Edition with the Surviving Manuscripts* (Manchester University Press, Manchester 1986)

56. J.W. Gibbs: *Elementary Principles in Statistical Mechanics* (Yale University Press, Hartford 1902, reprint Dover, New York 1960)

57. J.W. Gibbs: *Collected Papers, V I, Thermodynamics* (Longmans, Green & Co., London 1906, reprint Dover, New York 1961)

58. D.E. Goldman: J. Gen. Physiol. **27**, 365 (1943)

59. H. Goldstein: *Classical Mechanics*, 2nd edn. (Addison-Wesley, Reading 1980)

60. A.R.H. Goodwin, K.N. Marsh, W.A. Wakeham (eds): *Measurement of the Thermodynamic Properties of Single Phases, Experimental Thermodynamics Vol VI* (IUPAC Chemical Data Series No. 40) (Elsevier, Amsterdam Boston 2003).

61. I.S. Gradshteyn, I. M. Ryzhik: *Table of Integrals, Series, and Products* (Academic, San Diego 1980)

62. H.S. Green: *Information Theory and Quantum Physics* (Springer, Berlin Heidelberg 2000)

63. R. Haase: In: *Physical Chemistry: An Advanced Treatise*, ed. by W. Jost (Academic, New York 1971) p. 29

64. J.M. Haile: Fluid Phase Equilib. **26**, 103 (1986)

65. H. Haken: *Information and Self-Organization*, 2nd. edn. (Springer, Berlin Heidelberg 2000)

66. H.H. Hasegawa, D.J. Driebe: Phys. Rev. E **50**, 1781 (1994)

67. C.E. Hecht: *Statistical Thermodynamics and Kinetic Theory* (Freeman, New York 1990, reprint Dover, 1998)

68. H. von Helmholtz: Sitzber. kgl. preuss. Akad. Wiss. **1**, 22 (1882)

69. H. von Helmholtz: *Über die Erhaltung der Kraft* (Reimer, Berlin 1847)

70. C. S. Helrich, J. A. Schmucker, D. J. Woodbury: Biophys. J. **91**, 1116 (2006)

71. E.N. Hiebert: Reflections on the Origin and Verification of the Third Law of Thermodynamics. In: *Thermodynamics: History and Philosophy*, ed. by K. Martinás, L. Ropolyi, P. Szegedi (World Scientific, Singapore 1990)

72. B. Hille: *Ion Channels of Excitable Membranes*, 3rd ed. (Sinauer, Sunderland, Massachusetts 2001)

73. C.N. Hinshelwood and W.K. Hutchinson: Proc. Roy. Soc. Lon. **111A**, 380 (1926)
74. A.L. Hodgkin, B. Katz: J. Physiol. (London) **108**, 37 (1949)
75. G. Holton, S.G. Brush: *Physics, the Human Adventure* (Rutgers University Press, New Brunswick 2001)
76. I.K. Howard: J. Chem. Educ. **79**, 697 (2002)
77. K. Huang: *Statistical Mechanics*, 2nd edn. (John Wiley, New York 1987)
78. J. Huang, G. W. Feigenson: Biophys. J. **76**, 2142 (1999)
79. H. Ibach, H. Lüth: *Solid State Physics* (Springer, Berlin 1990)
80. E.T. Jaynes: The Gibbs Paradox. In: *Maximum Entropy and Bayesian Methods*, ed. by C. R. Smith, G. J. Erickson, P. O. Neudorfer (Kluwer, Dordrecht 1992) pp. 1–22
81. E.T. Jaynes: Phys. Rev. **106**, 620 (1957)
82. E.T. Jaynes: Phys. Rev. **108**, 171 (1957)
83. J. Jeans: *An Introduction to the Kinetic Theory of Gases* (Cambridge University Press, Cambridge 1940)
84. J.P. Joule, W. Thomson: Phil. Trans. Roy. Soc. Lon. **144**, 321 (1854)
85. J.P. Joule, W. Thomson: Phil. Trans. Roy. Soc. Lon. **152**, 579 (1862)
86. B. Katz: Arch. Sci. Physiol. **2**, 285 (1949)
87. J.H. Keenan, J.Chao, J. Kaye: *Gas Tables: Thermodynamic Properties of Air Products of Combustion and Component Gases, Compressible Flow Functions* (International Version) (Krieger, Melbourne 1992)
88. J.H. Keenan, F. G. Keyes, P.G. Hill, J.G. Moo: *Steam Tables : Thermodynamic Properties of Water Including Vapor, Liquid, and Solid Phases* (Krieger, Melbourne 1992)
89. J. Kestin: *A Course in Thermodynamics*, V I (McGraw-Hill, New York 1979)
90. A.I. Khinchin: *The Mathematical Foundations of Information Theory* (Dover, New York 1957)
91. A.I. Khinchin: *The Mathematical Foundations of Statistical Mechanics* (Dover, New York 1960)
92. J.G. Kirkwood: J. Chem. Phys. **3**, 300 (1935)
93. J.G. Kirkwood: Chem. Rev. **19**, 275 (1936)
94. M.J. Klein: The Beginnings of the Quantum Theory. In: *Proceedings of the International School of Physics "Enrico Fermi", Course LVII.* ed. by C. Weiner(Academic Press, New York 1977) pp. 1–39.
95. Yu. L. Klimontovich: *Statistical Physics* (Harwood, New York 1982)
96. D. Kondepudi, I. Prigogine: *Modern Thermodynamics* (Wiley, New York 1998)
97. S. Labík, J. Kolafa, A. Malijevský: Phys. Rev. E **71**, 21105 (2005)
98. K.J. Laidler, M.C. King: J. Phys. Chem. **87**, 2657 (1983)
99. L.D. Landau, E.M. Lifshitz: *Statistical Physics* (Addison-Wesley, Reading 1958)
100. P.T. Landsberg: 'The Laws of Thermodynamics'. In: Problems in Thermodynamics and Statistical Physics. ed. by P.T. Landsberg (Pion, London 1971) pp. 1–4 *Thermodynamics*. (Interscience, New York 1961)

101. P.T. Landsberg: *Thermodynamics and Statistical Mechanics* (Oxford University Press, Oxford 1978, reprint Dover, New York 1990)
102. E.W. Lemmon, M.L. Huber, M.O. McLinden: *NIST Reference Fluid Thermodynamic and Transport Properties-REFPROP-Version 8.0.* (National Institute of Standards and Technology, Boulder 2007).
103. D.S. Lemmons, C.M. Lund: Am. J. Phys. **67**, 1105 (1999)
104. G.N. Lewis, M. Randall (revised by K. S. Pitzer, L. Brewer): *Thermodynamics*, 2nd edn. (McGraw-Hill, New York 1961)
105. G.N. Lewis: Proc. Amer. Acad. Arts Sci. **37**, 49 (1901)
106. B. Mahon: *The Man Who Changed Everything: The Life of James Clerk Maxwell* (John Wiley, West Sussex 2003)
107. J.C. Maxwell: J. Chem. Soc. (London) **28**, 493 (1875)
108. J.C. Maxwell: *Theory of Heat* (Longmans, Green & Co., London 1888, reprint Dover, New York 2001)
109. J. C. Maxwell: On the Dynamical Theory of Gases. In: *The Scientific Papers of James Clerk Maxwell, V II*, ed. by W.D. Niven (Cambridge University Press, Cambridge 1980, reprint Dover, New York 1952)
110. J.R. Mayer: Liebig's Annalen der Chemie und Parmazie **42**, 239 (1842)
111. J.E. Mayer, M.G. Mayer: *Statistical Mechanics of Fluids* (Wiley, New York 1940)
112. A.D. McNaught, A. Wilkinson: *IUPAC Compendium of Chemical Terminology*, 2nd edn. (1997) (Royal Society of Chemistry, Cambridge 1997)
113. E. Merzbacher: *Quantum Mechanics*, 3rd ed. (Wiley, New York 1998)
114. H. Metiu: *Physical Chemistry: Statistical Mechanics* (Taylor and Francis, New York 2006)
115. N. Metropolis, A.N. Rosenbluth, M.N. Rosenbluth, A.H. Teller, E. Teller: J. Chem. Phys. **21**, 1087 (1953)
116. D. Mintzer: Transport Theory of Gases. In: *The Mathematics of Physics and Chemistry, V II*, ed. by H. Margenau, G.M. Murphy (D. van Nostrand, Princeton 1964) pp. 1–49
117. D.C. Montgomery, D.A. Tidman: *Plasma Kinetic Theory.* (McGraw-Hill, New York 1964).
118. C.E. Moore: *National Standard Reference Data Series (NSRDS) 3 Sec. 1–10, Selected Tables on Atomic Spectra*, (1965–1983)
119. W. Nernst: Z. Phys. Chem. **2**, 613 (1888)
120. W. Nernst: Sitzber. kgl. preuss. Akad. d. Wiss. V. **20**, 933 (1906)
121. G. Nicolis, I. Prigogine: *Self-organization in Nonequilibrium Systems* (Wiley, New York 1977)
122. G. Nicolis, I. Prigogine: *Exploring Complexity* (Freeman, New York 1989)
123. J.P. O'Connell, J.M. Haile: *Thermodynamics: Fundamentals for Applications* (Cambridge University Press, Cambridge 2005)
124. G.S. Ohm: *Die galvanische Kette mathematisch bearbeitet* (Riemann, Berlin 1827)
125. R. Omnès: *Understanding Quantum Mechanics* (Princeton University Press, Princeton 1999)

126. A. Pais: *Subtle is the Lord* (Oxford University Press, Oxford 1982)

127. P. Papon, J. Leblond, P.H.E. Meijer: *The Physics of Phase Transitions, Concepts and Applications* (Springer, Berlin Heidelberg 2006)

128. D. Park: *The How and the Why* (Princeton University Press, Princeton 1988)

129. W. Pauli: *Statistical Mechanics*, Pauli Lectures on Physics V IV (MIT Press, Cambridge 1973, reprint Dover, New York 2000)

130. M.J. Pilling, P.W. Seakins: *Reaction Kinetics* (Oxford University Press, Oxford 1995)

131. M. Planck: Ann. Phys. **4**, 553 (1901)

132. M. Planck: *Treatise on Thermodynamics* (Longmans, Green & Co., London 1927, reprint Dover, New York 1945)

133. I. Prigogine: *From Being to Becoming* (Freeman, New York 1980)

134. I. Prigogine: *The End of Certainty* (Free Press, New York 1997)

135. I. Prigogine: *Thermodynamics of Irreversible Processes* (Wiley Interscience, New York 1961)

136. H. Ramsperger: PNAS **13**, 849 (1927)

137. O. Redlich, J.N.S. Kwong: Chem. Rev. **44**, 233 (1949)

138. O. Redlich: *Thermodynamics: Fundamentals, Applications* (Elsevier, Amsterdam 1976)

139. L. Reichl: *A Modern Course in Statistical Physics*, 2nd edn. (Interscience, New York 1998)

140. J.R. Reitz, F.J. Milford, R.W. Christy: *Foundations of Electromagnetic Theory*, 4th edn. (Addison-Wesley, New York 1993)

141. M. Rukeyser:*Willard Gibbs*, (Doubleday, Garden City 1942, reprinted Ox Bow Press, Woodbridge 1988)

142. O. Sackur: Ann. Phys. **36**, 958 (1911)

143. O. Sakur: Ann. Phys. **40**, 67 (1913)

144. D.V. Schroeder: *An Introduction to Thermal Physics* (Addison Wesley Longman, San Francisco 2000)

145. F.W. Sears, G.L. Salinger: *Thermodynamics, Kinetic Teory, and Statistical Mechanics* (Addison-Wesley, Reading 1986)

146. C.E. Shannon: Bell Syst. Technol. J. **27**, 379 (1948)

147. C.E. Shannon: Bell Syst. Technol. J. **27**, 623 (1948)

148. C.E. Shannon, W. Weaver: *The Mathematical Theory of Communication* (University of Illinois Press, Urbana Chicago 1963)

149. A. Sommerfeld: *Partial Differential Equations, Lectures on Theoretical Physics, Vol I* (Academic Press, New York 1952)

150. A. Sommerfeld: *Thermodynamics and Statistical Mechanics, Lectures on Theroretical Physics, Vol V* (Academic Press, New York 1956)

151. T. Stonier: *Information and the Internal Structure of the Universe*, (Springer, London 1990)

152. D. ter Haar, H. Wergeland: *Elements of Thermodynamics* (Addison-Wesley, Reading1966)

153. H. Tetrode: Ann. Phys. **38**, 434 (1912)

154. M. Toda, R. Kubo, N. Satô: *Statistical Physics I* (Springer, Berlin Heidelberg 1983)
155. R.C. Tolman: *The Principles of Statistical Mechanics* (Oxford University Press, Oxford 1962).
156. J.D. van der Waals: Over de Continueteit van den Gas-en Vloeistoftoestand. Ph.D. Dissertation, Leiden University, Leiden (1873).
157. C.E. Vanderzee: J. Am. Chem. Soc. **74**, 4806 (1952)
158. W.G. Vincenti, C.H. Kruger, Jr: *Introduction to Physical Gas Dynamics* (John Wiley & Sons, New York 1965)
159. J.D. Walecka: *Introduction to General Relativity* (World Scientific, Singapore 2007)
160. C.Weiner ed.: *History of Twentieth Century Physics, Proceedings of the International School of Physics "EnricoFermi" Course LVII* (Academic, New York 1977)
161. Wigner, Eugene: CPAM **13**(1), 1 (1960).
162. Wilson, A.H. 1960, *Thermodynamics and Statistical Mechanics*, Cambridge University Press, London.
163. X. Yang: Ann. Rev. Phys. Chem. **58**, 433 (2007)
164. W. Yourgrau, S. Mandelstam: *Variational Principles in Dynamics and Quantum Theory*, 3rd ed. (Dover, New York 1979)

Index